Honda
XL/XR 80-200
Owners
Workshop
Manual

by Chris Rogers
With an additional Chapter on the 1981 to 1987 models
by Jeremy Churchill

Models covered
XL80S. 80 cc. US only 1979 to 1985
XR80. 80 cc. UK 1979 to 1980, US 1978 to 1984
XR80R. 80 cc. US only 1984 to 1987
XL100S. 99 cc. UK 1978 to 1983, US 1979 to 1985
XR100. 99 cc. US only 1981 to 1984
XR100R. 99 cc. US only 1984 to 1987
XL125S. 124 cc. UK 1978 to 1983, US 1978 to 1982, 1984
XL125R. 124 cc. UK only 1982 to 1987
XL185S. 180 cc. UK 1978 to 1982, US 1978 to 1983
XR185. 180 cc. US only 1979
XL200R. 195 cc. US only 1982 to 1984
XR200. 195 cc. UK 1980, US 1979 to 1984
XR200R. 195 cc. UK 1982 to 1983, US 1981 to 1983, 1986 to 1987

Note: This manual does not cover the 199 cc RFVC-engined (4-valve) XR200R model, sold in the US for 1984 to 1985

ABCDE

2

ISBN 978 1 85010 347 9

© Haynes Publishing 1990

Printed in the UK   (566-7P5)

British Library Cataloguing in Publication Data
A catalogue record for this book is available from the British Library

Library of Congress Control Number 86-82148

Haynes Publishing Group
Sparkford Nr Yeovil
Somerset BA22 7JJ   England

Haynes Publications, Inc
859 Lawrence Drive
Newbury Park
California 91320 USA

Acknowledgements

Our thanks are due to Paul Branson Motorcycles of Yeovil who supplied the machines featured in this manual.

We would also like to thank the Avon Rubber Company, who kindly supplied information and technical assistance on tyre fitting; NGK Spark Plugs (UK) Ltd for information on sparking plug maintenance and electrode conditions, and Renold Ltd for advice on chain care and renewal.

Last, but not least, thanks are due to all of those people at Sparkford who helped in the production of this manual. Particularly Alan Jackson and Tony Steadman who carried out the mechanical work and took the photographs respectively, and Mansur Darlington who edited the text.

About this manual

The purpose of this manual is to present the owner with a concise and graphic guide which will enable him to tackle any operation from basic routine maintenance to a major overhaul. It has been assumed that any work will be undertaken without the luxury of a well-equipped workshop and a range of manufacturer's service tools.

To this end, the machine featured in the manual was stripped and rebuilt in our own workshop, by a team comprising a mechanic, a photographer and the author. The resulting photographic sequence depicts events as they took place, the hands shown being those of the author and the mechanic.

The use of specialised, and expensive, service tools was avoided unless their use was considered to be essential due to risk of breakage or injury. There is usually some way of improvising a method of removing a stubborn component, provided that a suitable degree of care is exercised.

The author learnt his motorcycle mechanics over a number of years, faced with the same difficulties and using similar facilities to those encountered by most owners. It is hoped that this practical experience can be passed on through the pages of this manual.

Where possible, a well-used example of the machine is chosen for a workshop project, as this highlights any areas which might be particularly prone to give rise to problems. In this way, any such difficulties are encountered and resolved before the text is written, and the techniques used to deal with them can be incorporated in the relevant Section. Armed with a working knowledge of the machine, the author undertakes a considerable amount of research in order that the maximum amount of data can be included in the manual.

Each Chapter is divided into numbered Sections. Within these Sections are numbered paragraphs. Cross reference throughout the manual is quite straightforward and logical. When reference is made, 'See Section 6.10', it means Section 6, paragraph 10 in the same Chapter. If another Chapter were intended, the reference would read, for example, 'See Chapter 2, Section 6.10'. All the photographs are captioned with a section/paragraph number to which they refer and are relevant to the Chapter text adjacent.

Figures (usually line illustrations) appear in a logical but numerical order, within a given Chapter. Fig. 1.1 therefore refers to the first figure in Chapter 1.

Left-hand and right-hand descriptions of the machines and their components refer to the left and right of a given machine when the rider is seated normally.

Motorcycle manufacturers continually make changes to recommendations and specifications, and these, when notified, are incorporated into our manuals at the earliest opportunity.

Whilst every care is taken to ensure that the information in this manual is correct no liability can be accepted by the author or publishers for loss, damage or injury caused by any errors in, or omissions from, the information given.

Contents

The 1980 Honda XL100 S

The 1980 Honda XL185 S

The engine unit of the Honda XL125 S

Introduction to the Honda XL/XR 80 - 200 models

The present Honda empire has facilities which enable it to offer machines to cover every conceivable aspect of motorcycling.

Of all the machines described in this Manual, the XR models are designed and built for off-road use only; although not as powerful as their two-stroke rivals, they are much easier to ride and have encouraged many riders to take up serious competition work. Stripped of all but the bare essentials, they are lighter in weight than the XL models and are fitted with more powerful engines and better suspension components.

The XL models are designed as trail bikes; a class of motorcycle which became popular throughout the 1970s. Such machines will be used mainly on the road but are capable of off-road use should their owners feel so inclined. This is illustrated by the much more comprehensive equipment of these models, including fully-legal lighting etc, in conjunction with their high-level exhausts, large-diameter front wheels, semi-knobbly tyres and off-road styling.

The models described in this Manual can be sub-divided into two main groups. The XL/XR80 models are enlarged and refined versions of the XR75 model first seen in 1972; while the XL100 model employs a suitably enlarged version of the same basic engine/gearbox unit its cycle parts are largely those of the second main group, the XL/XR125, 185 and 200 models. These all use an enlarged and refined version of the same basic engine/gearbox unit which first appeared in the CB100 model of 1970. This has proved itself to be one of the most reliable and robust engine units in the motorcycle world, even when enlarged to twice its original size. The main modifications to the original design are the use of a detachable cylinder head cover and the substitution of CDI ignition for the original contact breaker system.

Since it is Honda's policy to refine each model during its production run, with modifications of varying degrees of importance being made at the beginning of the production year, it is essential to identify exactly your machine. To help the owner in this, the frame numbers are given below of each version of each model. Note that while UK models are identified by a suffix letter, US models are identified by the year of production. The approximate dates of importing are also given; note that these will not necessarily coincide with the date of sale.

UK models

XR80		
XR80-Z	HE01-5006322 to 5026984	May '79 to Feb '80
XL100 S		
XL100S-Z	HD01-5000056 to 5009069	Introduced Nov '78
XL125 S		
XL125S-Z	L125S-5004816 to 5038165	Aug '78 to Feb '79
XL125S-A	L125S-5100022 to 5133740	Introduced Feb '79
XL185S		
XL185S-Z	L185S-5000043 to 5019726	Aug '78 to Dec '79
XL185S-A	L185S-5100002 to 5103955	Introduced Jan '80
XR200		
XR200-A	ME02-5100012 to 5107049	Introduced Jan '80

US models

XL80 S		
1980	HD04-5000014 to 5013496	Oct '79 to Sept '80
XR80		
1979	HE01-5000012 to 5027474	Sept '78 to Sept '79
1980	HE01-5100003 to 5122777	Oct '79 to Sept '80
XL100 S		
1979	HD01-5002640 to 5012036	Jan to Sept '79
1980	HD01-5100009 to 5103211	Oct '79 to Sept '80
XL125 S		
1979	L125S-5008316 to 5027623	Sept '78 to Sept '79
1980	JD02-5000003 to 5006405	Oct '79 to Sept '80
XL185 S		
1979	L185S-5001103 to 5017934	Sept '78 to Sept '79
1980	MD02-5000003 to 5009422	Oct '79 to Sept '80
XR185		
1979	ME02-5000061 on	Jan to Sept '79
XR200		
1980	ME02-5100012 to 5107049	Oct '79 to Sept '80

Dimensions and weights

	XR80	XL80 S	XL100 S
Overall length	1700 mm (67.0 in)	1730 mm (68.1 in)	2050 mm (80.7 in)
Overall width	750 mm (29.5 in)	755 mm (29.7 in)	840 mm (33.1 in)
Overall height	1010 mm (39.8 in)	970 mm (38.2 in)	1080 mm (42.5 in)
Wheelbase	1140 mm (45.0 in)	1135 mm (44.7 in)	1310 mm (51.6 in)
Ground clearance (min)	195 mm (7.7 in)	195 mm (7.7 in)	280 mm (11.0 in)
Seat height	725 mm (28.5 in)	720 mm (28.3 in)	780 mm (30.7 in)
Dry weight	66.5 kg (146.3 lb)	72 kg (159 lb)	94.7 kg (208.9 lb)

	XL125 S	XL185 S	XR185	XR200
Overall length	2080 mm (81.9 in)	2080 mm (81.9 in)	2060 mm (81.1 in)	2060 mm (81.1 in)
Overall width	840 mm (33.1 in)	850 mm (33.5 in)	840 mm (33.1 in)	820 mm (32.3 in)
Overall height	1130 mm (44.5 in)	1130 mm (44.5 in)	1145 mm (45.1 in)	1145 mm (45.1 in)
Wheelbase	1315 mm (51.8 in)	1310 mm (51.6 in)	1325 mm (52.2 in)	1325 mm (52.2 in)
Ground clearance (min)	265 mm (10.6 in)	265 mm (10.6 in)	285 mm (11.2 in)	285 mm (11.2 in)
Seat height	820 mm (32.3 in)	820 mm (32.3 in)	860 mm (33.9 in)	860 mm (33.9 in)
Dry weight	104 kg (229 lb)	105.5 kg (233 lb)	98 kg (214.1 lb)	98 kg (214.1 lb)

Ordering spare parts

When ordering spare parts for the Honda X L and XR range of trail bikes covered in this manual, it is advisable to deal direct with an official Honda agent, who will be able to supply many of the items required ex-stock. It is advisable to get acquainted with the local Honda agent, and to rely on his advice when purchasing spares. He is in a better position to specify exactly the parts required and to identify the relevant spare part numbers so that there is less chance of the wrong part being supplied by the manufacturer due to a vague or incomplete description.

When ordering spares, always quote the frame and engine numbers in full, together with any prefixes or suffixes in the form of letters.

The frame number is found stamped on the left-hand side of the steering head, parallel with the forks. The engine number is stamped on the left-hand side of the crankcase, just forward of the gearchange lever shaft.

Use only parts of genuine Honda manufacture. A few pattern parts are available, sometimes at cheaper prices, but there is no guarantee that they will give such good service as the originals they replace. Retain any worn or broken parts until the replacements have been obtained; they are sometimes needed as a pattern to help identify the correct replacement when design changes have been made during a production run.

Some of the more expendable parts such as sparking plugs, bulbs, tyres, oils and greases etc., can be obtained from accessory shops and motor factors, who have convenient opening hours, and can often be found not far from home. It is also possible to obtain parts on a Mail Order basis from a number of specialists who advertise regularly in the motorcycle magazines.

Location of frame number

Location of engine number

Safety first!

Professional motor mechanics are trained in safe working procedures. However enthusiastic you may be about getting on with the job in hand, do take the time to ensure that your safety is not put at risk. A moment's lack of attention can result in an accident, as can failure to observe certain elementary precautions.

There will always be new ways of having accidents, and the following points do not pretend to be a comprehensive list of all dangers; they are intended rather to make you aware of the risks and to encourage a safety-conscious approach to all work you carry out on your vehicle.

Essential DOs and DON'Ts

DON'T start the engine without first ascertaining that the transmission is in neutral.

DON'T suddenly remove the filler cap from a hot cooling system – cover it with a cloth and release the pressure gradually first, or you may get scalded by escaping coolant.

DON'T attempt to drain oil until you are sure it has cooled sufficiently to avoid scalding you.

DON'T grasp any part of the engine, exhaust or silencer without first ascertaining that it is sufficiently cool to avoid burning you.

DON'T allow brake fluid or antifreeze to contact the machine's paintwork or plastic components.

DON'T syphon toxic liquids such as fuel, brake fluid or antifreeze by mouth, or allow them to remain on your skin.

DON'T inhale dust – it may be injurious to health (see *Asbestos* heading).

DON'T allow any spilt oil or grease to remain on the floor – wipe it up straight away, before someone slips on it.

DON'T use ill-fitting spanners or other tools which may slip and cause injury.

DON'T attempt to lift a heavy component which may be beyond your capability – get assistance.

DON'T rush to finish a job, or take unverified short cuts.

DON'T allow children or animals in or around an unattended vehicle.

DON'T inflate a tyre to a pressure above the recommended maximum. Apart from overstressing the carcase and wheel rim, in extreme cases the tyre may blow off forcibly.

DO ensure that the machine is supported securely at all times. This is especially important when the machine is blocked up to aid wheel or fork removal.

DO take care when attempting to slacken a stubborn nut or bolt. It is generally better to pull on a spanner, rather than push, so that if slippage occurs you fall away from the machine rather than on to it.

DO wear eye protection when using power tools such as drill, sander, bench grinder etc.

DO use a barrier cream on your hands prior to undertaking dirty jobs – it will protect your skin from infection as well as making the dirt easier to remove afterwards; but make sure your hands aren't left slippery. Note that long-term contact with used engine oil can be a health hazard.

DO keep loose clothing (cuffs, tie etc) and long hair well out of the way of moving mechanical parts.

DO remove rings, wristwatch etc, before working on the vehicle – especially the electrical system.

DO keep your work area tidy – it is only too easy to fall over articles left lying around.

DO exercise caution when compressing springs for removal or installation. Ensure that the tension is applied and released in a controlled manner, using suitable tools which preclude the possibility of the spring escaping violently.

DO ensure that any lifting tackle used has a safe working load rating adequate for the job.

DO get someone to check periodically that all is well, when working alone on the vehicle.

DO carry out work in a logical sequence and check that everything is correctly assembled and tightened afterwards.

DO remember that your vehicle's safety affects that of yourself and others. If in doubt on any point, get specialist advice.

IF, in spite of following these precautions, you are unfortunate enough to injure yourself, seek medical attention as soon as possible.

Asbestos

Certain friction, insulating, sealing, and other products – such as brake linings, clutch linings, gaskets, etc – contain asbestos. *Extreme care must be taken to avoid inhalation of dust from such products since it is hazardous to health.* If in doubt, assume that they *do* contain asbestos.

Fire

Remember at all times that petrol (gasoline) is highly flammable. Never smoke, or have any kind of naked flame around, when working on the vehicle. But the risk does not end there – a spark caused by an electrical short-circuit, by two metal surfaces contacting each other, by careless use of tools, or even by static electricity built up in your body under certain conditions, can ignite petrol vapour, which in a confined space is highly explosive.

Always disconnect the battery earth (ground) terminal before working on any part of the fuel or electrical system, and never risk spilling fuel on to a hot engine or exhaust.

It is recommended that a fire extinguisher of a type suitable for fuel and electrical fires is kept handy in the garage or workplace at all times. Never try to extinguish a fuel or electrical fire with water.

Note: *Any reference to a 'torch' appearing in this manual should always be taken to mean a hand-held battery-operated electric lamp or flashlight. It does **not** mean a welding/gas torch or blowlamp.*

Fumes

Certain fumes are highly toxic and can quickly cause unconsciousness and even death if inhaled to any extent. Petrol (gasoline) vapour comes into this category, as do the vapours from certain solvents such as trichloroethylene. Any draining or pouring of such volatile fluids should be done in a well ventilated area.

When using cleaning fluids and solvents, read the instructions carefully. Never use materials from unmarked containers – they may give off poisonous vapours.

Never run the engine of a motor vehicle in an enclosed space such as a garage. Exhaust fumes contain carbon monoxide which is extremely poisonous; if you need to run the engine, always do so in the open air or at least have the rear of the vehicle outside the workplace.

The battery

Never cause a spark, or allow a naked light, near the vehicle's battery. It will normally be giving off a certain amount of hydrogen gas, which is highly explosive.

Always disconnect the battery earth (ground) terminal before working on the fuel or electrical systems.

If possible, loosen the filler plugs or cover when charging the battery from an external source. Do not charge at an excessive rate or the battery may burst.

Take care when topping up and when carrying the battery. The acid electrolyte, even when diluted, is very corrosive and should not be allowed to contact the eyes or skin.

If you ever need to prepare electrolyte yourself, always add the acid slowly to the water, and never the other way round. Protect against splashes by wearing rubber gloves and goggles.

Mains electricity and electrical equipment

When using an electric power tool, inspection light etc, always ensure that the appliance is correctly connected to its plug and that, where necessary, it is properly earthed (grounded). Do not use such appliances in damp conditions and, again, beware of creating a spark or applying excessive heat in the vicinity of fuel or fuel vapour. Also ensure that the appliances meet the relevant national safety standards.

Ignition HT voltage

A severe electric shock can result from touching certain parts of the ignition system, such as the HT leads, when the engine is running or being cranked, particularly if components are damp or the insulation is defective. Where an electronic ignition system is fitted, the HT voltage is much higher and could prove fatal.

Routine maintenance

For information relating to 1981 — 1987 models, see Chapter 7

Periodic routine maintenance is a continuous process that commences immediately the machine is used. It must be carried out at specified mileage recordings or on a calendar date basis if the machine is not used regularly, whichever falls sooner. Maintenance should be regarded as an insurance policy, to help keep the machine in peak condition and to ensure long, trouble-free service. It has the additional benefit of giving early warning of any faults that may develop and will act as a regular safety check, to the obvious advantage of both rider and machine alike.

The various maintenance tasks are listed under their respective mileage and calendar headings, the model types being quoted where necessary. If further general information is required, it can be found within the manual under the pertinent Section heading in the relevant Chapter.

It should be remembered that the interval between the various maintenance tasks serves only as a guide. As the machine gets older or is used under particularly adverse conditions, it would be advisable to reduce the period between each check.

In order that the routine maintenance tasks are carried out with as much ease as possible, it is essential that a good selection of general workshop tools is available.

Included in the kit must be a range of metric ring or combination spanners, a selection of crosshead screwdrivers and at least one pair of circlip pliers.

Additionally, owing to the extreme tightness of most casing screws on Japanese machines, an impact screwdriver, together with a choice of large and small crosshead screw bits, is absolutely indispensable. This is particularly so if the engine has not been dismantled since leaving the factory.

Pre-ride inspection

All models
Complete items 3, 33, 35 and 36.
XL models only
Complete item 39.

Weekly or every 150 miles (250 km)

All models
Complete items 1, 2, 3, 11, 14 and 34.

Fortnightly or every 300 miles (500 km)

All models
Complete item 35.

Monthly or every 500 miles (800 km)

XL models only
Complete item 37.
XR models only.
Complete item 17.

Five weekly or every 600 miles (1000 km)

XL125 S models only
Complete items 4 and 5.

Two monthly or every 1250 miles (2000 km)

XL80 S and XL100 S models
Complete item 4.

Three monthly or every 1000 miles (1600 km)

XR models only
Complete items 4, 5, 6, 7, 8, 9, 10, 12, 13, 14, 15, 16, 18, 19, 20, 21, 23, 24, 25, 26, 27, 28, 30, 31, 32, 33, 36 and 38.

Three monthly or every 2000 miles (3200 km)

XL185 S models only
Complete items 4 and 5.

Four monthly or every 2500 miles (4000 km)

XL80 S and XL100 S models
Complete items 7, 8, 9, 11, 12, 13, 14, 15, 16, 17, 18, 19, 20, 21, 24, 25, 26, 27, 28, 30, 31, 32, 33, 38 and 39.
XL125 S models only
Complete all the items listed for the XL100 S models, less items 19 and 20.

Six monthly or every 4000 miles (6400 km)

XL185 S models only
Complete items 7, 8, 9, 10, 11, 12, 13, 14, 15, 16, 17, 18, 21, 24, 25, 26, 27, 28, 30, 31, 32, 33, 38 and 39.

Eight monthly or every 5000 miles (8000 km)

XL80 S and XL100 S models
Complete item 5.
XL125 S models only
Complete items 6 and 23.

Yearly or every 4000 miles (6400 km)

XR models only
Complete items 22 and 29.

Yearly or every 7500 miles (12000 km)

XL80 S, XL100 S and XL185 S models
Complete items 22, 23 and 29.
XL125 S models
Complete items 22 and 29.

Routine maintenance tasks

1 Safety check

Give the machine a close visual inspection, checking for loose nuts and fittings, frayed control cables etc. Check the tyres for damage, especially splitting on the sidewalls. Remove any stones or other objects caught between the treads. This is particularly important on the front tyre, where rapid deflation due to penetration of the inner tube will almost certainly cause total loss of control.

2 Legal check

Check that the tyres are not excessively worn or damaged and that the lights and speedometer function correctly. On models fitted with a horn and directional indicators, check that these components also function correctly. Note that if bulbs have to be renewed, then they must have the same rating as the original. Remember that if different wattage bulbs are used in the traffic indicators, the flashing rate will be altered.

3 Checking the engine/gearbox oil level

Unscrew the filler plug which is situated to the rear of the right-hand crankcase half; it will be noted that the plug incorporates a dipstick which should be wiped off using a clean, lint-free rag. Place the plug back in position, but do not screw it home; allow it to rest in position on the edge of the orifice. Remove the plug and note the level of the oil on the dipstick, which should be between the two level marks. Note that if the machine has been ridden recently, it should be allowed to stand for a few minutes to allow the oil clinging to the internal surfaces to drain down into the sump. The machine should be positioned upright on level ground before carrying out this check.

If required, replenish the oil reservoir with SAE 10W/40 engine oil. Do not run the engine with the oil level lower than the minimum level line and do not overfill the reservoir.

4 Changing the engine/gearbox oil

Before draining the oil from the engine/gearbox unit, start and run the engine until it is warm; this will help the oil to drain thoroughly. Obtain a container of at least 1.5 litre (2.64/3.17 Imp/US pint) capacity and place it beneath the engine unit to catch the old oil.

On XR80, XL80 S and XL100 S models remove the drain plug from the crankcase and allow the oil to drain thoroughly. Remove the filler plug to aid this. On all other models the oil must be drained through the oil filter screen chamber on the left-hand side of the crankcase, because no separate drain plug is fitted. To do this, remove the large hexagon-headed cap which retains the filter screen.

Refit the drain plug (where fitted) and filter assembly (where fitted), ensuring that the sealing washers are in good condition, and tighten them to the torque settings given on page 32. Note that on some models cleaning of the oil filter screen or centrifugal oil filter chamber should be carried out prior to refilling the engine with oil. See the maintenance schedule and the following two Sections.

Refill the crankcases to the recommended level on the dipstick; the amount of oil required is stated under the appropriate model heading in the Specifications under Quick Glance Maintenance. Refit the dipstick and start and run the engine for a few minutes, checking that there are no oil leaks at the disturbed components. Remove the dipstick and check that the oil level is at the upper level mark with the machine positioned upright on level ground. Top-up the oil, if necessary, and refit the dipstick.

Replenishing the engine/gearbox oil reservoir

Engine/gearbox oil drain plug

5 Cleaning the engine/gearbox oil filter screen

This procedure must be carried out with the engine/gearbox unit drained of oil.

XR80, XL80 S and XL100 S models

Access to the filter fltted to these machines is gained by first removing the kickstart lever and then disconnecting the brake lever so that it may be pivoted clear of the crankcase cover. Disconnect the clutch cable from the crankcase lever end by placing an open-ended spanner on the lever end and turning the lever towards the cable so that the cable nipple may be released from the lever and the threaded adjuster withdrawn from the crankcase bracket. Remove the right-hand crankcase cover by unscrewing the securing screws. This should be done evenly and in a diagonal sequence to avoid distortion to the cover.

With the crankcase cover removed, the oil filter, which takes the form of a flat gauze screen with a rubber surround, may be withdrawn from its housing in the base of the crankcase.

XL125 S, XL185 S, XR185 and XR200 models

The filter fitted to these machines is located behind a large hexagon-headed cap situated in the base of the left-hand crankcase. To gain proper access to this cap, remove the sumpguard from the machine by removing the two rear and one forward retaining bolts. Place a container beneath the cap in order to catch the small amount of oil contained within the filter housing. Remove the cap, followed by the spring and cylindrical filter screen. Clean the cap and inspect the O-ring for signs of damage or deterioration; if necessary, renew the O-ring. Finally, turn the engine over several times by operating the kickstart lever, so that any oil left in the filter housing is expelled.

All models

To clean the oil filter screen, wash the component in clean petrol or paraffin, using a small nylon-bristled brush to remove any traces of contamination. Place the filter in a well ventilated space to allow it to dry. Refitting the filter is a reversal of the removal procedure.

On the XR80, XL80 S and XL100 S models, ensure that the right-hand crankcase cover gasket is in good condition before the cover is refitted. Tighten the cover securing screws evenly and in a diagonal sequence to avoid distortion of the cover. Ensure that the clutch cable and rear brake are both correctly adjusted and that the kickstart lever is correctly positioned on its shaft.

Note that the forward sumpguard retaining bolt on the XL125 S, XL185 S, XR185 and XR200 models is locked in position by means of a tab washer. Ensure that the tabs of this washer are correctly drifted against the flats of the bolt head once the bolt has been fitted and tightened.

6 Cleaning the engine/gearbox centrifugal oil filter - XL125 S, XL185 S, XR185 and XR200 models

This procedure must be carried out with the engine/gearbox unit drained of oil.

Remove the kickstart lever and disconnect the brake lever so that it may be pivoted clear of the crankcase cover. Disconnect the clutch cable from the crankcase lever end, as described in the previous Section, and disconnect the decompressor cable from the rear of the clutch cover. Remove the sumpguard by unscrewing the one forward and two rear retaining bolts.

Access may now be gained to the filter by unscrewing the right-hand crankcase cover securing screws, evenly and in a diagonal sequence to avoid distortion to the cover, and removing the cover. Remove the three filter cover retaining screws and remove the cover. Carefully clean out the deposits which will have accumulated around the inner edge of the unit, noting that these may have become quite compacted and may need

scraping off. Wash each part out using a petrol soaked rag, and then dry them off.

Note the condition of the cover gasket and renew it, if considered necessary, before refitting the cover.

Refit the crankcase cover, after first checking the condition of the gasket and ensuring that the locating dowels are correctly positioned. Tighten the cover securing screws evenly and in a diagonal sequence. Refit the sumpguard and reconnect and correctly adjust the decompressor cable (except XL125 S), clutch cable and rear brake lever. Finally, refit the kickstart lever in the correct position on its shaft.

7 Checking and adjusting the valve clearances

It is important that the correct valve clearance is maintained to ensure the proper operation of the valve assemblies. A small amount of free play is designed into each valve train to allow for expansion of the various engine components. If the setting deviates greatly from that specified, a marked drop in performance will be evident. In the case of the clearance becoming too great, it will be found that valve operation will be noisy, and performance will drop off as a result of the valves not opening fully. If on the other hand, the clearance is too small, the valves may not close completely. This will not only causes loss of compression, but will also cause the exhaust valve to burn out very quickly. In extreme cases, the valve head may strike the piston crown, causing extensive damage to the engine. The clearances should be checked and adjusted with the engine cold, by the following method.

On the XR80, XL80 S and XL100 S models, access to the valve adjusters is obtained by removing the cylinder head cover; this is secured to the cylinder head by two special retaining bolts. It is also necessary on these models to expose the timing marks on the flywheel generator rotor by removing the left-hand crankcase cover.

On the XL125 S, XL185 S, XR185 and XR200 models, it is first necessary to remove the seat and fuel tank before gaining access to the valve adjusters by removing the two circular threaded covers from the cylinder head. Remove the two inspection caps from the left-hand crankcase cover to expose the flywheel generator rotor timing marks and the central rotor retaining nut.

To check and adjust the valve clearances, rotate the crankshaft anti-clockwise until the 'T' mark on the generator rotor aligns exactly with the index mark on the casing.

The engine should now be on top dead centre (TDC) on the compression stroke with both valves closed. To ensure this is so, inspect the valve mechanisms; if they are rocking when the 'T' mark is lined up, the engine is in the wrong position. Turn it over again until the 'T' mark lines up and the valves do not move when the rotor is moved slightly.

Using a feeler gauge, check the clearance between each rocker arm adjuster and the end of the valve stem with which it makes contact. The inlet and exhaust valves should both have a clearance identical to that given in the Specifications under Quick Glance Maintenance.

If the clearances are not correct, adjustment is made by slackening the locknut on the end of the rocker arm concerned and by turning the adjuster until the correct setting is obtained. Hold the adjuster still and tighten the locknut. Then recheck the setting before passing to the next valve.

On completion of the above procedure, refit all disturbed components noting the condition of any sealing rings and renewing them if thought necessary. Full details of fuel tank and seat removal and refitting are given in Chapters 2 and 4 respectively.

Note that on the XL185 S, XR185 and XR200 models, it is necessary to check the starter decompressor free play before the engine is started. Full details of this procedure are given under the appropriate Section heading in this Chapter.

Engine/gearbox oil filter screen (80 and 100)

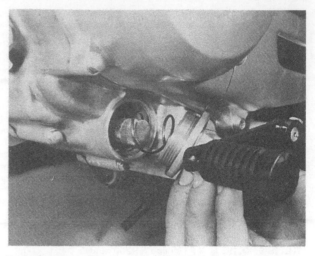
Engine/gearbox oil filter screen (125, 185 and 200)

Removing the engine/gearbox centrifugal oil filter cover

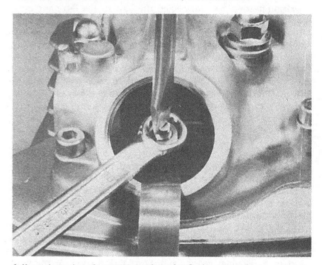
Adjust the valve clearance so that the feeler gauge is a sliding fit

8 Adjusting the cam chain tension

XR80, XL80 S and XL100 S models

Start by making fine adjustment of the cam chain tension at the adjuster situated on the left-hand side of the cylinder head. With the engine running, slacken the set plate bolt and turn the slotted adjuster anticlockwise to slacken the chain or clockwise to tension it, as shown in the accompanying figure (note the mid position shown by the punch mark).

If this method fails to take up sufficient chain slack, return the slotted adjuster to the mid position and secure the set plate bolt. Stop the engine and slacken the locknut of the main cam chain adjuster situated on the rear of the barrel, and slacken the central adjuster screw; adjustment will now be made automatically. Tighten the adjuster screw and secure with the locknut. Restart the engine and if necessary repeat the fine adjustment procedure.

XL125 S, XL185 S, XR185 and XR200 models

To adjust the cam chain tension on these models, start the engine and allow it to run at a fast tick-over. Prise the rubber boot from the tensioner pushrod housing and slacken the housing by applying a spanner to the large hexagon. The tension will adjust automatically. Retighten the housing. This will cause the two angled bushes to lock the tensioner pushrod in position, so maintaining the pre-set tension until adjustment is required again. Refit the rubber boot to the housing.

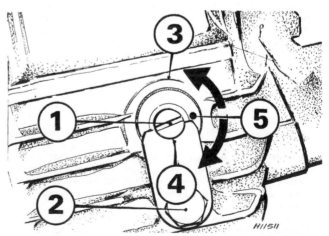
Cam chain fine adjustment tensioner — XR80 and XL100 S

1	Adjusting screw	4	Position where chain is tight
2	Set plate	5	Punch mark
3	Position where chain is slack		

Adjusting the cam chain tension (125, 185 and 200)

9 Adjusting the clutch

Clutch adjustment is correct if there is 10 - 20 mm (0.4 - 0.8 in) of free play at the handlebar lever end. The procedure for adjustment is as follows.

XR80, XL80 S and XL100 S models

Minor adjustments to the clutch cable free play on these models may be made by means of the adjuster and locknut at the handlebar lever housing. When making major adjustments, however, both the handlebar mounted adjuster and cable adjuster must be turned in all the way. With this done, loosen the adjusting screw locknut, situated at the base of the clutch cover projection just forward of the kickstart spindle. Turn the adjusting screw anti-clockwise until resistance is felt. Turn the screw clockwise 1/8 to ¼ of a turn. Hold the screw in position with the screwdriver and tighten the locknut. The cable adjuster may now be turned out until there is 25 mm (1.0 in) of free play at the handlebar lever end and the locknut then tightened. The handlebar mounted adjuster can be used for fine running adjustments. Ensure that the locknut is tightened.

XL125 S, XL185 S, XR185 and XR200 models

Major adjustments to clutch cable free play on these models is made by means of the adjuster passing through the bracket located at the base of the cylinder barrel. Adjustment is achieved simply by loosening the locknut, turning the adjuster the required amount and then retightening the locknut. The same principle should be applied to the handlebar mounted adjuster when fine adjustment is required.

10 Checking and adjusting the starter decompressor - XL185 S, XR185 and XR200 models

Check that the engine decompressor mechanism is adjusted correctly by carrying out the procedure listed in the following paragraph. It should be noted that excessive free play at the cylinder head lever will cause the starting procedure to be made more difficult, whereas insufficient free play may cause erratic idling and burning of the valves.

Adjustment of the decompressor mechanism should be carried out only after the exhaust valve clearance has been correctly adjusted. Align the T mark on the generator rotor with the index mark on the crankcase cover upper inspection hole, ensuring that the piston is on the compression stroke. Measure the amount of free play at the end of the cylinder head lever. This should be 1 - 2 mm (0.04 - 0.08 in). If necessary, adjust the amount of free play by loosening the locknut securing the cable adjuster and rotating the adjuster the necessary number

of turns to give the correct amount of free play. Tighten the locknut on completion of adjustment.

11 Draining the crankcase breather - USA models only

To drain the crankcase breather assembly, locate the end of the crankcase breather pipe, situated just behind the right-hand footrest, and remove the drain plug after having first released its retaining clip. This will allow any deposits retained within the pipe to drain away. Refit the plug and retain it in position with the clip once all deposits have drained from the pipe.

If the machine is ridden with the throttle continually in the fully-open position, or ridden through very humid conditions, the breather assembly must be drained more frequently. Note that a section of the drain tube is transparent; directly deposits are seen to appear in this section of tube, it is necessary to carry out the draining procedure.

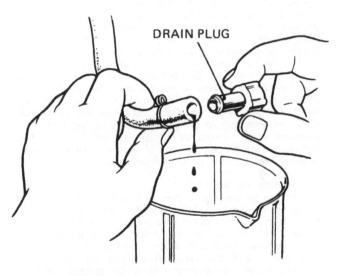

DRAIN PLUG

Draining the crankcase breather — USA models only

12 Adjusting the carburettor

The procedure which follows may be disregarded if the engine tick-over is correct and no roughness is evident at idle speeds.

Start the engine and run it until normal operating temperature is reached, turn the pilot jet screw inwards until the engine misfires or decreases in speed. Note the position of the screw, then turn it outwards until similar symptoms are observed. The screw should be set exactly between these two positions. This position is normally close to the number of turns, from the fully closed position, quoted in the Specifications Section of Chapter 2 for the particular model type. On machines fitted with pilot screw limiter caps the adjustment can only be made between the fixed limits.

If the idling speed is now too low or too high, adjust the idling speed (throttle stop) screw until engine will tick over at the recommended idling speed.

Note that these adjustments should always be made with the engine at normal operating temperature and with the air cleaner connected, otherwise a false setting will be obtained.

13 Checking and adjusting the carburettor choke

Check that the handlebar mounted choke knob can be moved up and down smoothly. If there are any signs of seizure in its movement then the operating cable should be inspected for damage; check also to see whether the cable is becoming trapped between any moving components whilst turning the handlebars from lock to lock.

With the choke knob pulled all the way out, check that the carburettor lever is in the fully up position. Push the knob fully down and check that there is 1 - 2 mm (1/16 - 1/8 in) of free play in the cable at the carburettor lever end. The amount of free play may be adjusted by loosening the carburettor-mounted cable clamp and moving the cable housing as required whilst holding the carburettor lever fully down. Retighten the clamp and recheck the choke for correct operation.

As well as moving smoothly throughout its operating range, the choke knob should stay where it is positioned. There is an adjuster located beneath the rubber cap under the choke knob which, when turned, will enable the correct amount of friction to be imposed on the choke knob shaft.

14 Checking the throttle operation

Check that the throttle twistgrip turns smoothly over the full operating range whilst opening the throttle and that it returns to the closed position automatically. This check should be carried out with the handlebars placed at full right-lock, full left-lock and in the central position.

Inspect the throttle cable for signs of damage, kinking or deterioration and check that there is no danger of it becoming trapped between moving components. Ensure that the throttle twistgrip is adjusted correctly by using the following procedure.

The throttle cable should be adjusted to give 2 - 6 mm (0.125 - 0.25 in) of free play at the twistgrip. Major adjustments may be made by means of the adjuster located beneath the rubber cap in the centre of the carburettor mixing chamber top. Note that to reach this adjuster, on all but the 80 cc models, it is first necessary to remove the seat and fuel tank. Fine adjustments may be made by loosening the locknut and turning the adjuster at the twistgrip end of the cable. On completion of adjustment, tighten both adjuster locknuts.

15 Cleaning the fuel filter

The fuel filter is located over the fuel tap stack. Removal of the tap from the fuel tank is therefore necessary before the filter may be removed from the tap for inspection and cleaning.

Before the fuel tap can be removed, it is first necessary to drain the tank. This is easily accomplished by removing the feed pipe from the carburettor float chamber and allowing the contents of the tank to drain into a clean receptacle, with the tap turned to the 'reserve' position. Alternatively, the tank can be removed and placed on one side, so that the fuel level is below the tap outlet.

Care must be taken not to damage the tank paintwork whilst doing this.

The tap unit is retained to the threaded boss on the tank by means of a gland nut. Undo this nut, remove the clip and fuel pipe from the fuel tap stub and withdraw the filter stack from the tank. The filter may now be removed and inspected for contamination and damage. It should be noted that this filter provides the only means of preventing particles of dirt or grit from reaching the carburettor and should therefore be in perfect condition; renew the filter if it is damaged in any way.

To clean the filter, wash it thoroughly in clean petrol, using a soft nylon-bristled brush to remove carefully any stubborn traces of contamination remaining. Place the filter in a well ventilated area to allow it to dry and inspect the seal located at the base of the stack for signs of deterioration or damage; renew it if considered necessary. Refitting of the filter and tap is a reversal of the removal procedure.

16 Inspecting the fuel feed pipe

The fuel feed pipe is made from thin walled synthetic rubber. This material will, after a long period of time, become hard and split, thereby causing a leakage of fuel. It is therefore necessary to inspect the pipe for any signs of deterioration and renew it if considered necessary.

It should be noted that the seal between the pipe ends and the fuel tap and carburettor unions is effected by an interference fit, with the pipe being held in position by a retaining clip. Any hardening of the pipe material will cause this seal to become less efficient, thus increasing the danger of a fuel leak at this point.

17 Cleaning the air filter element

XR80 and XL80 S models

To remove the air cleaner element fitted to these models, expose the three air cleaner case cover retaining nuts or screws by unclipping the right-hand side panel from its three frame locating grommets. Access may now be gained to the element by removing the three nuts and plain washers, releasing the cover to carburettor mouth retaining clip and pulling the cover away from the air cleaner case. The element is attached to the back of the cover.

XL100 S, XL125 S, XL185 S, XR185 and XR200 models

The air cleaner element fitted to these models is located behind the left-hand sidepanel and is fitted around a cylindrical holder which is held in position by a single central wingnut. The sidepanel may be removed by unscrewing the three cross-head retaining screws, thus allowing the panel to drop away from the air cleaner case. Care should be taken to ensure that the seal located in the sidepanel to case joining surface is not damaged during removal. It should be correctly positioned around the case mouth to avoid loss or damage during the following element cleaning procedure. The element and holder may now be removed by unscrewing the wingnut.

Where a primary air filter is fitted, this must be removed and cleaned as described in Chapter 2, Section 9.

All models

Remove the air cleaner element from its holder and wash it thoroughly in a non-flammable or high flash point solvent. Never use petrol or a low flash point solvent as it may well result in a fire or explosion once the engine is started. The element may then be squeezed gently (never wring) to remove most of the solvent and then allowed to dry thoroughly by standing it in a well ventilated area. Never wring out the element, because the foam will be damaged and lead to a requirement for premature replacement.

Once the element is properly dried, allow it to soak in SAE 80 - 90 gear oil. Gently squeeze out any excess oil and slide the element back over its housing. Refit the assembly in the air cleaner case using a reversal of the procedure used for removal, noting the following points. Check that the seal between the cover or sidepanel and air cleaner case is correctly located and undamaged. Where the element holder is marked with an arrow and the word TOP, ensure that the arrow points upward before tightening the retaining nut. Ensure that, on the XR80 and XL80 S machines, the carburettor to air cleaner cover connection is correctly made. Any air leaks allowed at this connection will affect the performance of the engine.

If the foam element becomes torn or perforated, it should be replaced without question. Never run the engine without the element connected to the carburettor because the carburettor is specially jetted to compensate for the addition of this component.

Refitting the air filter element (except 80 models)

18 Cleaning the exhaust spark arrester

Cleaning of the exhaust spark arrester and silencer baffles varies between the different types of machines and should be carried out as follows.

XR80 and XL80 S models

The spark arrester fitted to the exhaust system on these machines is detachable and may be removed for cleaning by unscrewing the three retaining screws at the silencer end. Clean any carbon deposits off the spark arrester with a wire brush and clean out the exhaust system by first removing the drain plug located in the exhaust pipe just under the right-hand sidepanel. The engine should now be started and revved approximately twenty times to clear the exhaust of any carbon build up.

XL100 S, XL125 S, XL185 S, XR185 and XR200 models

On these machines the spark arrester is a permanently fixed item. The system is cleaned of any carbon build up by removing the plate located underneath the rear of the silencer. An impact driver may be required to free the two crosshead retaining screws. If this tool is used, the silencer must be well supported to avoid any strain being placed upon the exhaust system mounting points.

With the plate and gasket removed, start the engine and allow it to idle. Place a thick pad of rag over the end of the silencer, thereby creating back pressure in the system, and rev the engine approximately twenty times to blow any carbon out of the silencer plate hole. Before refitting the plate, check the condition of the screws and gasket and renew if thought necessary.

All models

When carrying out the cleaning procedures it is strongly advised that the following safety precautions be observed. The machine should be run only in a well ventilated area. The area should be free of any combustible materials, due to there being an increased fire hazard once the spark arrester is removed. Always wear some form of eye protection to protect against dislodged particles of carbon. Wait until the exhaust system is cool to the touch before removing or refitting any components; this system becomes very hot during engine operation.

19 Checking and adjusting the contact breaker points - XR80, XL80 S and XL100 S models

To gain access to the contact breaker assembly, it is first necessary to remove the left-hand crankcase cover by unscrewing the five retaining screws. The contact breaker points may then be viewed through one of the slots in the flywheel generator rotor.

Remove the sparking plug and rotate the generator rotor slowly until the points can be viewed in their fully-open position through one of the rotor slots. Examine the contact faces. If they are dirty, burnt or pitted, they should be removed for renewal as described in Section 4 of Chapter 3. If the points are in sound condition, proceed with adjustment as described.

Contact breaker adjustment is carried out by loosening the single contact breaker locking screw slightly and moving the contact points nearer or further apart as the case may be, by levering with a screwdriver in the plate indentation provided.

Turn in the appropriate direction until the gap is within the range 0.3 to 0.4 mm (0.012 to 0.016 in) and then retighten the locking screw.

It is imperative that the points are open FULLY whilst this adjustment is made, or a false reading will result. Recheck the setting when the screw has been tightened; it is not unknown for the setting to move slightly during the retightening operation.

If adjustment has proved necessary, it is likely that the ignition timing will have been affected, and this should be checked. See the following Section for the relevant procedure.

Refit the crankcase cover by fitting and tightening the five retaining screws evenly and in a diagonal sequence. Ensure

the cover gasket is undamaged and correctly located before doing so. Refit the sparking plug.

Checking the contact breaker points gap (80 and 100)

20 Checking and re-setting the ignition timing

XR80, XL80 S and XL100 S models

In order to check the accuracy of the ignition timing on these models, it is first necessary to remove the flywheel generator rotor cover by unscrewing the five retaining screws.

It will be observed that the rotor is inscribed with two lines marked 'T' and 'F' and that there is a small pointer cast in the crankcase mating surface just below the centre line of the cylinder barrel.

In addition, there are two further parallel lines inscribed on the rotor. These indicate the point of full advance when they align with the fixed pointer.

If the ignition timing is correct, the 'F' line on the rotor will coincide exactly with the fixed pointer when the contact breaker points are just beginning to separate. The moment at which the contact points separate may be determined as follows.

Obtain an ordinary torch battery and bulb and three lengths of wire. Connect one end of a wire to the positive (+) terminal of the battery and one end of another wire to the negative (−) terminal of the battery. The negative lead may now be earthed to a point on the engine casing. Ensure the earth point on the casing is clean and that the wire is positively connected; a crocodile clip fastened to the wire is ideal. Take the free end of the positive lead and connect it to the bulb. The third length of wire may now be connected between the bulb and the black output lead from the generator. As the final connection is made, the bulb should light with the points closed.

Rotate the generator rotor anti-clockwise until the 'F' line on the rotor aligns with the fixed pointer. If the timing is correct the light should dim as both marks align thus indicating that the points have begun to separate.

If the ignition timing proves to be incorrect, adjust the contact breaker gap as described in the previous Section so that the light dims at the correct point.

If, after adjusting the contact breaker points to achieve the correct ignition timing, the gap is not within the recommended limits of 0.3 - 0.4 mm (0.012 - 0.016 in) the points should be renewed as a complete set.

It cannot be overstressed that optimum performance depends on the accuracy with which the ignition timing is set. Even a small error can cause a marked reduction in performance and the possibility of engine damage as the result of over-heating.

An alternative method of checking the ignition timing can be adopted, whilst the engine is running, using a stroboscopic lamp. When the light from the lamp is aimed at the timing marks it has the effect of 'freezing' the moving marks on the rotor in one position, and thus the accuracy of the timing can be seen. An additional advantage is also provided in that the correct and smooth functioning of the automatic advance-retard unit can be determined.

The stroboscopic lamp should be connected up as recommended by the lamp's manufacturer and the engine started. Up to about 1800 rpm the 'F' mark should be in line with the fixed index mark. As the speed is increased, the two parallel advance marks should be seen to move closer to the fixed index point. Note the rpm reading at which the initial movement of the marks occurs and compare it with that reading given in the Specifications at the beginning of Chapter 3.

Note also the rpm reading at which point full advance is reached (ie the parallel advance marks on the rotor are directly in line with the fixed index point) and compare it with that figure given in the Specifications.

Alterations to the ignition timing should be made in the normal way as described in the first paragraphs of this Section.

If it is found that the static ignition timing is correct but that full advance cannot be reached, or that progression to full advance is erratic, there is some indication that the mechanical automatic timing unit is not functioning correctly. Refer to Section 11 of Chapter 3 for details relating to examination of this component.

XL125 S, XL185 S, XR185 and XR200 models

These machines utilise an ignition system where the normal contact breaker unit is dispensed with; the ignition firing point being controlled entirely by electrical means. Because no mechanical components are used, and therefore wear is eliminated, the ignition timing should remain accurate almost indefinitely during normal usage.

Static timing of the system is covered in Section 49 of Chapter 1. If any components have been disturbed in the pulser generator unit then the static timing procedure must be carried out before proceeding further.

With the upper inspection cap removed from the flywheel generator cover, connect the strobe lamp into the system as recommended by the manufacturer of the lamp. Start the engine and aim the lamp at the index pointer and the rotor periphery.

At the engine idle speed of 1300 rpm, the 'F' mark on the generator rotor should be aligned with the index mark on the rotor cover. As the engine speed is allowed to increase, the 'F' mark will be seen to move away from the index mark. Note the rpm reading at which this initial movement occurs and compare it with that reading given in the Specifications at the beginning of Chapter 3 for the particular type of machine. Note also the rpm reading at which point full advance is reached (ie the parallel advance marks on the rotor are directly in line with the fixed index point) and compare it with that figure given in the Specifications. Any alterations to the ignition timing should be made by following the previously mentioned static timing procedure.

If it is found that the static ignition timing is correct but that full advance cannot be reached, or that progression to full advance is erratic, there is some indication that the mechanical automatic timing unit is not functioning correctly. Refer to Section 11 of Chapter 3 for details relating to examination of this component.

21 Cleaning and re-setting the sparking plug

Detach the sparking plug cap, and using the correct spanner, remove the sparking plug. Clean the electrodes using a wire brush followed by a strip of fine emery cloth or paper. Check the plug gap with a feeler gauge, adjusting it if necessary to within the range 0.6 - 0.7 mm (0.024 - 0.028 in). Make adjustments by bending the outer electrode, never the inner (central) electrode.

Before fitting the sparking plug, smear the threads with a graphited grease; this will aid subsequent removal.

22 Changing the front fork oil

To change the front fork oil, first remove the hexagon-headed plug from the top of each fork stanchion. Inspect the O-ring on each plug for signs of damage or deterioration and renew if necessary. Place a container beneath each fork leg and remove the drain plugs, located at the base of each leg to the rear of the spindle, to allow the oil to drain. Pump the fork legs up and down several times to expel any remaining fluid. Ensure that each drain plug sealing washer is in good condition and refit and tighten the plugs.

Each fork leg may now be refilled with the type and quantity of damping fluid listed in the Specifications under Quick Glance Maintenance. Refit the hexagon-headed plugs, complete with O-ring, to the top of each fork stanchion and tighten them to a torque of 1.5 - 3.0 kgf m (11 - 22 lbf ft) (except 80 cc models).

23 Checking and adjusting the steering head bearings

Check the steering head bearings for play by applying the front brake hard and seeing whether there is any movement as the machine is rocked backwards and forwards. If adjustment is necessary, slacken the top yoke to steering head retaining nut and make the necessary adjustment by tightening or loosening the slotted adjuster ring fitted to the steering stem immediately below the upper fork yoke. A C-spanner should be used to turn the ring.

Retighten the yoke retaining nut to lock the adjuster ring in position and re-check the steering head adjustment. Note that this nut should be torque loaded to 6.0 - 9.0 kgf m (43 - 65 lbf ft) and if a lockwasher is fitted beneath the nut, the tabs of the washer must be bent up against the flats of the nut so as to lock it in position.

Take care not to overtighten the adjuster ring, because this will place undue stress on the bearings and cause the machine to display a tendency to roll at low speeds. Do not confuse movement at the steering head with worn fork sliders.

24 Checking the swinging arm pivot bushes

Any wear in the swinging arm pivot bushes may be detected by placing blocks under the sumpguard or footrest assembly so as to raise the rear wheel clear of the ground, and pulling and pushing on the fork ends in a horizontal direction with one hand whilst holding the frame firmly in the other. Ensure the machine is blocked securely to prevent it toppling during this test. Any play in the bushes will be greatly magnified by the leverage effect of the fork legs. No play at all should be detected during this test. If play is apparent then the swinging arm assembly should be removed from the machine and the bushes renewed. Full details of bush removal, examination and refitting may be found in Section 12 of Chapter 4.

25 Lubricating the control cables

Before carrying out the following procedure, check to see if the control cables fitted to the machine are of the nylon lined type. If they are found to be nylon lined then they should not be lubricated because the oil may cause the nylon to swell, thereby causing total cable seizure.

Lubricate the control cables thoroughly with motor oil or an all-purpose oil. A good method of lubricating the cables is shown in the accompanying illustration, using a plasticine funnel. This method has the disadvantage that the cables usually need removing from the machine. An hydraulic cable oiler which pressurises the lubricant overcomes this problem.

It is advisable, in order to prevent any drying up of the cables in the intervals between carrying out the above listed procedure, to apply regularly a few drops of motor oil to the exposed lengths of inner cable of each control cable.

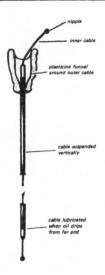

Lubricating a control cable

Examine the prop stand rubber for wear

26 Examining and lubricating the instrument drive cables

It is advisable to detach the speedometer and tachometer drive cables from time to time in order to check whether they are adequately lubricated and whether the outer cables are compressed or damaged at any point along their run. A jerky or sluggish movement at the instrument head can often be attributed to a cable fault.

To grease the cable, uncouple both ends and withdraw the inner cable. (On some model types this may not be possible, in which case a badly seized cable will have to be renewed as a complete assembly).

After removing any old grease, clean the inner cable with a petrol soaked rag and examine the cable for broken strands or other damage. Do not check the cable for broken strands by passing it through the fingers and palm of the hand, this may well cause a painful injury if a broken strand snags the skin. It is best to wrap a piece of rag around the cable and pull the cable through it, any broken strands will snag on the rag.

Re-grease the cable with high melting point grease, taking care not to grease the last six inches closest to the instrument head. If this precaution is not observed, grease will work into the instrument and immobilise the sensitive movement.

If the cable breaks, it is usually possible to renew the inner cable alone, provided the outer cable is not damaged or compressed at any point along its run. Before inserting the new inner cable, it should be greased in accordance with the instructions given in the preceding paragraph. Try to avoid tight bends in the run of the cable because this will accelerate wear and make the instrument movement sluggish.

27 Examining the prop stand

Check that the prop stand pivot bolt is secure and well lubricated and that the spring is in good condition and not over-stretched. If the stand does not retract properly or extends whilst the machine is in motion, a serious accident is more or less inevitable.

On those XL models fitted with a rubber pad at the base of the stand, Honda recommend that the pad be renewed once it is worn to the wear line marked on the rearmost face of the pad. The new pad should be marked 'BELOW 259 LB ONLY', no other type of pad is suitable. Removal of the pad is achieved simply by unscrewing the retaining nut and bolt and withdrawing the pad from its housing. Refitting the pad is a reversal of this process, ensuring the arrow on the pad points down-

wards and faces outwards away from the machine. The pad should also be renewed if the rubber has deteriorated in any way.

28 Examining the wheel rims and spokes

Examination of the wheel rim and spokes is identical for both the front and rear wheels. Position blocks underneath the sumpguard plate so that the wheel to be inspected is raised clear of the ground. Spin the wheel and check the rim alignment; this should be no more than 2.0 mm (0.1 in) out of true. Small irregularities in alignment can be corrected by tightening the spokes in the affected area although a certain amount of experience is necessary to prevent over-correction. Any flats in the wheel rim will be evident at the same time. These are more difficult to remove and in most cases it will be necessary to have the wheel rebuilt on a new rim. Apart from the effect on stability, a flat will expose the tyre bead and walls to greater risk of damage if the machine is run with a deformed wheel.

Check for loose and broken spokes. Tapping the spokes is the best guide to tension. A loose spoke will produce a quite different sound and should be tightened by turning the nipple in an anti-clockwise direction. Always check for run out by spinning the wheel again. If the spokes have to be tightened by an excessive amount, it is advisable to remove the tyre and tube as detailed in Section 16 of Chapter 5. This will enable the protruding ends of the spokes to be ground off, thus preventing them from chafing the inner tube and causing punctures.

29 Inspecting and lubricating the wheel bearings

With both wheels removed from the machine, the wheel bearings should be removed from their wheel housings and inspected in accordance with the information given in Sections 6 and 11 of Chapter 5. Full details of wheel and bearing removal and refitting are also given in Chapter 5.

30 Adjusting the front brake

If adjustment of the front brake is correct, there should be 10 - 20 mm (0.39 - 0.75 in) on XL80S models, 20 - 30 mm (0.75 - 1.23 in) on all other models, of free play at the end of the handlebar lever before the brake begins to operate. Major adjustments should be made by turning the cable lower adjuster, whereas any minor adjustments may then be made by turning the adjuster at the handlebar lever bracket. Lock both adjusters in position by tightening the locknuts and check the brake for correct operation by spinning the wheel and applying the brake lever. There should be no indication of the brake binding as the wheel is spun. If the brake shoes are heard to be brushing against the surface of the wheel drum, back off the cable adjuster slightly until all indication of binding disappears.

31 Adjusting the rear brake

If the adjustment of the rear brake is correct, the rear brake pedal will have 20 - 30 mm (0.75 - 1.25 in) of free play at the footplate end of the lever before the brake begins to operate.

The length of travel is controlled by the adjuster at the end of the brake operating rod, close to the brake operating arm. If the nut is turned clockwise, the amount of travel is reduced and vice-versa. Always check that the brake is not binding after adjustments have been made by raising the rear wheel clear of the ground and rotating it whilst listening for the brake shoes brushing against the surface of the wheel drum. If necessary, back off the adjuster slightly until all indication of binding disappears.

Note that it may be necessary to re-adjust the height of the stop lamp switch if the pedal height has been changed to any marked extent. The body of the switch is threaded, so that it can be raised or lowered, after the locknuts have been slackened. If the stop lamp lights too soon, the switch should be lowered and vice-versa.

32 Checking brake shoe wear

Apply the front and rear brakes in turn and check to see if the arrow on the indicator plate located beneath the brake backplate lever aligns with the corresponding triangle mark on the backplate casting. If this is so, the brake shoes must be renewed because they will have worn beyond their acceptable service limit. Full details on dismantling, examining and re-assembling the brake assemblies are given in Sections 4 and 10 of Chapter 5.

33 Inspecting the brake operating components

Inspect both front and rear brake operating levers for free and unobstructed range of movement. If either lever shows signs of seizure it must be lubricated at its pivot point. Both levers have a specified degree of free play allowed before their respective brakes begin to operate, this should be set by following the instructions given in the Sections on brake adjustment.

The front brake operating cable must be inspected for signs of damage, deterioration, seizure or kinking; check also that it is not in danger of becoming trapped between moving components. It must be remembered that the front brake provides the main method of stopping the machine and any damage to the cable, or lack of lubrication, will cause the efficiency of the brake to deteriorate rapidly. Full details of control cable lubrication may be found under the relevant heading in this Chapter.

Check all connections throughout the braking system for incorrect assembly and looseness. If necessary, retighten any connections and ensure that they are correctly locked in position. Check also the large return spring at the lower end of the front brake cable for damage and renew it if thought necessary. The same applies to the rear brake operating rod spring. Finally, inspect all protective gaiters for deterioration or splitting of the rubber and renew if necessary.

34 Checking the tyre pressures

Check the tyre pressures with a gauge that is known to be accurate. It is worthwhile purchasing a pocket gauge for this purpose, as the gauges on garage forecourts cannot always be relied upon as being accurate. The readings should not be taken after the machine has been used, as the tyres will have become warm. This will have caused the pressure to increase, giving a false reading.

The recommended tyre pressures are:

	XR80	XL80 S and XL100 S	XL125 S and XL185 S	XR185 and XR200
Front	18 psi (1.2 kg/ sq cm)	21 psi (1.5 kg/ sq cm)	21 psi (1.5 kg/ sq cm)	14 psi (1.0 kg/ sq cm)
Rear	20 psi (1.4 kg/ sq cm)	21 psi (1.5 kg/ sq cm)	21 psi (1.5 kg/ sq cm)	14 psi (1.0 kg/ sq cm)

35 Final drive chain adjustment and lubrication

XR80 and XL80 S models

With the rear wheel raised clear of the ground and the gear lever placed in the neutral position, depress the chain at a mid-point between the sprockets on its lower run and measure the amount of slack. If the chain tension is correct, this slack should measure 20 mm (0.75 in).

XL100 S, XL125 S and XL185 S models

Position the machine on its side stand and select neutral. Select a point on the chain midway between the sprockets on its lower run and move the chain up and down at this point. The amount of slack present in the chain should measure 25 - 35 mm (1.0 - 1.4 in) for XL100 S models or 30 - 40 mm (1.25 - 1.60 in) for XL125 S and XL185 S models.

XR185 models

Place the machine on its side stand and select neutral. Before inspecting the chain tension, check the degree of wear on the chain tensioner slider block. If wear extends to the wear line on the slider then the block should be renewed.

Note the chain tension gauge marks on the chain guard plate just behind and to the rear of the chain tensioner. If the upper surface of the chain aligns with the upper chain mark on the gauge, then it should be adjusted down to align with the bottom chain mark.

If the gauge marks have become obliterated during the life of the machine, then the following check should be made. Measure the distance between the bottom of the swinging arm fork and the top of the chain at a point on the chain run just forward of the tensioner slider block. If this distance is less than the service limit of 21 mm (0.9 in), then it should be corrected to 31 mm (1.25 in) by adjusting the chain tension.

XR200 models

The method of examining chain tension on this machine is similar to that given for the XR185 model, except that the chain tension gauge on the chain guard plate takes the form of two rivets. The chain should be adjusted so that the upper surface of the chain aligns with the upper edge of the lower rivet. The chain must be adjusted before its upper surface reaches the upper edge of the upper rivet. There are no alternative measurements given because the rivets are a permanent fixture.

All models

The method of drive chain adjustment is the same for all the model types covered in this Manual and is as follows. Remember that a chain rarely wears evenly and the tightest point in the chain should be found by rotating the rear wheel, before adjustment is made.

Remove the split-pin from the wheel spindle nut and loosen the nut just enough to allow the wheel to be drawn backwards by means of the two drawbolt adjusters.

Always adjust the drawbolts an equal amount in order to preserve wheel alignment. The fork ends are clearly marked with a series of horizontal lines above or below the adjusters, to provide a simple, visual check. If desired, wheel alignment can be checked by running a plank of wood parallel to the machine, so that it touches the side of the rear tyre. If wheel alignment is correct, the plank will be equidistant from each side of the front wheel tyre, when tested on both sides of the rear wheel. It will not touch the front wheel tyre because this tyre is of smaller cross section.

On completion of chain adjustment, tighten the wheel spindle nut to the torque figure given in the Specifications Section of Chapter 5 and fit a new split-pin. Check that the rear wheel rotates freely and check the rear brake pedal free play.

Do not run the chain overtight to compensate for uneven wear. A tight chain will place excessive stresses on the gearbox and rear wheel bearings, leading to their early failure. It will also absorb a surprising amount of power.

After checking the chain tension, ensure that the chain is well lubricated because lack of oil will accelerate the rate of wear of both chain and sprockets. If the chain is found to be dry, the application of engine oil will act as a temporary expedient. A recommended alternative to this is the aerosol type chain lubricant; this type of lubricant is less likely to be flung off the moving chain. More details of chain lubrication and inspection may be found in Section 14 of Chapter 5.

36 Examining the final drive chain tensioner - XR185 and XR200 models

Check that the tensioner slider block has not worn to the line inscribed on its rearmost face. If the block has worn beyond this limit, it should be renewed by removing the two retaining screws attaching it to the tensioner arm.

Ensure the tensioner arm pivots freely about its attachment point on the swinging arm fork and is held against the chain by the return spring. Inspect this spring for signs of fatigue or failure and renew it if necessary. If the tensioner arm shows signs of seizing on its pivot, then the arm should be detached from the swinging arm fork bracket by removing the single flange bolt. Inspect the bolt, collar and oil seal and renew any item that is worn or damaged. Clean and re-lubricate the pivot assembly, and check that the tensioner arm now pivots freely and is returned by the spring to its position against the chain.

It should be noted that it is essential that the chain tensioner operates correctly at all times. Should it fail or seize on its pivot, then the chain tension will be lost, resulting in the chain being allowed to run off the rear wheel sprocket, possibly with disastrous consequences.

37 Checking the battery electrolyte level - except XR models

On the XL100 S, XL125 S and XL185 S models, the battery electrolyte level may be checked through the cut-out in the battery cover situated to the rear of the right-hand sidepanel. The electrolyte level must be maintained between the UPPER and LOWER level marks on the battery. If this level is low, unclip the sidepanel and battery cover and unscrew the battery filler caps. Carefully add distilled water, using a plastic funnel or small syringe, until the electrolyte level rises to the UPPER mark. Refit the battery filler caps, screwing them in firmly, refit the battery cover and the sidepanel.

Details of how to deal with acid spillage from the battery, or damage to the battery casing causing a leakage of electrolyte, may be found in Section 6 of Chapter 6.

38 Checking and adjusting the headlamp beam alignment

XL80 S, XL100 S, XL125 S and XL185 S models

Beam height on these models is effected by pivoting the headlamp about the two mounting bolts. These bolts should be loosened slightly before any adjustment is carried out and retightened on completion. Horizontal beam adjustment is provided by the adjuster screw which passes through the headlamp rim at the 9 o'clock position.

XR185 and XR200 models

Beam height on the combined headlamp/number plate unit fitted to these models is effected by screwing in or out the adjuster screw located directly beneath the headlamp lens. This serves to pivot the headlamp unit about its hinged mounting bracket. No provision for horizontal beam adjustment is made.

All models

It should be noted that UK lighting regulations stipulate that the lighting system must be arranged so that the light will not dazzle a person standing in the same horizontal plane as the vehicle at a distance greater than 25 feet from the lamp, whose eye level is not less than 3 feet 6 inches above that plane. It is easy to approximate this setting by placing the machine 25 feet away from a wall, on a level road, and setting the dipped beam height so that it is concentrated at the same height as the distance from the centre of the headlamp to the ground. The rider must be seated normally during this operation and also the pillion passenger, if one is carried regularly.

For the US, the headlamp beam should be adjusted as specified by the State laws and regulations.

39 Checking and adjusting the rear brake stop lamp switch

Check to ensure that when the rear brake pedal is depressed to the point where the brake begins to engage, the stop lamp illuminates. If this is not the case, the stop lamp switch must be adjusted and, if necessary, tested in accordance with the instructions given in Section 18 of Chapter 6.

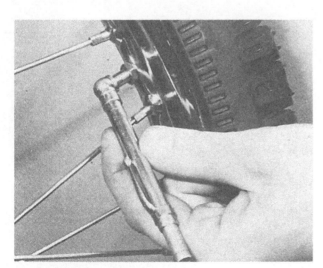

Checking the tyre pressures

Maintain the battery electrolyte level between the UPPER and LOWER marks

Quick glance maintenance adjustments and capacities

Engine oil capacity

   XR80 and XL80 S .... 0.9 litre (1.58/1.90 Imp/US pint)

   XL100 S:

      At oil change .... 0.9 litre (1.58/1.90 Imp/US pint)

      At filter cleaning .... 1.0 litre (1.76/2.11 Imp/US pint)

   All others:

      At oil change .... 0.9 litre (1.58/1.90 Imp/US pint)

      At filter cleaning .... 1.1 litre (1.94/2.32 Imp/US pint)

Front fork capacity (per leg)

   XR80 .... 118 cc (4.14/3.98 Imp/US fl oz)

   XL80 S .... 112 cc (4.00/3.85 Imp/US fl oz)

   XL100 S .... 140 cc (4.93/4.73 Imp/US fl oz)

   XL125 S and XL185 S .... 155 cc (5.46/5.24 Imp/US fl oz)

   XR185 and XR200 .... 170 cc (5.98/5.75 Imp/US fl oz)

Valve clearances (cold)

	Inlet	Exhaust
XR185 and XR200	0.05 mm (0.002 in)	0.08 mm (0.003 in)
All others	0.05 mm (0.002 in)	0.05 mm (0.002 in)

Sparking plug gap .... 0.6 - 0.7 mm (0.024 - 0.028 in)

Contact breaker gap

   XR80, XL80 S and XL100 S only .... 0.3 - 0.4 mm (0.012 - 0.016 in)

Tyre pressures

	Front	Rear
XR80	18 psi (1.2 kg/cm^2)	20 psi (1.4 kg/cm^2)
XL80 S, XL100 S, XL125 S and XL185 S	21 psi (1.5 kg/cm^2)	21 psi (1.5 kg/cm^2)
XR185 and XR200	14 psi (1.0 kg/cm^2)	14 psi (1.0 kg/cm^2)

Recommended lubricants

Component	Lubricant
Engine/gearbox unit	SAE 10W/40
Front forks	Automatic transmission fluid (ATF)
Wheel bearings	High melting point grease
Swinging arm	High melting point grease
Steering head ball races	High melting point grease
Final drive chain	Aerosol chain lubricant or special chain grease such as Linklyfe
Prop stand pivot	Multi-purpose grease
Brake pedal pivot	Multi-purpose grease
Speedometer drive gear	High melting point grease
Brake camshaft	High melting point grease
Throttle twistgrip	Multi-purpose grease

Tools and working facilities

The first priority when undertaking maintenance or repair work of any sort on a motorcycle is to have a clean, dry, well-lit working area. Work carried out in peace and quiet in the well-ordered atmosphere of a good workshop will give more satisfaction and much better results than can usually be achieved in poor working conditions. A good workshop must have a clean flat workbench or a solidly constructed table of convenient working height. The workbench or table should be equipped with a vice which has a jaw opening of at least 4 in (100 mm). A set of jaw covers should be made from soft metal such as aluminium alloy or copper, or from wood. These covers will minimise the marking or damaging of soft or delicate components which may be clamped in the vice. Some clean, dry, storage space will be required for tools, lubricants and dismantled components. It will be necessary during a major overhaul to lay out engine/gearbox components for examination and to keep them where they will remain undisturbed for as long as is necessary. To this end it is recommended that a supply of metal or plastic containers of suitable size is collected. A supply of clean, lint-free, rags for cleaning purposes and some newspapers, other rags, or paper towels for mopping up spillages should also be kept. If working on a hard concrete floor note that both the floor and one's knees can be protected from oil spillages and wear by cutting open a large cardboard box and spreading it flat on the floor under the machine or workbench. This also helps to provide some warmth in winter and to prevent the loss of nuts, washers, and other tiny components which have a tendency to disappear when dropped on anything other than a perfectly clean, flat, surface.

Unfortunately, such working conditions are not always available to the home mechanic. When working in poor conditions it is essential to take extra time and care to ensure that the components being worked on are kept scrupulously clean and to ensure that no components or tools are lost or damaged.

A selection of good tools is a fundamental requirement for anyone contemplating the maintenance and repair of a motor vehicle. For the owner who does not possess any, their purchase will prove a considerable expense, offsetting some of the savings made by doing-it-yourself. However, provided that the tools purchased meet the relevant national safety standards and are of good quality, they will last for many years and prove an extremely worthwhile investment.

To help the average owner to decide which tools are needed to carry out the various tasks detailed in this manual, we have compiled three lists of tools under the following headings: *Maintenance and minor repair*, *Repair and overhaul*, and *Specialized*. The newcomer to practical mechanics should start off with the simpler jobs around the vehicle. Then, as his confidence and experience grow, he can undertake more difficult tasks, buying extra tools as and when they are needed. In this way, a *Maintenance and minor repair* tool kit can be built-up into a *Repair and overhaul* tool kit over a considerable period of time without any major cash outlays. The experienced home mechanic will have a tool kit good enough for most repair and overhaul procedures and will add tools from the specialized category when he feels the expense is justified by the amount of use these tools will be put to.

It is obviously not possible to cover the subject of tools fully here. For those who wish to learn more about tools and their use there is a book entitled *Motorcycle Workshop Practice Manual* (Bk no 1454) available from the publishers of this manual.

As a general rule, it is better to buy the more expensive, good quality tools. Given reasonable use, such tools will last for a very long time, whereas the cheaper, poor quality, item will wear out faster and need to be renewed more often, thus nullifying the original saving. There is also the risk of a poor quality tool breaking while in use, causing personal injury or expensive damage to the component being worked on.

For practically all tools, a tool factor is the best source since he will have a very comprehensive range compared with the average garage or accessory shop. Having said that, accessory shops often offer excellent quality tools at discount prices, so it pays to shop around. There are plenty of tools around at reasonable prices, but always aim to purchase items which meet the relevant national safety standards. If in doubt, seek the advice of the shop proprietor or manager before making a purchase.

The basis of any toolkit is a set of spanners. While open-ended spanners with their slim jaws, are useful for working on awkwardly-positioned nuts, ring spanners have advantages in that they grip the nut far more positively. There is less risk of the spanner slipping off the nut and damaging it, for this reason alone ring spanners are to be preferred. Ideally, the home mechanic should acquire a set of each, but if expense rules this out a set of combination spanners (open-ended at one end and with a ring of the same size at the other) will provide a good compromise. Another item which is so useful it should be considered an essential requirement for any home mechanic is a set of socket spanners. These are available in a variety of drive sizes. It is recommended that the $\frac{1}{2}$-inch drive type is purchased to begin with as although bulkier and more expensive than the $\frac{3}{8}$-inch type, the larger size is far more common and will accept a greater variety of torque wrenches, extension pieces and socket sizes. The socket set should comprise sockets of sizes between 8 and 24 mm, a reversible ratchet drive, an extension bar of about 10 inches in length, a spark plug socket with a rubber insert, and a universal joint. Other attachments can be added to the set at a later date.

Maintenance and minor repair tool kit

Set of spanners 8 – 24 mm
Set of sockets and attachments
Spark plug spanner with rubber insert – 10, 12, or 14 mm as appropriate
Adjustable spanner
C-spanner/pin spanner
Torque wrench (same size drive as sockets)
Set of screwdrivers (flat blade)
Set of screwdrivers (cross-head)
Set of Allen keys 4 – 10 mm
Impact screwdriver and bits
Ball pein hammer – 2 lb
Hacksaw (junior)
Self-locking pliers – Mole grips or vice grips
Pliers – combination
Pliers – needle nose
Wire brush (small)
Soft-bristled brush
Tyre pump
Tyre pressure gauge
Tyre tread depth gauge
Oil can
Fine emery cloth
Funnel (medium size)
Drip tray
Grease gun

Set of feeler gauges
Brake bleeding kit
Strobe timing light
Continuity tester (dry battery and bulb)
Soldering iron and solder
Wire stripper or craft knife
PVC insulating tape
Assortment of split pins, nuts, bolts, and washers

Repair and overhaul toolkit

The tools in this list are virtually essential for anyone undertaking major repairs to a motorcycle and are additional to the tools listed above. Concerning Torx driver bits, Torx screws are encountered on some of the more modern machines where their use is restricted to fastening certain components inside the engine/gearbox unit. It is therefore recommended that if Torx bits cannot be borrowed from a local dealer, they are purchased individually as the need arises. They are not in regular use in the motor trade and will therefore only be available in specialist tool shops.

Plastic or rubber soft-faced mallet
Torx driver bits
Pliers — electrician's side cutters
Circlip pliers — internal (straight or right-angled tips are available)
Circlip pliers — external
Cold chisel
Centre punch
Pin punch
Scriber
Scraper (made from soft metal such as aluminium or copper)
Soft metal drift
Steel rule/straight edge
Assortment of files
Electric drill and bits
Wire brush (large)
Soft wire brush (similar to those used for cleaning suède shoes)
Sheet of plate glass
Hacksaw (large)
Valve grinding tool
Valve grinding compound (coarse and fine)
Stud extractor set (E-Z out)

Specialized tools

This is not a list of the tools made by the machine's manufacturer to carry out a specific task on a limited range of models. Occasional references are made to such tools in the text of this manual and, in general, an alternative method of carrying out the task without the manufacturer's tool is given where possible. The tools mentioned in this list are those which are not used regularly and are expensive to buy in view of their infrequent use. Where this is the case it may be possible to hire or borrow the tools against a deposit from a local dealer or tool

hire shop. An alternative is for a group of friends or a motorcycle club to join in the purchase.

Valve spring compressor
Piston ring compressor
Universal bearing puller
Cylinder bore honing attachment (for electric drill)
Micrometer set
Vernier calipers
Dial gauge set
Cylinder compression gauge
Vacuum gauge set
Multimeter
Dwell meter/tachometer

Care and maintenance of tools

Whatever the quality of the tools purchased, they will last much longer if cared for. This means in practice ensuring that a tool is used for its intended purpose; for example screwdrivers should not be used as a substitute for a centre punch, or as chisels. Always remove dirt or grease and any metal particles but remember that a light film of oil will prevent rusting if the tools are infrequently used. The common tools can be kept together in a large box or tray but the more delicate, and more expensive, items should be stored separately where they cannot be damaged. When a tool is damaged or worn out, be sure to renew it immediately. It is false economy to continue to use a worn spanner or screwdriver which may slip and cause expensive damage to the component being worked on.

Fastening systems

Fasteners, basically, are nuts, bolts and screws used to hold two or more parts together. There are a few things to keep in mind when working with fasteners. Almost all of them use a locking device of some type; either a lock washer, lock nut, locking tab or thread adhesive. All threaded fasteners should be clean, straight, have undamaged threads and undamaged corners on the hexagon head where the spanner fits. Develop the habit of replacing all damaged nuts and bolts with new ones.

Rusted nuts and bolts should be treated with a rust penetrating fluid to ease removal and prevent breakage. After applying the rust penetrant, let it 'work' for a few minutes before trying to loosen the nut or bolt. Badly rusted fasteners may have to be chiseled off or removed with a special nut breaker, available at tool shops.

Flat washers and lock washers, when removed from an assembly should always be replaced exactly as removed. Replace any damaged washers with new ones. Always use a flat washer between a lock washer and any soft metal surface (such as aluminium), thin sheet metal or plastic. Special lock nuts can only be used once or twice before they lose their locking ability and must be renewed.

If a bolt or stud breaks off in an assembly, it can be drilled out and removed with a special tool called an E-Z out. Most dealer service departments and motorcycle repair shops can perform this task, as well as others (such as the repair of threaded holes that have been stripped out).

Chapter 1 Engine, clutch and gearbox

For information relating to 1981 — 1987 models, see Chapter 7

Contents

Specifications

Engine

	XR80	XR185	XR200
Type	Air cooled, single cylinder, overhead camshaft, 4-stroke		
Bore	47.5 mm (1.870 in)	63.0 mm (2.48 in)	65.5 mm (2.57 in)
Stroke	45.0 mm (1.771 in)	57.8 mm (2.28 in)	57.8 mm (2.28 in)
Capacity	79.7 cc (4.85 cu in)	180.2 cc (11.01 cu in)	195.0 cc (11.80 cu in)
Compression ratio	9.7 : 1	10.0 : 1	10.0 : 1
Maximum horsepower	—	18.0 BHP @ 9000 rpm	19.0 BHP @ 9000 rpm
Maximum torque	—	1.58 kg m (11.49 lb ft) @ 7500 rpm	1.65 kg m @ 7000 rpm 11.49 lb ft @ 7500 rpm

Cylinder barrel

	XR80	XR185	XR200
Standard bore	47.50 - 47.51 mm (1.8700 - 1.8704 in)	63.00 - 63.01 mm (2.4803 - 2.4807 in)	65.50 - 65.51 mm (2.5787 - 2.5791 in)
Service limit	47.6 mm (1.874 in)	63.10 mm (2.484 in)	65.6 mm (2.58 in)
Taper limit	0.1 mm (0.004 in)	0.1 mm (0.004 in)	0.1 mm (0.004 in)
Ovality limit	0.1 mm (0.004 in)	0.1 mm (0.004 in)	0.1 mm (0.004 in)
Barrel to head face distortion limit	0.1 mm (0.004 in)	0.1 mm (0.004 in)	0.1 mm (0.004 in)
Cylinder bore to piston clearance	0.010 - 0.045 mm (0.0004 - 0.0018 in)	0.015 - 0.045 mm (0.0006 - 0.0018 in)	0.015 - 0.045 mm (0.0006 - 0.0018 in)
Service limit	0.2 mm (0.008 in)	0.1 mm (0.004 in)	0.1 mm (0.004 in)

Piston

	XR80	XR185	XR200
Outside diameter	47.465 - 47.490 mm (1.8686 - 1.8697 in)	62.955 - 62.985 mm (2.4785 - 2.4797 in)	65.465 - 65.485 mm (2.577 - 2.578 in)
Service limit	47.40 mm (1.866 in)	62.90 mm (2.476 in)	65.4 mm (2.575 in)
Gudgeon pin OD	12.994 - 13.0 mm (0.5116 - 0.5118 in)	14.994 - 15.0 mm (0.5903 - 0.5906 in)	14.994 - 15.0 mm (0.5903 - 0.5906 in)
Service limit	12.98 mm (0.511 in)	14.96 mm (0.589 in)	14.96 mm (0.589 in)
Gudgeon pin hole bore	13.002 - 13.008 mm (0.5119 - 0.5121 in)	15.002 - 15.008 mm (0.5906 - 0.5909 in)	15.002 - 15.008 mm (0.5906 - 0.5909 in)
Service limit	13.06 mm (0.514 in)	15.04 mm (0.592 in)	15.04 mm (0.592 in)
Gudgeon pin to piston clearance	0.002 - 0.014 mm (0.0001 - 0.0006 in)	0.002 - 0.014 mm (0.0001 - 0.0006 in)	0.002 - 0.014 mm (0.0001 - 0.0006 in)
Service limit	0.03 mm (0.001 in)	0.02 mm (0.001 in)	0.02 mm (0.001 in)

Piston rings

	XR80	XR185 and 200
Ring to groove clearance:		
Top	0.015 - 0.050 mm (0.0006 - 0.0020 in)	0.015 - 0.050 mm (0.0006 - 0.0020 in)
Second	0.015 - 0.045 mm (0.0006 - 0.0018 in)	0.015 - 0.045 mm (0.0006 - 0.0018 in)
Service limit:		
Top	0.15 mm (0.006 in)	0.09 mm (0.004 in)
Second	0.15 mm (0.006 in)	0.09 mm (0.004 in)
End gap:		
Top and second	0.15 - 0.35 mm (0.006 - 0.014 in)	0.20 - 0.40 mm (0.008 - 0.016 in)
Oil scraper	0.30 - 0.90 mm (0.01 - 0.04 in)	0.20 - 0.90 mm (0.010 - 0.04 in)

Service limit:

Top and second ...	...	...	...	...	...	...	0.5 mm (0.02 in)	0.5 mm (0.02 in)
Oil scraper ...	...	...	...	...	...	...	1.1 mm (0.04 in)	—

Valves

Valve stem OD:									
Inlet ...	...	...	...	...	...	...	...	5.450 - 5.465 mm (0.2146 - 0.2152 in)	5.450 - 5.465 mm (0.2146 - 0.2152 in)
Exhaust ...	...	...	...	...	...	...		5.430 - 5.445 mm (0.2138 - 0.2144 in)	5.430 - 5.445 mm (0.2138 - 0.2144 in)
Service limit:									
Inlet ...	...	...	...	...	...	...	...	5.42 mm (0.2134 in)	5.42 mm (0.2134 in)
Exhaust ...	...	...	...	...	...	...		5.40 mm (0.213 in)	5.40 mm (0.213 in)
Valve guide ID:									
Inlet ...	...	...	...	...	...	...	...	—	5.475 - 5.485 mm (0.2156 - 0.2159 in)
Exhaust ...	...	...	...	...	...	...		—	5.475 - 5.485 mm (0.2156 - 0.2159 in)
Service limit:									
Inlet ...	...	...	...	...	...	...	...	—	5.50 mm (0.217 in)
Exhaust ...	...	...	...	...	...	...		—	5.50 mm (0.217 in)
Stem to guide clearance:									
Inlet ...	...	...	...	...	...	...	...	0.010 - 0.035 mm (0.0004 - 0.0014 in)	0.010 - 0.035 mm (0.0004 - 0.0014 in)
Exhaust ...	...	...	...	...	...	...		0.030 - 0.055 mm (0.0012 - 0.0022 in)	0.030 - 0.055 mm (0.0012 - 0.0022 in)
Service limit:									
Inlet ...	...	...	...	...	...	...	...	0.08 mm (0.0031 in)	0.12 mm (0.005 in)
Exhaust ...	...	...	...	...	...	...		0.10 mm (0.004 in)	0.14 mm (0.006 in)
Valve face width (inlet and exhaust)	...	...	...	...	...		—	1.7 mm (0.07 in)
Service limit ...	...	...	...	...	...	...		—	2.0 mm (0.08 in)
Valve seat width (inlet and exhaust)	...	...	...	...			1.0 mm (0.0394 in)	1.2 mm (0.05 in)
Service limit ...	...	...	...	...	...	...		1.5 mm (0.06 in)	1.5 mm (0.06 in)

Valve springs

Free length:									
Inner ...	...	...	...	...	...	...	...	28.05 mm (1.1043 in)	39.4 mm (1.55 in)
Outer ...	...	...	...	...	...	...	...	33.8 mm (1.3307 in)	45.5 mm (1.79 in)
Service limit:									
Inner ...	...	...	...	...	...	...	...	27.0 mm (1.063 in)	35.5 mm (1.40 in)
Outer ...	...	...	...	...	...	...	...	32.7 mm (1.2874 in)	41.0 mm (1.61 in)

Valve timing at 1 mm lift

Inlet valve opens ...	...	...	...	...	...	...	...	8° BTDC	15° BTDC
Exhaust valve opens ...	...	...	...	...	...	...		40° BBDC	45° BBDC
Inlet valve closes ...	...	...	...	...	...	...		40° ABDC	45° ABDC
Exhaust valve closes ...	...	...	...	...	...	...		8° ATDC	15° ATDC

Valve clearance (cold)

Inlet ...	...	...	...	...	...	...	...	0.05 mm (0.002 in)	0.05 mm (0.002 in)
Exhaust ...	...	...	...	...	...	...	...	0.05 mm (0.002 in)	0.08 mm (0.003 in)

Camshaft

Cam height:									
Inlet ...	...	...	...	...	...	...	...	27.677 - 27.717 mm (1.0896 - 1.0974 in)	31.675 - 31.875 mm (1.2470 - 1.2549 in)
Exhaust ...	...	...	...	...	...	...		27.540 - 27.586 mm (1.0833 - 1.0861 in)	31.279 - 31.479 mm (1.2315 - 1.2393 in)
Service limit:									
Inlet ...	...	...	...	...	...	...	...	27.5 mm (1.0827 in)	31.55 mm (1.242 in)
Exhaust ...	...	...	...	...	...	...		27.36 mm (1.0772 in)	31.25 mm (1.230 in)
Journal OD:									
Right-hand ...	...	...	...	...	...	...		—	19.967 - 19.980 mm (0.7861 - 0.7866 in)
Left-hand ...	...	...	...	...	...	...		—	33.959 - 33.975 mm (1.3370 - 1.3376 in)
Service limit:									
Right-hand ...	...	...	...	...	...	...		—	19.90 mm (0.784 in)
Left-hand ...	...	...	...	...	...	...		—	33.90 mm (1.335 in)
Camshaft bush ID ...	...	...	...	...	...		—	20.005 - 20.026 mm (0.7876 - 0.7884 in)

Service limit	—	20.05 mm (0.789 in)
Cylinder head left-hand bearing ID	—	34.000 - 34.025 mm (1.3386 - 1.3396 in)
Service limit	—	34.05 mm (1.341 in)
Camshaft to bearing clearance service limit	—	0.1 mm (0.004 in)
Camshaft bush to camshaft clearance service limit	—	0.1 mm (0.004 in)

Rockers

Rocker arm bore ID	—	12.000 - 12.018 mm (0.4724 - 0.4730 in)
Service limit	—	12.05 mm (0.474 in)
Rocker arm shaft OD	—	11.995 - 11.997 mm (0.4722 - 0.4723 in)
Service limit	—	11.93 mm (0.470 in)
Rocker arm to shaft clearance	0.013 - 0.037 mm (0.0005 - 0.0015 in)	0.005 - 0.041 mm (0.0002 - 0.0016 in)
Service limit	0.1 mm (0.0039 in)	0.08 mm (0.003 in)

Camshaft chain

Guide wear service limit	1.0 mm (0.0394 in)	—
Tensioner wear service limit	1.0 mm (0.0394 in)	—

Crankshaft

Connecting rod small-end ID	13.016 - 13.034 mm (0.5124 - 0.5132 in)	15.010 - 15.028 mm (0.5909 - 0.5917 in)
Service limit	13.04 mm (0.5134 in)	15.06 mm (0.593 in)
Connecting rod big-end axial clearance:		
Standard	0.10 - 0.35 mm (0.004 - 0.014 in)	0.05 - 0.30 mm (0.002 - 0.012 in)
Service limit	0.50 mm (0.02 in)	0.80 mm (0.032 in)
Crankshaft runout:		
Standard	0 - 0.035 mm (0.0014 in) For right-hand end 0 - 0.020 mm (0.0008 in) For chain sprocket boss end	0 - 0.008 mm (0 - 0.0003 in)
Service limit	0.085 mm (0.0033 in) For right-hand end 0.070 mm (0.0027 in) For chain sprocket boss end	0.05 mm (0.002 in)

	XR80	XR185	XR200
Clutch			
Type		Wet multiplate	
Spring free length	26.1 mm (1.03 in)	35.5 mm (1.40 in)	37.9 mm (1.49 in)
Service limit	24.0 mm (0.95 in)	32.5 mm (1.28 in)	34.7 mm (1.37 in)
Friction plate thickness	2.8 - 2.9 mm (0.1102 - 0.1142 in)	2.9 - 3.0 mm (0.11 - 0.12 in)	2.9 - 3.0 mm (0.11 - 0.12 in)
Service limit	2.5 mm (0.0984 in)	2.6 mm (0.10 in)	2.6 mm (0.10 in)
Friction plate maximum warpage	0.20 mm (0.008 in)	0.20 mm (0.008 in)	0.20 mm (0.008 in)
Plain plate maximum warpage	0.20 mm (0.008 in)	0.20 mm (0.008 in)	0.20 mm (0.008 in)
Gearbox			
Type	5-speed constant mesh	6-speed constant mesh	6-speed constant mesh
Primary reduction ratio	4.437 : 1	3.333 : 1	3.333 : 1 (21/70)
Gear ratios:			
1st	2.692 : 1	3.083 : 1	2.769 : 1 (13/36)
2nd	1.823 : 1	1.941 : 1	1.941 : 1 (17/33)
3rd	1.400 : 1	1.450 : 1	1.450 : 1 (20/29)
4th	1.130 : 1	1.130 : 1	1.130 : 1 (23/26)
5th	0.960 : 1	0.923 : 1	0.923 : 1 (26/24)
6th	—	0.785 : 1	0.785 : 1 (28/22)
Final reduction ratio	3.285 : 1 (46/14)	3.866 : 1	3.846 : 1 (13/50)
Selector fork bore ID	12.000 - 12.018 mm (0.4724 - 0.4731 in)	12.000 - 12.018 mm (0.4724 - 0.4731 in)	12.000 - 12.018 mm (0.4724 - 0.4731 in)

	XL80 S / XL100 S / XL125 S (col 1)		
Service limit	12.05 mm (0.474 in)	12.05 mm (0.474 in)	12.05 mm (0.474 in)
Selector fork shaft OD	11.976 - 11.994 mm (0.4715 - 0.4722 in)	11.976 - 11.994 mm (0.4715 - 0.4722 in)	11.976 - 11.994 mm (0.4715 - 0.4722 in)
Service limit	11.96 mm (0.471 in)	11.96 mm (0.471 in)	11.96 mm (0.471 in)
Selector fork claw end thickness	4.93 - 5.00 mm (0.1941 - 0.1969 in)	4.93 - 5.00 mm (0.1941 - 0.1969 in)	4.93 - 5.00 mm (0.1941 - 0.1969 in)
Service limit	4.5 mm (0.177 in)	4.5 mm (0.177 in)	4.5 mm (0.177 in)
Selector fork guide pin to drum groove clearance	0.05 - 0.20 mm (0.0020 - 0.0079 in)	—	—
Service limit	0.30 mm (0.0118 in)	—	—

Kickstart

Spindle OD	17.940 - 17.968 mm (0.7063 - 0.7074 in)	19.959 - 19.980 mm (0.7858 - 0.7866 in)	19.959 - 19.980 mm (0.7858 - 0.7866 in)
Service limit	17.86 mm (0.703 in)	19.90 mm (0.783 in)	19.90 mm (0.783 in)
Pinion ID	18.020 - 18.041 mm (0.7094 - 0.7103 in)	20.000 - 20.021 mm (0.7874 - 0.7882 in)	20.000 - 20.021 mm (0.7874 - 0.7882 in)
Service limit	18.07 mm (0.711 in)	20.05 mm (0.789 in)	20.05 mm (0.789 in)

Engine

	XL80 S	XL100 S	XL125 S	XL185 S
Type	Air cooled, single cylinder, overhead camshaft, 4-stroke			
Bore	47.5 mm (1.870 in)	53.0 mm (2.087 in)	56.5 mm (2.22 in)	63.0 mm (2.48 in)
Stroke	45.0 mm (1.177 in)	45.0 mm (1.177 in)	49.5 mm (1.95 in)	57.8 mm (2.28 in)
Capacity	79.7 cc (4.85 cu in)	99.2 cc (6.06 cu in)	124 cc (7.57 cu in)	180 cc (10.98 cu in)
Compression ratio	9.4 : 1	9.4 : 1	9.4 : 1	9.2 : 1
Maximum horsepower	—	9.7 BHP @ 9500 rpm	12.0 BHP @ 9500 rpm	14.3 BHP @ 8000 rpm
Maximum torque	—	0.75 kg m (5.42 lb ft) @ 8000 rpm	0.98 kg m (7.09 lb ft) @ 8000 rpm	1.43 kg m (10.34 lb ft) @ 6500 rpm

Cylinder barrel

	XL80 S	XL100 S	XL125 S	XL185 S
Standard bore	47.50 - 47.51 mm (1.8701 - 1.8705 in)	53.00 - 53.01 mm (2.0866 - 2.0870 in)	56.50 - 56.51 mm (2.2244 - 2.2248 in)	63.00 - 63.01 mm (2.4803 - 2.4807 in)
Service limit	47.61 mm (1.8744 in)	53.10 mm (2.091 in)	56.60 mm (2.228 in)	63.01 mm (2.484 in)
Taper limit	0.05 mm (0.002 in)	0.1 mm (0.004 in)	0.1 mm (0.004 in)	0.1 mm (0.004 in)
Ovality limit	0.05 mm (0.002 in)	0.1 mm (0.004 in)	0.1 mm (0.004 in)	0.1 mm (0.004 in)
Barrel to head face distortion limit ...	0.1 mm (0.004 in)	0.1 mm (0.004 in)	0.1 mm (0.004 in)	0.1 mm (0.004 in)
Cylinder bore to piston clearance	0.010 - 0.045 mm (0.0004 - 0.0018 in)	0.010 - 0.040 mm (0.0004 - 0.0016 in)	0.010 - 0.040 mm (0.0004 - 0.0016 in)	0.015 - 0.045 mm (0.0006 - 0.0018 in)
Service limit	0.1 mm (0.004 in)	0.1 mm (0.004 in)	0.1 mm (0.004 in)	0.1 mm (0.004 in)

Piston

Outside diameter	...	...	...	...	47.465 - 47.490 mm (1.8687 - 1.8697 in)	52.970 - 52.990 mm (2.0854 - 2.0862 in)	56.470 - 56.490 mm (2.2232 - 2.2401 in)	62.965 - 62.985 mm (2.4789 - 2.4797 in)
Service limit ...	...	...	...	...	47.40 mm (1.8661 in)	52.90 mm (2.083 in)	56.40 mm (2.220 in)	62.90 mm (2.476 in)
Gudgeon pin OD	...	...	...	...	12.994 - 13.000 mm (0.5115 - 0.5118 in)	13.994 - 14.000 mm (0.5509 - 0.5906 in)	14.994 - 15.000 mm (0.5903 - 0.5906 in)	14.994 - 15.000 mm (0.5903 - 0.5906 in)
Service limit ...	...	...	...	...	12.98 mm (0.5110 in)	13.96 mm (0.550 in)	14.96 mm (0.589 in)	14.96 mm (0.589 in)
Gudgeon pin hole bore	...	...	...		13.002 - 13.008 mm (0.5118 - 0.5121 in)	14.002 - 14.008 mm (0.5513 - 0.5515 in)	15.002 - 15.008 mm (0.5906 - 0.5909 in)	15.002 - 15.008 mm (0.5906 - 0.5909 in)
Service limit ...	...	...	...	...	13.055 mm (0.5139 in)	14.04 mm (0.553 in)	15.04 mm (0.592 in)	15.04 mm (0.592 in)
Gudgeon pin to piston clearance	...	...			0.08 mm (0.003 in)	0.002 - 0.014 mm (0.0001 - 0.0006 in)	0.002 - 0.014 mm (0.0001 - 0.0006 in)	0.002 - 0.014 mm (0.0001 - 0.0006 in)
Service limit ...	...	...	...	...	0.03 mm (0.001 in)	0.02 mm (0.001 in)	0.02 mm (0.001 in)	0.02 mm (0.001 in)

Piston rings

								XL80 S	XL100 S	XL125 S and 185 S
Ring to groove clearance:										
Top	...	...	...	...	...	...	...	0.015 - 0.050 mm (0.0006 - 0.0020 in)	0.015 - 0.050 mm (0.0006 - 0.0020 in)	0.015 - 0.050 mm (0.0006 - 0.0020 in)
Second	...	...	...	...	...	...	...	0.015 - 0.045 mm (0.0006 - 0.0018 in)	0.015 - 0.045 mm (0.0006 - 0.0018 in)	0.015 - 0.045 mm (0.0006 - 0.0018 in)
Oil ...	...	...	...	...	...	...	...	0.035 - 0.560 mm (0.0014 - 0.0221 in)	0.035 - 0.180 mm (0.0014 - 0.0071 in)	—
Service limit:										
Top	...	...	...	...	...	...	...	0.1 mm (0.004 in)	0.1 mm (0.004 in)	0.09 mm (0.004 in)
Second	...	...	...	...	...	...	...	0.1 mm (0.004 in)	0.1 mm (0.004 in)	0.09 mm (0.004 in)
Oil ...	...	...	...	...	...	...	...	—	—	—
End gap:										
Top and second ...	...	...	...	...	...	...		0.15 - 0.35 mm (0.006 - 0.014 in)	0.15 - 0.35 mm (0.006 - 0.014 in)	0.15 - 0.35 mm (0.006 - 0.014 in)
Oil scraper	...	...	...	...	...	...		0.30 - 0.90 mm (0.012 - 0.035 in)	0.20 - 0.90 mm (0.008 - 0.035 in)	0.20 - 0.90 mm (0.008 - 0.035 in)
Service limit:										
Top and second ...	...	...	...	...	...	...		0.50 mm (0.02 in)	0.50 mm (0.02 in)	0.50 mm (0.02 in)
Oil scraper	...	...	...	...	...	...		—	—	—

Valves

								XL80 S	XL100 S	XL125 S and 185 S
Valve stem OD:										
Inlet	...	...	...	...	...	...	...	5.450 - 5.465 mm (0.2145 - 0.2151 in)	5.450 - 5.465 mm (0.2145 - 0.2151 in)	5.450 - 5.465 mm (0.2146 - 0.2152 in)
Exhaust	...	...	...	...	...	...	...	5.430 - 5.445 mm (0.2137 - 0.2143 in)	5.430 - 5.445 mm (0.2137 - 0.2143 in)	5.430 - 5.445 mm (0.2138 - 0.2144 in)

Service limit:			
Inlet	5.42 mm (0.213 in)	5.42 mm (0.213 in)	5.42 mm (0.213 in)
Exhaust	5.40 mm (0.213 in)	5.40 mm (0.213 in)	5.40 mm (0.213 in)
Valve guide ID:			
Inlet and exhaust...	5.475 - 5.485 mm (0.2155 - 0.2159 in)	5.475 - 5.485 mm (0.2155 - 0.2159 in)	5.475 - 5.485 mm (0.2156 - 0.2159 in)
Service limit:			
Inlet and exhaust...	5.50 mm (0.217 in)	5.50 mm (0.217 in)	5.50 mm (0.217 in)
Stem to guide clearance:			
Inlet	0.010 - 0.035 mm (0.0004 - 0.0013 in)	0.010 - 0.035 mm (0.0004 - 0.0013 in)	0.010 - 0.035 mm (0.0004 - 0.0013 in)
Exhaust	0.030 - 0.055 mm (0.0012 - 0.0022 in)	0.030 - 0.055 mm (0.0012 - 0.0022 in)	0.030 - 0.055 mm (0.0012 - 0.0022 in)
Service limit:			
Inlet	0.08 mm (0.003 in)	0.08 mm (0.003 in)	0.12 mm (0.005 in)
Exhaust	0.10 mm (0.004 in)	0.10 mm (0.004 in)	0.14 mm (0.006 in)
Valve face width (inlet and exhaust)	1.1 - 1.5 mm (0.04 - 0.06 in)	1.1 - 1.5 mm (0.04 - 0.06 in)	1.7 mm (0.067 in)
Service limit	1.8 mm (0.07 in)	1.8 mm (0.07 in)	2.0 mm (0.078 in)
Valve seat width (inlet and exhaust)	1.0 mm (0.04 in)	1.0 mm (0.04 in)	1.2 mm (0.047 in)
Service limit	1.5 mm (0.06 in)	1.5 mm (0.06 in)	1.5 mm (0.059 in)

Valve springs

Free length:			
Inner	28.05 mm (1.10 in)	28.05 mm (1.10 in)	39.4 mm (1.55 in)
Outer	34.80 mm (1.37 in)	34.80 mm (1.37 in)	45.5 mm (1.79 in)
Service limit:			
Inner	27.6 mm (1.09 in)	27.6 mm (1.09 in)	35.5 mm (1.40 in)
Outer	33.7 mm (1.33 in)	33.7 mm (1.33 in)	41.0 mm (1.61 in)

Valve timing at 1 mm lift

Inlet valve opens	5° ATDC	10° BTDC	10° BTDC
Exhaust valve opens	15° BBDC	40° BBDC	40° BBDC
Inlet valve closes	15° ABDC	35° ABDC	40° ABDC
Exhaust valve closes	5° BTDC	5° ATDC	10° ATDC

Valve clearance (cold)

Inlet	0.05 mm (0.002 in)	0.05 mm (0.002 in)	0.05 mm (0.002 in)
Exhaust	0.05 mm (0.002 in)	0.05 mm (0.002 in)	0.05 mm (0.002 in)

Camshaft

Cam height:			
Inlet	27.005 mm (1.0632 in)	27.950 mm (1.1004 in)	31.675 - 31.875 mm (1.2470 - 1.2549 in)
Exhaust	27.005 mm (1.0632 in)	27.866 mm (1.0971 in)	31.279 - 31.479 mm (1.2315 - 1.2393 in)
Service limit:			
Inlet	26.0 mm (1.03 in)	27.8 mm (1.09 in)	31.55 mm (1.242 in)
Exhaust	26.0 mm (1.03 in)	27.7 mm (1.09 in)	31.25 mm (1.230 in)
Journal OD:			
Right-hand	19.950 - 19.968 mm (0.7854 - 0.7861 in)	19.950 - 19.968 mm (0.7854 - 0.7861 in)	19.967 - 19.980 mm (0.7861 - 0.7866 in)

	XL80 S / col1	XL100 S / col2	XL125 S and XL185 S / col3
Left-hand	19.950 - 10.068 mm (0.7854 - 0.7861 in)	19.950 - 19.908 mm (0.7854 - 0.7861 in)	33.959 - 33.975 mm (1.3370 - 1.3376 in)
Service limit:			
Right-hand	19.90 mm (0.783 in)	19.90 mm (0.783 in)	19.90 mm (0.784 in)
Left-hand	19.90 mm (0.783 in)	19.90 mm (0.783 in)	33.90 mm (1.335 in)
Camshaft bush ID	—	—	20.005 - 20.026 mm (0.7876 - 0.7884 in)
Service limit	—	—	20.05 mm (0.789 in)
Cylinder head left-hand bearing ID	—	—	34.000 - 34.025 mm (1.3386 - 1.3396 in)
Service limit	—	—	34.05 mm (1.341 in)
Camshaft holder ID	20.008 - 20.063 mm (0.7877 - 0.7899 in)	20.008 - 20.063 mm (0.7877 - 0.7899 in)	—
Service limit	20.15 mm (0.793 in)	20.15 mm (0.793 in)	—

Rockers

	XL80 S	XL100 S	XL125 S and XL185 S
Rocker arm bore ID	10.000 - 10.015 mm (0.3937 - 0.3943 in)	10.000 - 10.015 mm (0.3937 - 0.3943 in)	12.000 - 12.018 mm (0.4724 - 0.4730 in)
Service limit	10.1 mm (0.40 in)	10.1 mm (0.40 in)	12.05 mm (0.474 in)
Rocker arm shaft OD	9.978 - 9.987 mm (0.3930 - 0.3931 in)	9.978 - 9.987 mm (0.3930 - 0.3931 in)	11.977 - 11.995 mm (0.4715 - 0.4722 in)
Service limit	9.91 mm (0.390 in)	9.91 mm (0.390 in)	11.93 mm (0.470 in)
Rocker arm to shaft clearance	0.013 - 0.037 mm (0.0005 - 0.0014 in)	0.013 - 0.037 mm (0.0005 - 0.0014 in)	0.005 - 0.041 mm (0.0002 - 0.0016 in)
Service limit	0.19 mm (0.008 in)	0.08 mm (0.003 in)	0.08 mm (0.003 in)

Crankshaft

	XL80 S	XL100 S	XL125 S and XL185 S
Maximum runout	0.1mm (0.004 in)	0.1 mm (0.004 in)	0.05 mm (0.002 in)
Connecting rod small-end ID	13.016 - 13.034 mm (0.5124 - 0.5131 in)	14.012 - 14.030 mm (0.5517 - 0.5524 in)	15.010 - 15.028 mm (0.5909 - 0.5917 in)
Service limit	13.08 mm (0.5150 in)	14.05 mm (0.553 in)	15.06 mm (0.593 in)
Connecting rod big-end clearance:			
Axial	0.10 - 0.35 mm (0.0039 - 0.0138 in)	0.10 - 0.35 mm (0.0039 - 0.0138 in)	0.05 - 0.30 mm (0.002 - 0.012 in)
Service limit	0.6 mm (0.024 in)	0.6 mm (0.024 in)	0.8 mm (0.032 in)
Radial	0 - 0.008 mm (0 - 0.0003 in)	0 - 0.008 mm (0 - 0.0003 in)	0 - 0.008 mm (0 - 0.0003 in)
Service limit	0.05 mm (0.002 in)	0.05 mm (0.002 in)	0.05 mm (0.002 in)

Clutch

	XL80 S	XL100 S	XL125 S and XL185 S
Type		Wet, multiplate	
Spring free length	26.1 mm (1.03 in)	28.4 mm (1.12 in)	33.2 mm (1.31 in) 35.5 mm (1.40 in)

Service limit	24.0 mm (0.95 in)	26.3 mm (1.04 in)	30.0 mm (1.18 in)	32.5 mm (1.28 in)
Friction plate thickness	3.0 mm (0.12 in)	3.0 mm (0.12 in)	2.9 - 3.0 mm (0.11 - 0.12 in)	2.9 - 3.0 mm (0.11 - 0.12 in)
Service limit	2.6 mm (0.102 in)	2.6 mm (0.102 in)	2.6 mm (0.102 in)	2.6 mm (0.102 in)
Friction plate maximum warpage	0.20 mm (0.01 in)	0.20 mm (0.01 in)	0.20 mm (0.01 in)	0.20 mm (0.01 in)
Plain plate maximum warpage	0.20 mm (0.01 in)	0.20 mm (0.01 in)	0.20 mm (0.01 in)	0.20 mm (0.01 in)

Gearbox

Type	4-speed constant mesh	5-speed constant mesh	6-speed constant mesh	5-speed constant mesh
Primary reduction ratio	4.437 : 1	4.437 : 1	3.333 : 1	3.333 : 1
Gear ratios:				
1st	3.083 : 1	3.083 : 1	3.083 : 1	2.769 : 1
2nd	1.882 : 1	1.882 : 1	1.941 : 1	1.722 : 1
3rd	1.333 : 1	1.400 : 1	1.400 : 1	1.272 : 1
4th	1.042 : 1	1.130 : 1	1.130 : 1	1.000 : 1
5th	—	0.923 : 1	0.923 : 1	0.777 : 1
6th	—	—	0.785 : 1	—
Final reduction ratio	2.467 : 1	3.000 : 1 (45/15)	4.000 : 1	3.500 : 1
Selector fork bore ID	12.000 - 12.018 mm (0.4724 - 0.4731 in)	12.000 - 12.018 mm (0.4724 - 0.4731 in)	12.000 - 12.018 mm (0.4724 - 0.4731 in)	12.000 - 12.018 mm (0.4724 - 0.4731 in)
Service limit	12.05 mm (0.474 in)	12.05 mm (0.474 in)	12.05 mm (0.474 in)	12.05 mm (0.474 in)
Selector fork shaft OD	11.976 - 11.994 mm (0.4715 - 0.4722 in)	11.976 - 11.994 mm (0.4715 - 0.4722 in)	11.976 - 11.994 mm (0.4715 - 0.4722 in)	11.976 - 11.994 mm (0.4715 - 0.4722 in)
Service limit	11.96 mm (0.471 in)	11.96 mm (0.471 in)	11.96 mm (0.471 in)	11.96 mm (0.471 in)
Selector fork claw end thickness	4.93 - 5.00 mm (0.1941 - 0.1969 in)	4.93 - 5.00 mm (0.1941 - 0.1969 in)	4.93 - 5.00 mm (0.1941 - 0.1969 in)	4.93 - 5.00 mm (0.1941 - 0.1969 in)
Service limit	4.7 mm (0.19 in)	4.7 mm (0.19 in)	4.5 mm (0.18 in)	4.5 mm (0.18 in)

Kickstart

	XL80 S	XL100 S	XL125 S and 185 S
Spindle OD	17.940 - 17.968 mm (0.7063 - 0.7074 in)	17.959 - 17.980 mm (0.7070 - 0.7078 in)	19.959 - 19.980 mm (0.7858 - 0.7866 in)
Service limit	17.86 mm (0.703 in)	17.88 mm (0.704 in)	19.90 mm (0.783 in)
Pinion ID	18.020 - 18.041 mm (0.7094 - 0.7103 in)	18.020 - 18.041 mm (0.7094 - 0.7103 in)	20.000 - 20.021 mm (0.7874 - 0.7882 in)
Service limit	18.07 mm (0.711 in)	18.07 mm (0.711 in)	20.05 mm (0.789 in)

Torque wrench settings

	XR80, XL80 S and XL100 S kgf m (lbf ft)	XL125 S, XL185 S, XR185 and XR200 kgf m (lbf ft)
Cylinder head 8 mm nuts	—	1.8 - 2.0 (13 - 14)
Cylinder head cover 6 mm bolts	0.8 - 1.2 (5.8 - 8.7)	0.8 - 1.2 (5.8 - 8.7)
Camshaft holder retaining nuts	1.8 - 2.0 (13 - 14)	—
Camshaft sprocket bolts	1.0 - 1.4 (7.0 - 10.0)	0.8 - 1.2 (5.8 - 8.7)
Generator rotor centre nut or bolt	6.5 - 7.5 (47 - 54)	4.0 - 5.0 (29 - 36)
Clutch centre nut	—	4.0 - 5.0 (29 - 36)
Centrifugal oil filter centre nut	—	4.0 - 5.0 (29 - 36)

Primary drive pinion retaining nut...	...	...	...	...	3.5 - 4.5 (25 - 33)	—
Oil drain plug	...	...	...	...	2.0 - 3.0 (14 - 22)	—
Oil filter gauze screen cap	...	...	...	...	—	1.0 - 2.0 (7 - 14)
Engine to frame mounting bolt nuts	...	...	...	...	1.8 - 2.3 (13 - 16)	2.7 - 3.3 (20 - 24) - 8 mm
						3.0 - 4.0 (22 - 29) - 10 mm
CDI pulser rotor retaining bolt	...	...	...	...	—	0.8 - 1.2 (6 - 9)
Gearchange pedal clamp bolt	...	...	...	...	0.8 - 1.2 (6 - 9)	0.8 - 1.2 (6 - 9)
Kickstart lever clamp bolt	...	...	...	...	0.8 - 1.2 (6 - 9)	0.8 - 1.2 (6 - 9)

1 General description

The Honda range of trail bikes in the 80 cc to 200 cc capacity class employs two basic designs of single-cylinder, air-cooled, 4-stroke engine. Both represent a familiar design amongst small Honda motorcycles and have proved extremely reliable and robust over several years of manufacture in many different machine applications. The engine types are divided between the various model types as follows: XR80, XL80 S and XL100 S models; and XL125 S, XL185 S, XR185 and XR200 models.

Both units are of all-alloy construction, employing vertically split crankcases which house both the crankshaft assembly and the gear clusters. The cylinder head and cylinder barrel are also of light alloy, the latter incorporating a steel liner in which the cylinder bore is machined.

In common with virtually all the 4-stroke engines in the Honda range and with Japanese engine design in general, these engine units feature an overhead camshaft to operate the valve mechanism. The overhead camshaft actuates the two valves, one inlet and one exhaust, by means of rockers that bear directly on the camshaft. The camshaft is driven by a small, but strong, endless chain which passes up through a cast-in tunnel on the left-hand side of the cylinder barrel and head. The advantage of the chain-driven overhead camshaft arrangement to operate the valves, instead of the traditional pushrod system, is that higher engine revolutions may be sustained without risking damage. This is particularly beneficial in a small 4-stroke engine where useful power is obtainable only at high engine speeds. That the system is effective can be judged from the high recommended engine speeds given for these units.

Lubrication is provided by a small trochoid oil pump feeding the major engine components. The lubricating oil is contained in the lower portion of the crankcase which forms a combined sump and an oil bath for the gearbox components. Two filters are included in the lubrication system, one of the centrifugal type (except XR80, XL80 S and XL100 S models) and the other in the form of a wire mesh screen.

A flywheel generator is mounted on the left-hand end of the crankshaft and is fully enclosed behind a detachable crankcase side cover. The clutch is mounted on the right-hand side of the engine and is also enclosed behind a detachable side cover along with the oil pump assembly.

2 Operations with engine/gearbox in frame

It is not necessary to remove the engine unit from the frame in order to carry out the following servicing operations:
1 Removal and refitting of the cylinder head cover
2 Removal and refitting of the cylinder head (XR80 and XL80 S models only)
3 Removal and refitting of the cylinder barrel and piston (XR80 and XL80 S models only)
4 Removal and refitting of the camshaft and rockers
5 Removal and refitting of the clutch assembly
6 Removal and refitting of the flywheel generator
7 Removal and refitting of the oil pump and filters
8 Removal and refitting of the contact breaker or CDI unit assembly

When several operations need to be undertaken simultaneously, it will probably be advantageous to remove the complete engine unit from the frame, an operation that should take approximately one hour, working at a leisurely pace. This will give the advantage of better access and more working space.

3 Operations with engine/gearbox removed from frame

Certain operations can be accomplished only if the complete engine unit is removed from the frame. This is because it is necessary to separate the crankcases to gain access to the parts concerned, or because there is insufficient clearance to withdraw parts after they have been slackened or freed from their normal location. These operations include:
1 Removal and refitting of the main bearings
2 Removal and refitting of the crankshaft assembly
3 Removal and refitting of the gear cluster, selectors and gearbox main bearings
4 Removal and refitting of the cylinder head, cylinder barrel and piston (note previous Section for XR80 and XL80 S models)

4 Method of engine/gearbox removal

As mentioned previously, the engine and gearbox are built in unit and it is necessary to remove the unit complete in order to gain access to either component. Separation is accomplished after the engine/gearbox unit has been removed from the frame and refitting cannot take place until the crankcases have been reassembled. When the crankcases are separated, the gearbox internals will also be exposed. There is no means of working on the engine without disturbing the gearbox, or on the gearbox without first dismantling the engine.

5 Removing the engine/gearbox unit

1 Removal of the engine/gearbox unit will be made much easier by raising the machine to an acceptable working height and thus preventing the discomfort of squatting or kneeling down to work on the various component parts. Raising the machine may be achieved by using either a purpose built lift or a stout table or by building a platform from substantial planks and concrete blocks.
2 With the machine placed on the work surface and supported on its propstand, remove the sumpguard. On 80 cc models the guard may be removed by first unscrewing the two retaining bolts at the rear of the guard and then loosening the forward mounting bolt located at the base of the frame front tube. This will allow the guard to be pivoted down and away from the engine. On all other models it is necessary to remove the two rear and one forward retaining bolts before detaching the guard from the machine. It should be noted that before attempting to remove the forward retaining bolt, the tabs on the lockplate should be knocked back away from the bolt head flats.
3 Place the machine in neutral and mark the position of the gear lever in relation to that of the shaft to ensure correct

positioning of the lever during refitting. Remove the gear lever by loosening the clamp bolt and pulling the lever from its shaft.

4 Place a clean container of at least one litre capacity beneath the engine unit. Remove the crankcase drain plug and allow the oil to drain from the engine unit. Clean the drain plug and inspect the condition of the sealing ring; renew the ring if considered necessary. When the oil has finished draining, refit and tighten the drain plug. Draining the oil from the engine is best carried out after the engine has been run and is warm; the oil will be thinner and so drain more quickly.

5 On all except the 80 and 100 cc models, move the container with the engine oil to the underside of the left-hand crankcase cover and remove the hexagon-headed plug from the crankcase. A small additional quantity of oil will drain off when the plug is removed and the spring and filter withdrawn. Clean the plug and note the condition of the O-ring. If the ring is damaged in any way or has become hard or flattened, then it must be renewed. Place the plug, filter and spring aside after having first cleaned the filter gauze in petrol.

6 Remove the seat by releasing the two mounting bolts. These are located one either side towards the rear of the seat. The seat may then be lifted up and rearwards off its forward mounting point. Loss of the two mounting bolts may be avoided by relocating them in the frame/seat attachment points.

7 Remove the fuel tank by first turning the petrol tap to the 'Off' position and releasing the fuel pipe retaining clip. This will allow the pipe to be pulled off the stub at the base of the tap. Careful use of a small screwdriver may be necessary to help ease the pipe off the stub. Once the pipe is detached, allow any fuel in the pipe to drain into a small clean container. Remove the tank breather pipe (where fitted) by pulling it off the tank filler cap centre stub. The pipe may be fastened to the stub by means of a spring clip, in which case this clip must first be removed. The tank may now be detached from the frame by unscrewing the single retaining bolt at the rear of the tank and pulling the tank up and rearwards off its front mounting rubbers.

8 Remove the right-hand sidepanel by pulling it outwards off its frame attachment points. On 80 cc models, the left-hand sidepanel may be removed by the same method. On all other models the panel may be released by unscrewing the three locating screws.

9 After carefully placing the seat, tank and sidepanels in a storage space where they are likely to be safe from damage, proceed to remove the exhaust system.

XR80 and XL80 S models

On these models the exhaust system is a one-piece unit and may easily be detached from the machine after removal of the right-hand sidepanel and rear shock absorber. Unscrew and remove the two pipe to cylinder head retaining nuts. These should be released evenly to avoid any risk of distortion to the pipe joint plate. Earlier exhaust systems may then be removed by unscrewing the single retaining bolt situated at a mid-point in the system length. Later systems have two retaining bolts, one of which is situated at a mid-point in the system length and the other being situated just behind the seat mounting bolt. Removal of the rearmost bolt is simple and straightforward but it may, however, be necessary to remove the heat guard from the front half of the system in order to gain access to the head of the Allen bolt. Once the machine has been used over a period of time, the screws attaching this guard will become seized in their threads and some force will be required to remove them; an impact driver is ideal for this purpose, although the pipe must be well supported before removal is attempted.

XL100 S, XL125 S, XL185 S, XR185 and XR200 models

The exhaust system fitted to these models is a two-piece unit, the front pipe being attached to the silencer assembly by a clamp retained in position by an Allen bolt. When removing the engine unit, it is only necessary to remove the front pipe. This is achieved by first unscrewing the two pipe to cylinder head retaining nuts. Loosen the nuts evenly, to avoid any risk of distortion to the pipe joint plate. Release the clamp by undoing the Allen bolt and pull the pipe forward, twisting it slightly to aid removal once the pipe is clear of the cylinder head. It may be found necessary to remove the heat guard from the pipe in order to give better access to the head of the clamp bolt. Details of guard removal are given in the above paragraph for 80 cc models.

All models

10 Disconnect the clutch cable from the lever at the right-hand crankcase cover end by placing an open-ended spanner on the lever end and turning the lever towards the cable. The cable nipple may then be released from the lever and the threaded adjuster withdrawn from the crankcase bracket. On 80 and 100 cc models, remove the crankcase bracket by unscrewing the retaining nut. Loosely refit the nut and lockwasher to avoid loss. On 125, 185 and 200 cc models, release the cable from the forward engine plate cable clamp and allow the cable to hang clear in front of the engine.

11 Disconnect the tachometer drive cable (where fitted) from the right-hand crankcase cover boss by removing the crosshead retaining screw and pulling the cable end out of the cover recess. Refit the screw to avoid loss. Release the cable from the forward engine plate cable clamp and allow the cable to hang clear of the engine.

12 Unclip the crankcase breather pipe retaining clip and pull the pipe from the crankcase stub. Pull the HT lead suppressor cap from the sparking plug and unclip the HT lead from the cylinder head. Tuck the lead out of the way on the frame top tube.

13 Gain access to the final drive sprocket by removing its guard. On the 80 and 100 cc models, the guard is part of the flywheel generator cover. Removal of this cover is achieved by unscrewing the five crosshead retaining screws. All other models have a guard which is separate from the generator cover and is removed by unscrewing two retaining screws. In all cases, an impact driver will be needed to free the screws.

14 Undo and remove the two final drive sprocket locking plate retaining bolts and remove the plate. The sprocket may now be drawn off the splined shaft complete with the drive chain. It may be considered beneficial, however, at this stage, to remove the chain from the machine so that it may be cleaned and inspected for wear and damage. Removal of the chain is achieved by removing the spring clip on the connecting link with the use of a pair of pointed nose pliers; the connecting link may then be withdrawn thus allowing the chain to separate.

15 On machines fitted with a battery, isolate the battery from the electrical system by disconnecting the lead to the battery negative (–) terminal. It is advisable, when removing the engine unit, also to remove the battery from the machine so that it may be serviced as described in Chapter 6 of this Manual.

16 Locate the wiring harness connection block, normally situated near the frame downtube just forward of the air filter cover, and pull apart the two halves thus disconnecting the flywheel generator from the rest of the electrical system. If necessary, detach the generator leads from the frame downtube by releasing the clip.

17 On machines equipped with CDI ignition, trace the leads from the CDI pulse generator unit to the rubber cover located on the frame top tube just above the cylinder head. Pull back this cover to expose the bullet connections and unclip the necessary wires to disconnect the CDI unit. Pull the wires out of the cover and allow them to hang loose from the CDI unit.

18 On models equipped with conventional contact breaker ignition, the leads from the points are disconnected when separating the two halves of the flywheel generator to main harness connection block.

19 Detach the carburettor from the cylinder head by unscrewing the inlet stub to cylinder head securing nuts. These two nuts should be slackened evenly to avoid any risk of distortion to

the inlet stub mating face. If it is considered necessary to remove the carburettor at this point for any servicing or cleaning procedures then the following dismantling sequence should be followed, otherwise the carburettor may be left in position whilst the engine is removed from the machine.

20 Undo the carburettor to airbox rubber hose intake clamp. Unscrew and remove the carburettor top and withdraw the throttle valve assembly. Position this assembly where it will not be damaged during engine removal. On carburettors fitted with a cable operated choke, undo the choke cable clamp crosshead screw located on the left-hand side of the carburettor body and detach the cable from the carburettor lever. The carburettor may now be withdrawn from the frame complete with inlet stub and drain, fuel and breather hoses. When disconnecting the carburettor, ensure that any gaskets or sealing rings are retained, inspected for deterioration and renewed if necessary. Proceed to remove the engine from the frame as follows.

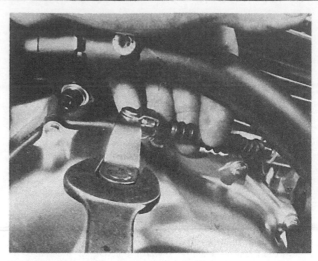

5.10 Disconnect the cable from the clutch lever

XR80 and XL80 S models

21 Remove the two remaining bolts and washers to release the footrest assembly from the underside of the machine. The machine must be held upright and the weight taken off the propstand because the propstand pivot point is part of the footrest assembly. Once the footrest assembly is removed, support the machine by means of a block placed underneath the frame just forward of the rear wheel.

22 Remove the engine front mounting plates, along with the sumpguard, by unscrewing and removing the four nuts, bolts and plain washers.

23 With the engine supported by an assistant, remove the two remaining engine mounting bolts located at the rear of the engine. Lift the engine clear of the frame.

XL100 S, XL125 S, XL185 S, XR185 and XR200 models

24 On models fitted with a decompressor, it is first necessary to disconnect the decompressor cable clamp from the rear of the clutch cover, followed by the cable inner from the cam follower lever. Move the cable clear of the engine. Unscrew and remove the three upper engine to frame plate retaining nuts, withdraw the three bolts and remove the plates. Remove the nuts from the engine to frame rear mounting bolts. Withdraw the rear lower mounting bolt. Some difficulty may be experienced in manoeuvring a spanner past the footrest attachment in order to undo the rear lower nut, but with a little patience the job is easily achieved.

5.11 Disconnect the tachometer drive cable by removing the retaining screw

25 At this point the engine will still be held in the frame by the four bolts passing through the engine front mounting plate and by the single rear mounting bolt. Carry out a final check around the engine to ensure all cables, electrical wires, etc have been moved clear of the engine where possible. Obtain a pad (a small block of wood or rubber, or a folded piece of rag is ideal) and place it within reach of the engine. It is advisable to obtain the aid of an assistant before proceeding any further with engine removal because the machine's propstand is not really suitable for supporting the machine during the final engine removal procedure. The machine should be either held steady or supported by a block placed under the footrest mounting brackets during the following procedure.

26 Remove the four forward mounting plate bolts and detach the plate from the frame tube. Allow the engine to pivot forwards, placing the pad between the cylinder head and frame front tube so that damage to either is avoided.

27 With the engine supported, remove the remaining rear mounting bolt. The spacer tube (where fitted) will become detached as the bolt is drifted out. Note the position of this tube for reference when refitting the engine. With the engine moved free from its mounting points, lift it out sideways from the frame. In practice, the engine was lifted out from the right-hand side of the machine.

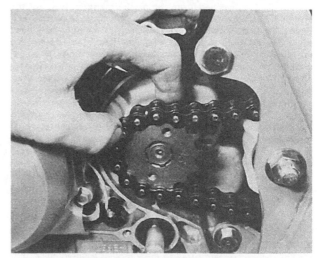

5.14 The sprocket may be drawn off the splined shaft complete with the drive chain

5.16 Pull apart the connection block to disconnect the flywheel generator

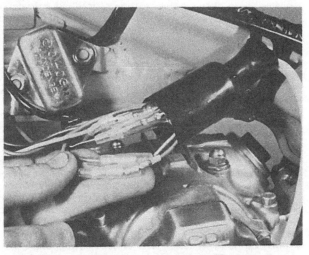

5.17 Expose and disconnect the wires to the CD1 pulse generator (except 80 and 100 models)

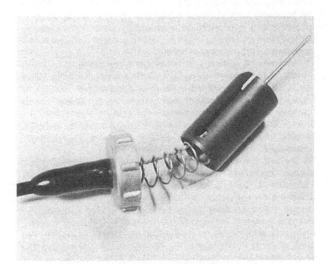

5.20 Unscrew the carburettor top and withdraw the throttle valve assembly

5.24 Some difficulty may be experienced in locating a spanner on the engine rear lower mounting nut (except 80)

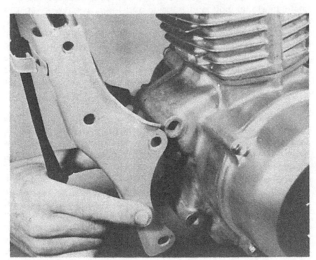

5.26 Detach the engine front mounting plate

6 Dismantling the engine/gearbox unit: general

1 Before commencing work on the engine unit, the external surfaces should be cleaned thoroughly. A motorcycle engine has very little protection from road grit and other foreign matter, which will find its way into the dismantled engine if this simple precaution is not taken. One of the proprietary cleaning compounds, such as 'Gunk' or 'Jizer' can be used to good effect, particularly if the compound is worked into the film of oil and grease before it is washed away. Special care is necessary when washing down to prevent water from entering the now exposed parts of the engine unit.
2 Never use undue force to remove any stubborn part unless specific mention is made of this requirement. There is invariably good reason why a part is difficult to remove, often because the dismantling operation has been tackled in the wrong sequence.
3 Mention has already been made of the benefits of owning an impact driver. Most of these tools are equipped with a standard ½ inch drive and an adaptor which can take a variety of screwdriver bits. It will be found that most engine casing screws will need jarring free due both to the effects of assembly by power tools and an inherent tendency for screws to become pinched in alloy castings. If an impact screwdriver is not

available, it is often possible to use a crosshead screwdriver fitted with a T handle as a substitute. An example of this would be the crosshead screwdriver and the box spanner, both from the machine's standard tool kit, fitted together, to form a T-handle attachment.

4 A cursory glance over many machines of only a few years' use, will almost invariably reveal an array of well-chewed screw heads. Not only is this unsightly, it can also make emergency repairs impossible. It should also be borne in mind that there are a number of types of crosshead screwdrivers which differ in the angle and design of the driving tangs. To this end, it is always advisable to ensure that the correct tool is available to suit a particular screw.

5 Before commencing dismantling, make arrangements for storing separately the various sub-assemblies and ancillary components, to prevent confusion on reassembly. Where possible, replace nuts and washers on the studs or bolts from which they were removed and refit nuts, bolts and washers to their components. This too will facilitate straightforward reassembly.

6 Identical sub-assemblies, such as valve springs and collets or rocker arms and pins etc should be stored separately, to prevent accidental transposition and to enable them to be fitted in their original locations.

7 Dismantling the engine/gearbox unit: removing the overhead camshaft and cylinder head

1 As stated earlier, it is possible to remove the overhead camshaft and rocker assembly without removing the engine from the frame; the procedure for the different models being as follows.

XR80, XL80 S and XL100 S models

2 Remove the cylinder head cover by unscrewing the two securing bolts.

3 Remove the two bolts that hold the sprocket to the camshaft. It should be noted that these bolts differ and their positions in relation to the sprocket should be marked for reference when refitting.

4 Obtain a length of wire, detach the sprocket from the cam chain and prevent the chain from falling into the crankcase by suspending it with the length of wire. The sprocket may now be removed from the cylinder head.

5 Remove the camshaft holder by unscrewing the four securing nuts, evenly and in a diagonal sequence to avoid any risk of distortion. The camshaft may now be lifted out of the cylinder head bearing recesses.

6 Remove the cam chain adjuster retaining bolt and plate and pull the adjuster out of the cylinder head. Unscrew and remove the one remaining cylinder head to barrel securing bolt, located on the left-hand side of the engine, and lift the cylinder head from the barrel. Do not use excessive force to achieve this and on no account attempt to lever the cylinder head from the barrel. A sharp tap with a wooden block and mallet should suffice to free the head.

7 At this point, note the location of the cylinder head to cover and the cylinder head to cylinder barrel locating dowels. There should be two dowels between each of the mating faces. One of the cylinder barrel dowels should be fitted with an O-ring; the position of this O-ring should be noted.

8 The cam chain guide blade may now be pulled from its location at the front of the cam chain tunnel.

XL125 S, XL185 S, XR185 and XR200 models

9 On these models it is not necessary to remove the cylinder head cover in order to remove the camshaft. It is however,

necessary to remove the CDI pulser generator assembly to enable the camshaft sprocket and then the camshaft itself to be removed.

10 Remove the cap from the centre of the flywheel generator cover to expose the end of the crankshaft. Remove the smaller cap above it to expose the timing marks and rotate the crankshaft anti-clockwise to align the T mark on the generator rotor with the index mark on the edge of the hole. The piston should now be at top dead centre (TDC) on the compression stroke. Rotation of the crankshaft is achieved by placing a socket or box spanner on the generator rotor centre nut. To check whether or not the piston is on the compression stroke, remove the sparking plug and place a finger over the hole. Compression will be felt as the piston rises.

11 Expose the CDI unit by unscrewing the two cover retaining screws and lifting the cover from the cylinder head. Check that the mark on the baseplate aligns with the corresponding mark on the unit housing. If not, or no marks are present, mark the position of the baseplate in relation to the housing for reference when refitting the assembly.

12 Remove the pulser generator/baseplate assembly by unscrewing the two crosshead retaining screws and detach the electrical lead rubber grommet from the housing recess. The rotor may now be removed from the end of the camshaft by unscrewing the retaining bolt. Take care to keep the bolt and plain washer together; it is a good idea to ensure they are not separated or lost by loosely refitting them to the camshaft on completion of the camshaft removal procedure.

13 Withdraw the dowel pin from the camshaft end, to allow removal of the unit housing, by using a pair of pointed nose pliers. Unscrew the two hexagon-headed bolts and remove the housing from the cylinder head. Check the condition of the housing O-ring and gasket and renew them if necessary. The housing/camshaft oil seal should also be inspected for damage. Any damage to this seal is usually indicated by an oil leak; if undamaged, the seal may be left in position.

14 Store the complete CDI unit assembly in a dry clean place, covered to protect it from any contamination by dirt or oil.

15 Check that the O mark on the cam sprocket aligns with the index mark on the cylinder head cover with the piston at TDC. Unscrew the two bolts from the centre of the sprocket, so that the sprocket is released from the camshaft flange. The two bolts may well be tight. A ring spanner should be applied to the alternator rotor centre nut to prevent engine rotation whilst the two sprocket bolts are slackened and removed.

16 If the machine is relatively new, there may not be enough slack in the cam chain to enable the sprocket to be pulled clear of the cam flange. To overcome this, unscrew the cam chain tensioner bolt and housing and insert a screwdriver into the hole. Push downwards on the tensioner plunger to release the tension on the tensioner blade and thus the chain. If a top-end only overhaul is envisaged, the cam chain must be prevented from dropping down into the crankcase when the sprocket is removed. To prevent this, secure the chain with a suitable length of stiff wire. Lift the sprocket off the end of the camshaft and, by wriggling it gently, pull the camshaft free from the head.

17 To remove the cylinder head cover slacken evenly, in the reverse sequence given in Fig. 1.25, the four Allen bolts and the four domed nuts securing the cover to the cylinder head. This procedure will minimise any risk of the cover becoming distorted.

18 The cover may now be lifted off the cylinder head to expose the valve gear. Note that there are two dowels to locate the cover correctly and no gasket is required at the jointing surface. Note also that the rocker assembly will be detached as the head cover is lifted away, because the assembly will still be fitted inside the cover. The rocker assembly can be left in situ in the cover until the examination and renovation stage as described in Section 30.

19 To remove the cylinder head, unscrew the one long bolt fitted to the left-hand side of the head. Finally, slacken and remove the last remaining bolt in the head, fitted to the rear of

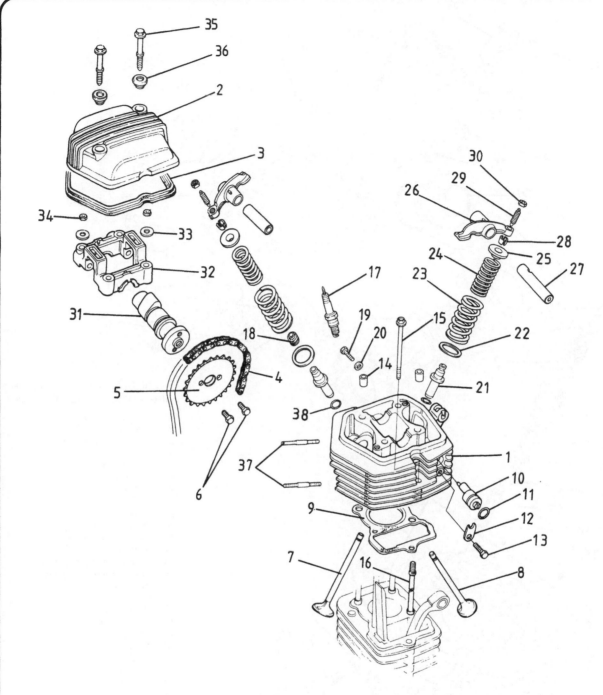

Fig. 1.1 Cylinder head, camshaft and valves — XR80 and XL100 S

1	Cylinder head	14	Dowel	27	Rocker arm spindle — 2 off
2	Cylinder head cover	15	Bolt	28	Collet — 4 off
3	Seal	16	Stud	29	Adjusting screw —2 off
4	Cam chain	17	Sparking plug	30	Locknut — 2 off
5	Camshaft sprocket	18	Seal (— 2 off, later models)	31	Camshaft
6	Bolt — 2 off	19	Bolt	32	Camshaft holder
7	Inlet valve	20	Washer	33	Washer
8	Exhaust valve	21	Valve guide — 2 off	34	Nut
9	Cylinder head gasket	22	Lower spring seat — 2 off	35	Bolt
10	Adjusting screw	23	Outer spring — 2 off	36	Rubber grommet
11	O-ring	24	Inner spring — 2 off	37	Stud — 2 off
12	Set plate	25	Upper spring seat — 2 off	38	O-ring — 2 off
13	Bolt	26	Rocker arm — 2 off		

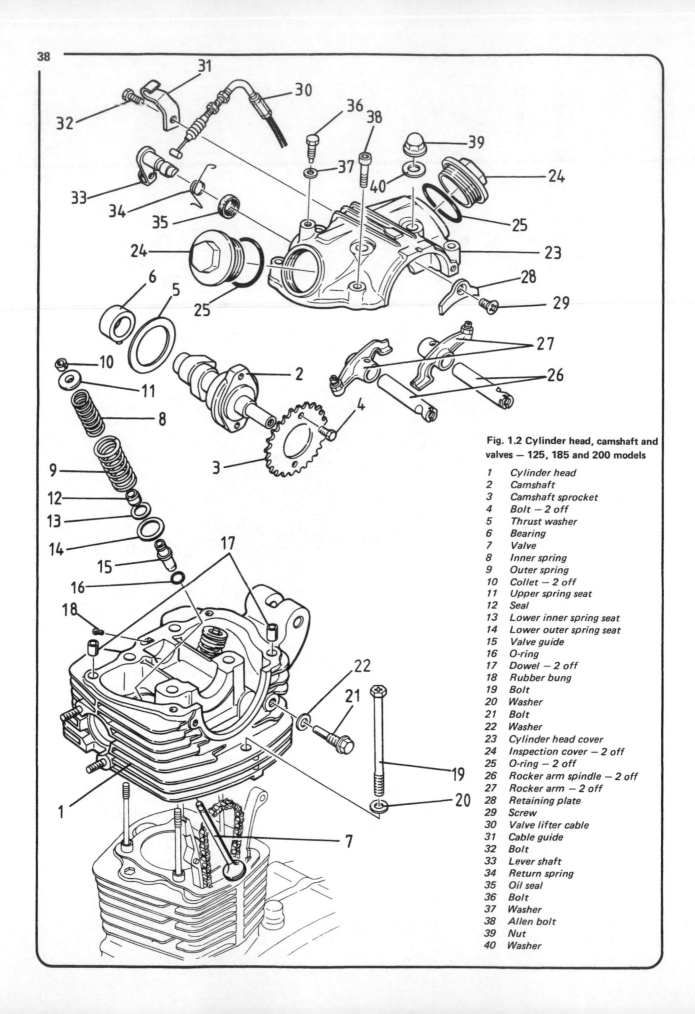

Fig. 1.2 Cylinder head, camshaft and valves — 125, 185 and 200 models

1	Cylinder head
2	Camshaft
3	Camshaft sprocket
4	Bolt — 2 off
5	Thrust washer
6	Bearing
7	Valve
8	Inner spring
9	Outer spring
10	Collet — 2 off
11	Upper spring seat
12	Seal
13	Lower inner spring seat
14	Lower outer spring seat
15	Valve guide
16	O-ring
17	Dowel — 2 off
18	Rubber bung
19	Bolt
20	Washer
21	Bolt
22	Washer
23	Cylinder head cover
24	Inspection cover — 2 off
25	O-ring — 2 off
26	Rocker arm spindle — 2 off
27	Rocker arm — 2 off
28	Retaining plate
29	Screw
30	Valve lifter cable
31	Cable guide
32	Bolt
33	Lever shaft
34	Return spring
35	Oil seal
36	Bolt
37	Washer
38	Allen bolt
39	Nut
40	Washer

the left-hand side of the head. This is the bolt that secures the top end of the rear cam chain tensioner guide blade.

20 The cylinder head can now be lifted away and placed to one side to await further attention. Make a note of the position and sizes of the dowels fitted over three of the four studs. One of these is fitted with an O-ring and acts as an oil feed passage. If the head appears to be resisting attempts on its removal, it is probably the head gasket sticking to the mating surfaces of the head and barrel. Tap the head lightly with a soft-headed mallet in order to jar it free. On no account attempt to lever the cylinder head and barrel apart because this will only cause damage to the alloy castings and, in extreme cases, distortion to the mating faces.

21 Note the fitting of a small rubber bung in a small oil passage located between the two right-hand stud holes on the upper mating surface of the cylinder head. Take great care not to lose this item because it performs an important function in directing the supply of oil around the cylinder head.

22 The cam chain tensioner guide blade, fitted to the front of the cam chain tunnel, may now be pulled up out of the tunnel if further dismantling is required.

7.3 Note that the camshaft sprocket securing bolts differ (80 and 100)

7.6a Remove the cam chain adjuster retaining bolt and plate...

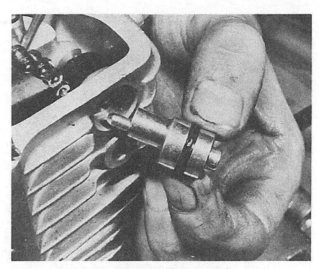

7.6b ... and withdraw the adjuster (80 and 100)

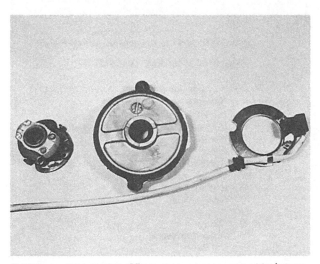

7.12 Store the complete CD1 pulser generator assembly (where fitted) in a dry, clean place

8 Dismantling the engine/gearbox unit: removing the cylinder barrel and piston

1 If it is intended that the engine be dismantled fully, then the camshaft chain may be allowed to drop into the crankcase. If however, it is intended that only the barrel be removed, then the chain must be prevented from falling into the depths of the crankcase. A length of stiff wire or a metal rod, ie a screwdriver blade of suitable length, will suffice.

2 There·are no further retaining bolts, and the barrel should now lift off. If the barrel appears to be sticking to the crankcase apply a few light taps with a soft-headed mallet; this should be sufficient to dislodge it.

3 Ease the cylinder barrel gently upwards, sliding it along the holding down studs. Take care to support the piston and rings as it emerges from the cylinder bore, otherwise there is risk of damage or ring breakage. If the crankcases are not to be separated, it is advisable to pack the crankcase mouth with clean rag before the piston is withdrawn from the bore, in case the piston rings have broken. This will prevent sections of broken

ring from falling into the crankcase. Note the positioning of, and remove the two locating dowels fitted to two of the holding down studs.

4 Prise one of the gudgeon pin circlips out of position, then press the gudgeon pin out of the small-end eye through the piston boss. If the pin is a tight fit, it may be necessary to warm the piston so that the grip on the gudgeon pin is released. A rag soaked in warm water and wrapped around the piston should suffice. The piston may be detached from the connecting rod once the gudgeon pin is clear of the small-end eye.

5 If the gudgeon pin is still a tight fit after warming the piston, it can be lightly tapped out of position with a hammer and soft metal drift. **Do not** use excess force and make sure the connecting rod is supported during this operation, or there is a risk of it bending.

6 With the piston free of the connecting rod, remove the second circlip and fully withdraw the gudgeon pin from the piston. Place the piston and gudgeon pin aside for further attention. On no account reuse the circlips, they should be discarded and new ones fitted during rebuilding.

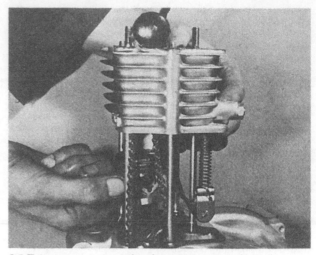

8.3 Take care to support the piston as it emerges from the cylinder bore

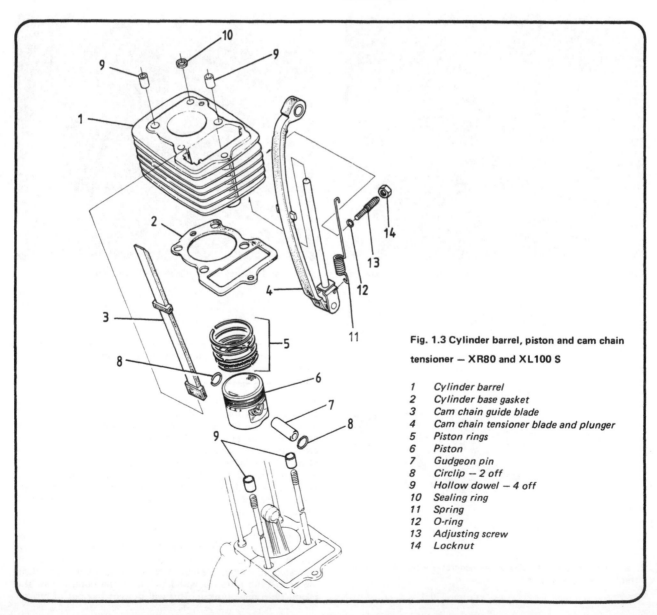

Fig. 1.3 Cylinder barrel, piston and cam chain tensioner — XR80 and XL100 S

1   Cylinder barrel
2   Cylinder base gasket
3   Cam chain guide blade
4   Cam chain tensioner blade and plunger
5   Piston rings
6   Piston
7   Gudgeon pin
8   Circlip — 2 off
9   Hollow dowel — 4 off
10   Sealing ring
11   Spring
12   O-ring
13   Adjusting screw
14   Locknut

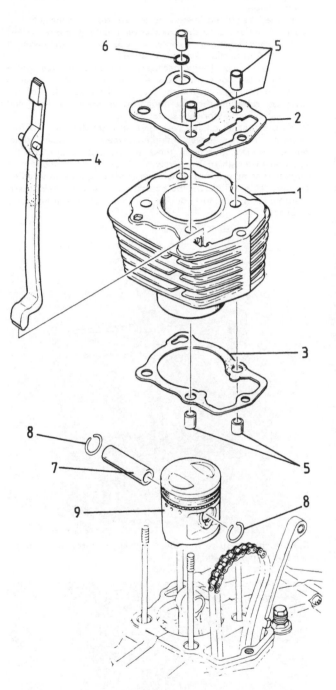

Fig. 1.4 Cylinder barrel and piston — 125, 185 and 200 models

1	Cylinder barrel
2	Cylinder head gasket
3	Cylinder base gasket
4	Cam chain guide blade
5	Hollow dowel — 5 off
6	O-ring
7	Gudgeon pin
8	Circlip — 2 off
9	Piston

9 Dismantling the engine/gearbox unit: removing the flywheel generator

XR80, XL80 S and XL100 S models

1 Position the engine on the work surface with the left-hand side uppermost. The generator cover will have already been removed to allow access to the final drive sprocket when removing the engine from the frame.

2 Lock the engine as described in paragraph 7 of this Section and remove the rotor securing nut (and washer, if fitted).

3 A Honda flywheel extractor tool is normally required to pull off the rotor as it is a keyed taper fit. If the extractor is not available and a puller is used on the outside as an alternative, note that the inner edge of the rotor is chamfered which may cause the puller to slip when the pressure is applied. On the machine that we dismantled, the rotor was extremely tight on the shaft and it was found necessary to use the set up shown in the accompanying photograph. Two shaped pieces of metal were clamped to the rotor by joining two Jubilee clips together. A puller was fitted onto the shaped metal pieces and a piece of chain was fixed round the legs to stop them spreading. After applying penetrating oil to the end of the shaft and applying some considerable force to the puller, the rotor eventually came off the shaft taper. Common sense should prevail in how much pressure is used to free the rotor. If the rotor proves particularly stubborn, seek the advice of a Honda Service Agent before any damage is done.

4 Remove the stator plate by unscrewing the two retaining screws (three on the XL100 S model), separating the electrical leads and rubber grommet from the crankcase recess and lifting the assembly away from the machine. Note that on the XL100 S model, the lead from the neutral switch must also be disconnected.

XL125 S, XL185 S, XR185 and XR200 models

5 With the engine lying on its right-hand side, suitably supported by wooden blocks, remove the four hexagon-headed screws that retain the generator cover to the crankcase and lift off the cover. Note the condition of the O-ring fitted to the cover and renew it if any signs of damage or deterioration are found. In practice, it was found that the four generator cover retaining screws were extremely tight. It is advisable to use a socket or box spanner with a tommy bar extension when slackening these screws; check that the socket or box spanner is an extremely good fit on the screw head to avoid any risk of damage to either the screw head or alloy castings by the tool slipping. It may be found that the lead from the generator is clamped to the crankcase at a point just above the final drive sprocket; if this is found to be so then the crosshead screw retaining the clamp must be slackened and the wire released from the clamp before attempting to remove fully the cover.

6 The stator assembly is fitted to the inside of the generator cover and may be removed, if thought to be defective, by unscrewing the two or four retaining screws.

7 Before removal of the rotor can be attempted, the crankshaft must be prevented from rotating. This is achieved by passing a close-fitting bar through the small-end eye and allowing the ends of the bar to rest on wooden blocks placed on each side of the crankcase mouth. Never allow the ends of the bar to come into direct contact with the jointing face. If the rotor is to be removed with the engine in the frame, crankshaft rotation may be prevented by selecting top gear and applying the rear brake.

8 Remove the rotor retaining bolt and plain washer. The rotor is a tapered fit on the crankshaft end and is located by a Woodruff key; it therefore requires pulling from position. The rotor boss is threaded internally to take the special Honda service tool No 07733 - 0010000 or 07933 - 2000000. If this tool cannot be acquired, then it is possible to use the rear wheel spindle as a puller because the thread in the boss is identical to that of the spindle. Clean the spindle threads of all oil and road dirt and thread the end into the rotor boss until it is finger-tight. Further tighten the bolt with a spanner to draw the rotor

off its taper. Should the rotor prove stubborn, sharply tap the end of the spindle to jar the rotor free. On no account strike the rotor itself as this can easily damage the unit or cause some of its magnetism to become lost. The same principles as stated above should be adopted when using the service tool.

All models
9 After rotor removal, prise the Woodruff key from the tapered crankshaft end. Ensure it is not lost, by storing it somewhere safe until reassembly.

---

10 Dismantling the engine/gearbox unit: removing the cam chain and cam chain tensioner

---

XR80, XL80 S and XL100 S models
1 The cam chain tensioner assembly fitted to these models may be removed from the cylinder barrel after the cylinder barrel itself has been removed from the engine unit.
2 Loosen the cam chain tensioner screw locknut and remove both the screw and locknut from the rear of the cylinder barrel. It is now possible to withdraw the tensioner blade and tensioner plunger and spring from the base of the cylinder barrel. Remove the O-ring from the cylinder barrel recess and renew it.

XL125 S, XL185 S, XR185 and XR200 models
3 The cam chain tensioner assembly fitted to these models is located behind the flywheel generator rotor. It is therefore necessary to remove the rotor, as described in the previous Section, before the following dismantling procedure may be carried out.
4 Remove the cam chain tensioner arm by unscrewing the single retaining bolt. Unscrew and remove the cam chain adjusting bolts from the crankcase and withdraw the two collars from their location. There should be a rubber protective cap fitted over the adjusting assembly. Inspect the condition of the O-ring and sealing washer and renew them if necessary before storing the assembly. The tensioner plunger and blade may now be withdrawn from the crankcase.

---

11 Dismantling the engine/gearbox unit: removing the right-hand crankcase cover - XR80, XL80 S and XL100 S models

---

1 When removing the crankcase cover with the engine in the frame, slacken the kickstart lever clamp bolt and pull the lever

clear of its shaft. Disconnect the clutch and tachometer cables from the cover by following the procedure given in Section 5 of this Chapter.
2 It may be found necessary to remove the sumpguard in order to gain access to the lower cover retaining screws. Details of sumpguard removal for the various models are given in paragraph 2 of Section 5 of this Chapter.
3 It may also be found necessary to depress the rear brake lever in order to gain access to the lower cover retaining screws. If necessary, allow the lever to drop well clear of the cover by releasing the operating rod at the drum lever end and unclipping the stoplight switch spring.
4 Ensure that all oil has been drained from the engine before removing the nine cover retaining screws. An impact driver will be needed to release these screws and because of the variation in length of the screws, it is advisable to make up a cardboard template into which the screws may be inserted on removal. This will ensure the screws are refitted in their correct positions and not lost during the rest of the dismantling sequence.

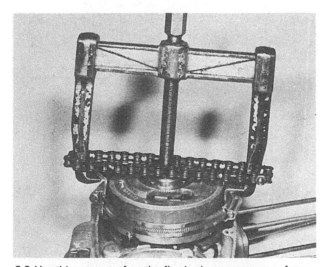

9.3 Use this set-up to free the flywheel generator rotor from the shaft taper (80 and 100)

9.8 The rear wheel spindle may be used as a puller to remove the flywheel generator rotor (except 80 and 100)

10.4a On all except 80 and 100 models, remove the cam chain tensioner arm by unscrewing the single securing bolt ...

10.4b ... remove the protective cap to expose the cam chain adjuster ...

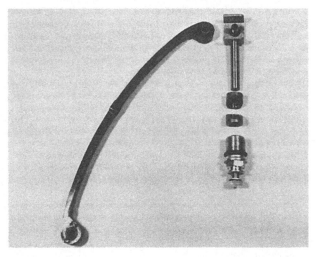

10.4c ... and remove, inspect and store the tensioner assembly

12 Dismantling the engine/gearbox unit: removing the centrifugal oil filter - XL125 S, XL185 S, XR185 and XR200 models

1 When removing the filter unit with the engine in the frame, the crankcase must first be drained of oil and the kickstart lever, clutch cable and tachometer cable disconnected. Details of disconnecting the clutch and tachometer cables from the clutch cover may be found in Section 5 of this Chapter. The kickstart lever is removed by loosening its clamp bolt and pulling it from the shaft. It may also be found necessary to depress the rear brake lever in order to gain access to the cover retaining screw situated below the kickstart lever shaft. If necessary allow the lever to drop well clear of the cover by releasing the operating rod at the drum lever end and unclipping the stoplight switch spring. On models fitted with a decompressor, remove the decompressor cable clamp from the rear of the cover and disconnect the cable end from the cam follower lever. It may also be necessary, on some earlier models, to remove the sumpguard as described in paragraph 2 of Section 5 of this Chapter.

2 It was found that the cover retaining screws were very tight and needed some force to loosen them. Ensure that the spanner used is a very good fit on the screw heads and that a good leverage is obtainable on the spanner. Loosen the screws evenly and in a diagonal sequence to avoid any risk of distortion to the cover. It is a good idea to make up a cardboard template prior to removing the screws because they are of varying lengths. This will ensure that the screws are refitted in their correct positions in the cover and are not lost during the rest of the dismantling sequence.

3 With the engine in the frame, prevent the crankshaft from turning by placing the machine in top gear and applying the rear brake. With the engine removed, the method of stopping crankshaft rotation is the same as that given in paragraph 7 of Section 9 of this Chapter.

4 With the clutch cover removed from the engine unit, slacken and remove the three screws which secure the oil filter cover. As the cover is lifted away, place some rag beneath the unit to catch the residual oil which will be caught inside the filter assembly.

5 The inner half of the filter housing is retained by a slotted nut which will require the use of a peg spanner to release it. This tool is available as a Honda service tool. If this tool cannot be obtained, it is possible to fabricate a suitable tool from a length of thick-walled tubing. Refer to the accompanying illustration for details, cutting away the segments shown with a hacksaw to leave four tangs. If the machine is to be regarded

as a long term purchase, it may be considered worthwhile spending some time with a file to obtain a good fit. The end can then be heated to a cherry red colour and quenched in oil to harden the tangs. An axial hole can be drilled to accept a tommy bar.

6 Lock the crankshaft by whichever method is appropriate. The securing nut can then be removed, and the inner housing pulled off the crankshaft. Note that the special washer, fitted behind the nut, is marked 'outside' for reference during reassembly.

13 Dismantling the engine/gearbox unit: removing the clutch assembly and primary drive pinion

1 Lift out the clutch pushrod, then remove the four bolts which secure the clutch thrust plate, unscrewing them in a diagonal sequence until the clutch spring pressure is released. Lift off the thrust plate and the four clutch springs and place them to one side.

2 On the XR80, XL80 S and XL100 S models, remove the circlip which retains the outer clutch centre. All other models have a locknut and washer in place of this circlip. Before attempting to remove the locknut, stop the crankshaft from rotating by following the procedures given in paragraph 3 of Section 12 of this Chapter. The locknut may be removed by using the tool described in paragraph 5 of Section 12 of this Chapter.

3 Lift the clutch centre out of the clutch drum, followed by the friction and plain plates. The inner clutch centre may now be removed, followed by the splined washer and clutch drum housing.

4 On the XR80, XL80 S and XL100 S models, the primary drive gear is retained on the crankshaft splines by a lockwasher and nut. Lock the crankshaft in position and unscrew the nut. Remove and discard the lockwasher, replacing it with a new item on reassembly, and remove the pressure relief valve assembly followed by the pinion and collar.

5 On all other models, the primary drive gear may simply be slid off the crankshaft end splines.

14 Dismantling the engine/gearbox unit: removing the oil pump and tachometer drive gear

1 The oil pump is retained by two screws which pass through the inner pinion casing into the crankcase. It will be noted that

the front half of the pinion casing has two holes in it. A screw-driver can be passed through these, and the corresponding holes in the pinion, in order to release the two crosshead retaining screws. Note that these may well be tight and may need the services of an impact driver to release them. The pump can then be lifted away as an assembly.

2 Note that there are two O-rings that mate up with the underside of the pump body. These should be removed and renewed during reassembly, to obviate the risk of oil leakage.

3 If it is found necessary to remove the tachometer worm drive gear, the pump unit may be split by removing the two cover retaining screws. The tachometer gear may then be withdrawn along with the oil pump drive gear and the two gears separated. This procedure may be carried out with the pump unit either fitted or removed from the machine.

15 Dismantling the engine/gearbox unit: removing the gear-change shaft and mechanism

1 The gearchange shaft runs in a bore through the crankcase, emerging on the left-hand side of the engine. Once the gear-change pedal is removed from the left-hand end of the shaft, the shaft and gearchange mechanism may be removed as follows.

XR80, XL80 S and XL100 S models
2 Remove the index arm retaining bolt and remove the index arm with the spring still attached.

3 Remove the index plate by unscrewing the central retaining bolt. Withdraw the four operating pins from the end of the gearchange drum.

4 The gearchange spindle assembly may now be withdrawn from the crankcase, taking care to retain the gearchange arm spring located behind the arm itself.

XL125 S, XL185 S, XR185 and XR200 models
5 Withdraw the gearchange shaft from the crankcase. Access may now be gained to the detent mechanism, the two compon-ent parts of which (the star-shaped camplate and the drum stopper arm) are both retained by one 6 mm bolt each. Remove the camplate fixing bolt and lift off the camplate, noting that it is located by a small pin, which should be removed to prevent loss. Use a pair of pliers or a magnetic screwdriver to aid removal of the pin.

6 The drum stopper arm (detent mechanism) is similarly retained. Release the tension on the stopper arm return spring gradually by slackening the pivot bolt. Remove the stopper arm and its return spring.

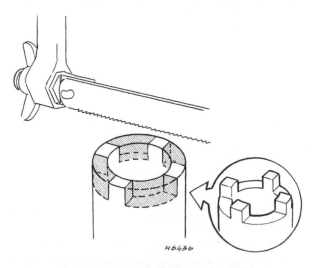

H6436

Fig. 1.5 Home made filter nut peg-spanner

12.4 On all except 80 and 100 models remove the centrifugal oil filter cover ...

12.5a ... to expose the filter housing retaining nut ...

12.5b ... which may be removed by using a specially manufactured tool

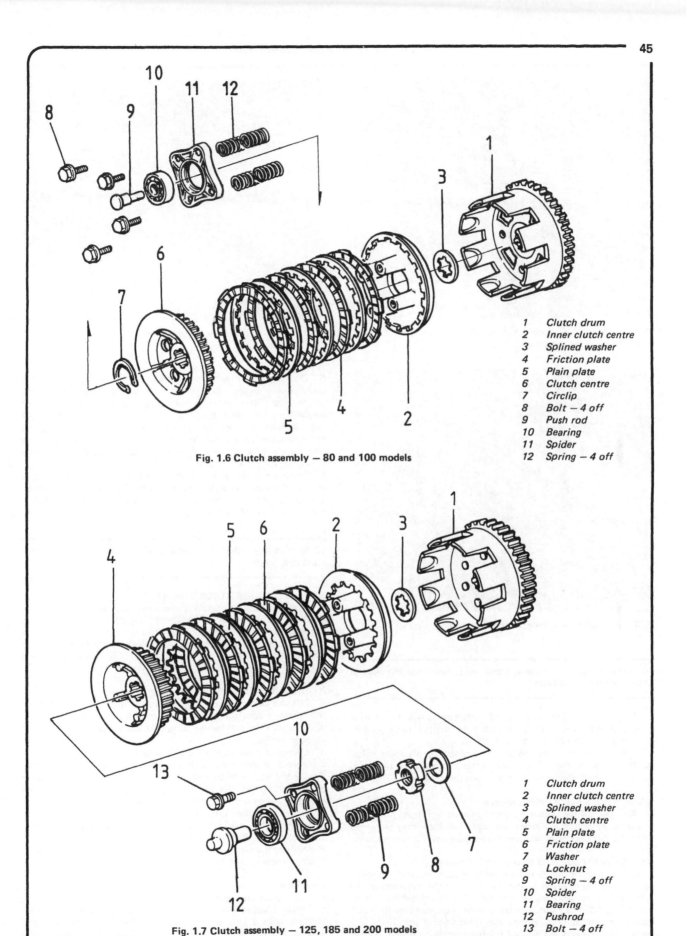

Fig. 1.6 Clutch assembly — 80 and 100 models

1 Clutch drum
2 Inner clutch centre
3 Splined washer
4 Friction plate
5 Plain plate
6 Clutch centre
7 Circlip
8 Bolt — 4 off
9 Push rod
10 Bearing
11 Spider
12 Spring — 4 off

Fig. 1.7 Clutch assembly — 125, 185 and 200 models

1 Clutch drum
2 Inner clutch centre
3 Splined washer
4 Clutch centre
5 Plain plate
6 Friction plate
7 Washer
8 Locknut
9 Spring — 4 off
10 Spider
11 Bearing
12 Pushrod
13 Bolt — 4 off

15.2 On 80 and 100 models, remove the index arm and spring by unscrewing the retaining bolt ...

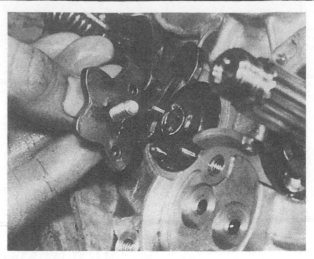

15.3 ... remove the index plate and pins ...

All models
5 Cleaning the filter now will enable it to be refitted straight-away during engine reassembly. Blow any particles of dirt etc out of the filter using a compressed air-line. If this is not available, wash the filter in a clean solvent such as petrol. If any particles, perhaps of rubber, adhere to the filter, agitate the solvent with a small stiff-bristled brush. Take care not to damage the filter. If any damage is apparent in the gauze construction, the filter must be renewed.
6 When using petrol for washing purposes, take extreme care as petrol vapour is highly inflammable. Cleaning should preferably be carried out in the open air or in well ventilated surroundings away from naked flames.

17 Dismantling the engine/gearbox unit: separating the crankcase halves

XR80, XL80 S and XL100 S models
1 With the collar and spring removed from the kickstart lever shaft and all previously mentioned components removed from the engine unit, there are only two screws left holding the crankcase halves together. Remove these two screws with an impact driver, supporting the engine unit on suitable wooden blocks with the right-hand casing half uppermost.

XL125 S, XL185 S, XR185 and XR200 models
2 Lift the cam chain off the crankshaft sprocket and withdraw the chain through the crankcase mouth. Support the engine unit on its side, left-hand casing half uppermost, using suitable wooden blocks.
3 Loosen the ten crankcase half securing screws, evenly and in a diagonal sequence to avoid distortion of the mating surfaces. Ensure the socket or box spanner used to loosen these screws is of a good fit on the screw heads, as once the heads become rounded, removal of the screws will be very difficult.
4 Turn the engine unit over and remove the single bolt located in front of the crankcase mouth. As well as holding the crankcase halves together, this bolt also retains the clutch cable guide. Reposition the engine unit with the left-hand casing half uppermost and remove the ten securing screws.

All models
5 The upper crankcase half should now lift away quite easily, the main bearing remaining attached to the crankshaft. Light tapping with a soft faced mallet on the crankshaft end and around the crankcase joining faces should aid removal if the crankcase half proves stubborn.

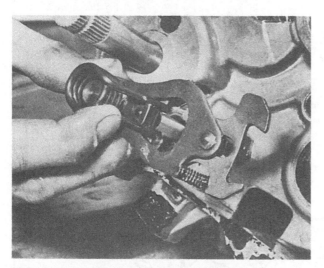

15.4 ... and withdraw the gear change spindle assembly

16 Dismantling the engine/gearbox unit: removing the neutral switch and gauze oil filter

1 Although the above mentioned components will not prevent crankcase separation, it is advisable to remove them as this will need to be done prior to reassembly.
2 The neutral indicator switch is retained by a rubber collar on the inside of the outer casing. After this has been removed, the switch body can be withdrawn from the crankcase.

XR80, XL80 S and XL100 S models
3 The gauze filter fitted to these models is retained in a slot in the lower right-hand crankcase half. The filter may be pulled from the slot once the right-hand crankcase cover is removed.

XL125 S, XL185 S, XR185 and XR200 models
4 If the gauze filter fitted to these models was not removed earlier from its position in the lower part of the left-hand crankcase half, it should now be removed. It is fitted behind a large hexagon-headed cap nut, similar to those used to cover the valve adjustment mechanism. Check the condition of the O-ring seal on the housing cap nut and of the filter retaining spring. Renew these components if necessary.

6   Never drive a wedge, such as the point of a screwdriver, between the mating faces of the crankcase in an attempt to force the crankcase halves apart. This will cause irreparable damage to the mating faces, necessitating an expensive repair or replacement.

18 Dismantling the engine/gearbox unit: removing the crankshaft assembly

1   The use of steel inserts in each crankcase half means that the main bearings are a light sliding fit and should offer little resistance during crankshaft removal. The bearings should therefore remain in position on the crankshaft.
2   Note that on the XR80, XL80 S and XL100 S models, the camshaft chain will need to be removed from the crankshaft sprocket and pulled clear of the engine unit before the crankshaft can be removed from the left-hand crankcase half.

16.3 The gauze oil filter fitted to the 80 and 100 models is retained in the right-hand crankcase

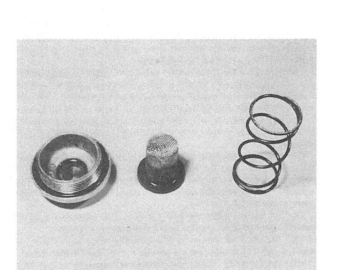

16.4 Check the condition of the cap nut O-ring and filter retaining spring (125, 185 and 200)

18.1 The main bearings should remain in position on the crankshaft

19 Dismantling the engine/gearbox unit: removal of the gearbox components and kickstart mechanism

1   The kickstart shaft assembly may be removed from the crankcase half as a complete unit. On the XL125 S, XL185 S, XR185 and XR200 models, it will however be necessary to unscrew the stop bolt before doing so. Remove the decompressor cam, spring and seat if fitted.
2   Remove the gearbox components by first withdrawing the selector fork shaft and removing the selector forks. Refit the forks on the shaft in the order in which they were removed and place the assembly to one side.
3   Withdraw both mainshaft and layshaft assemblies as a complete unit. Note the positions of the shims fitted to the shaft ends and place the complete assembly to one side. If the assembly is not to be dismantled, it is advisable to secure the gear clusters, selector forks and shaft in their correct relative positions with elastic bands before proceeding further.
4   Withdraw the gearchange selector drum from the crankcase, noting the neutral indicator switch rotor fitted to the end boss.

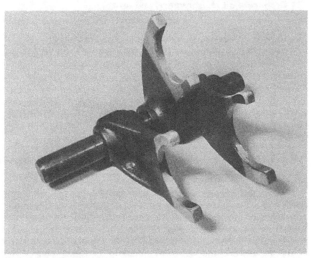

19.2 Refit the selector forks on their shaft in the order in which they were removed

19.4 Note the neutral indicator switch rotor fitted to the gear-change selector drum

20 Examination and renovation: general

1   Before examining the parts of the dismantled engine unit for wear, it is essential that they should be cleaned thoroughly. Use a paraffin/petrol mix to remove all traces of old oil and sludge that may have accumulated within the engine.
2   Examine the crankcase castings for cracks or other signs of damage. If a crack is discovered, it will require professional repair.
3   Examine carefully each part to determine the extent of wear, if necessary checking with the tolerance figures listed in the Specifications Section of this Chapter, or accompanying the text.
4   Use a clean, lint-free rag for cleaning and drying the various components, otherwise there is risk of small particles obstructing the internal oilways.
5   Should any studs or internal threads require repair, now is the appropriate time to attend to them. Where internal threads are stripped or badly worn, it is preferable to use a thread insert, rather than tap oversize. Most dealers can provide a thread reclaiming service by the use of Helicoil thread inserts. They enable the original component to be re-used.

21 Crankcase oil seals: removal and refitting

1   It may be necessary, before removing the bearings contained in either of the crankcase halves, to remove the oil seals from their respective housings in the castings. These seals are easily removed by prising them out of position with the flat of a screwdriver. Great care should be taken, however, to ensure that the alloy in or around the seal housings is not damaged during this operation. Equal care should be taken when fitting the new seals; use a soft wooden block to drift them into position.

22 Crankshaft and gearbox main bearings: removal and refitting

1   If wear is apparent in the crankshaft main bearings, then the bearings must be removed from the crankshaft by using a puller of the correct type. On no account should the crankshaft assembly be subject to severe force, such as hammer blows, in an attempt to remove the bearings. Any such force will serve only to distort the assembly, bearing in mind that it is of pressed construction.
2   It should also be noted that if the crankshaft main bearings are defective, then the big-end bearing is probably also in poor

condition. Taking all these factors into account it is considered advisable to take the crankshaft assembly to an official Honda Service Agent who will be able to give an expert opinion as to the condition of the assembly and who will also have the special tools required to replace the big-end and bearings readily at hand.
3   It should be noted that some models are fitted with a crankshaft cam chain sprocket of such a design that it is not possible to employ a standard type of puller in order to remove the sprocket from the shaft. It is therefore necessary to remove the main bearing situated behind the sprocket in order to draw the sprocket itself off the shaft. Before disturbing the cam chain drive sprocket note the remarks on sprocket positioning given in Section 29.
4   If it is not possible to obtain the services of a Honda Service Agent and the special bearing puller is not available, then it may be possible to free the main bearings from the crankshaft by using the following procedure.
5   Insert two thin steel wedges, one on each side of the bearing, and with these clamped in a vice hit the end of the crankshaft squarely with a rawhide mallet in an attempt to drive the crankshaft through the bearing. When the bearing has moved the initial amount, it should be possible to insert a conventional two or three legged sprocket puller, to complete the drawing-off action. Great care should be taken when using this technique. If the bearings prove stubborn, abandon the attempt, and wait until a suitable puller can be obtained.
6   The gearbox bearings are of a relatively light fit in the crank-case castings and can therefore be removed by using a tube of suitable diameter or a soft metal drift and hammer. With the bearing that cannot be drifted out in this manner, owing to one side being blanked off, the only practical solution is to employ the use of a slide hammer as shown in the photograph accompanying this text.
7   Warming the crankcases will aid removal of the bearings. Care should be taken not to overdo this because distortion of the castings may occur, a slight rise in temperature being all that is necessary. Immersing the casing in boiling water is the recommended method.
8   Ensure, before applying any force to the crankcase halves, that they are well supported around the bearing housing.
9   Both crankshaft and gearbox bearings may be fitted by using a length of tube, or a socket, as a drift with which to tap them into position. With the crankshaft bearings, ensure the diameter of the tube is the same as that of the inner race of the bearings. With the gearbox bearings, the diameter of the tube used should be equal to that of the outer race of the bearings. It is essential that the component to which the bearing is to be fitted is well supported during the fitting procedure.

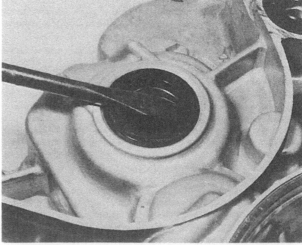

21.1a Care should be taken when removing the crankcase oil seals

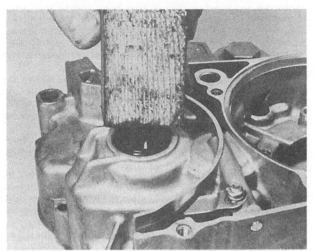

21.1b Tap the oil seals into position with a soft wooden block

22.6 It is necessary to use a slide hammer to remove the bearing fitted in the blind crankcase recess

5   The crankshaft main bearings are of the journal ball type. If wear is evident in the form of play or if the bearings feel rough as they are rotated, then renewal of the bearings is necessary.

22.9 Use a length of tube or a socket to refit the bearings

23 Examination and renovation: big-end and main bearings

1   Failure of the big-end bearing is invariably accompanied by a knock from within the crankcase that progressively becomes worse. Some vibration will also be experienced. There should be no vertical play in the big-end bearing after the old oil has been washed out. If even a small amount of play is evident, the bearing is due for replacement. Do not run the machine with a worn big-end bearing, otherwise there is risk of breaking the connecting rod or crankshaft.
2   A certain amount of side movement of the connecting rod is intentional. Refer to the Specifications at the beginning of this Chapter and measure the big-end axial clearance by inserting a feeler gauge between the flywheel inner face and the face of the big-end eye boss.
3   It is not possible to separate the flywheel assembly in order to replace the bearing because the parallel sided crankpin is pressed into the flywheels. Big-end repair should be entrusted to an official Honda Service Agent who will have the necessary repair or replacement facilities.
4   Failure of the main bearings is usually evident in the form of an audible rumble from the bottom end of the engine, accompanied by vibration. The vibration will be most noticeable through the footrests.

24 Examination and renovation: gudgeon pin, small-end and piston bosses

1   The fit of the gudgeon pin in both the small-end eye of the connecting rod, and in the piston bosses should be checked. In the case of the small-end eye, slide the pin into position and check for wear by moving the pin up and down. The pin should be a light sliding fit with no discernible radial play. If play is detected, it will almost certainly be the small-end eye which has worn rather than the gudgeon pin, although in extreme cases, the latter may also have become worn. The connecting rod is not fitted with a bush type of small-end bearing, and consequently a new connecting rod will have to be fitted if worn. This is not a simple job, as the flywheels must be parted to fit the new component, and is a job for a Honda Service Agent. It should be borne in mind that if the small-end has worn, it is likely that the big-end bearing will require attention.
2   Check the fit of the gudgeon pin in the piston. This is normally a fairly tight fit, and it is not unusual for the piston to have to be warmed slightly to allow the pin to be inserted and removed. After considerable mileages have been covered, it is possible that the bosses will have become enlarged. If this proves to be the case, it will be necessary to renew the piston to effect a cure. It is worth noting, as an aid to diagnosis, that wear in the above areas is characterised by a metallic rattle when the engine is running.

25 Examination and renovation: piston and piston rings

1   If a rebore is necessary, the existing piston and rings can be disregarded because they will be replaced with their oversize equivalents as a matter of course.
2   Remove all traces of carbon from the piston crown, using a soft scraper to ensure the surface is not marked. Finish off by polishing the crown, with metal polish, so that carbon does not adhere so easily in the future. Never use emery cloth.
3   Piston wear usually occurs at the skirt or lower end of the piston and takes the form of vertical streaks or score marks on the thrust side. There may also be some variation in the thickness of the skirt.

4   The piston ring grooves may also become enlarged in use, allowing the piston rings to have greater side float. If the clearance exceeds that given in Specifications for any given ring, then the piston is due for renewal. It is unusual for this amount of wear to occur on its own.

5   Piston ring wear is measured by removing the rings from the piston and inserting them in part of the cylinder bore which has not been subject to wear. Ideally, the rings should be inserted in the cylinder bore approximately 0.5 inch from the bottom of the bore, using the crown of the piston to locate them squarely. Measure the end gap with a feeler gauge; if it exceeds the figure given in the Specifications at the beginning of this Chapter, the rings require renewal.

6   Remove the piston rings by pushing the ends apart with the thumbs whilst gently easing each ring from its groove. Great care is necessary throughout this operation because the rings are brittle and will break easily if overstressed. If the rings are gummed in their grooves, three strips of tin can be used, to ease them free, as shown in the accompanying illustration.

7   Examine the working surface of each piston ring. If discoloured areas are evident, the ring should be renewed because these areas indicate the blow-by of gas. Check that there is not a build-up of carbon on the back of the ring or in the piston ring groove, which may cause an increase in the radial pressure. A portion of broken ring affords the best means of cleaning out the piston ring grooves.

8   The rings may be refitted to the piston by carefully pulling the ends apart just enough to allow the rings to pass over the piston crown and into their respective grooves. Always fit the rings with the marked (T, R or N) surface uppermost, taking care not to confuse the top ring with the second ring. When fitted, the rings should rotate freely in their grooves. Space the ring end gaps 120 degrees apart. The oil ring side rails must have their end gaps positioned 20 mm (0.8 in) or more away from the spacer ring end gap. The assembly may now be placed to one side ready for refitting to the connecting rod.

26 Examination and renovation: cylinder barrel

1   The usual indications of a badly worn cylinder barrel and piston are excessive oil consumption and piston slap, a metallic rattle that occurs when there is little or no load on the engine. If the top of the bore of the cylinder barrel is examined carefully, it will be found that there is a ridge on the thrust side, the depth of which will vary according to the amount of wear that has taken place. This marks the limit of travel of the uppermost piston ring.

2   Measure the bore diameter just below the ridge at the top of the barrel, at a mid-point in the bore and at the bottom of the bore. Measurements should be made with an internal micrometer in line with the gudgeon pin axis and a second measurement made at right angles to the first measurement. If the difference between two readings taken at different levels exceeds 0.1 mm (0.004 in), it is necessary to have the cylinder rebored and an oversize piston and rings fitted.

3   If an internal micrometer is not available, it is possible to determine the amount of bore wear by inserting the piston, without rings, so that it is just below the ridge at the top of the bore. Measure the distance between the bore wall and the side of the piston. Move the piston down to the bottom of the bore and repeat the measurement. Doing this, and subtracting the lesser measurement from the greater, will give the difference between the bore diameter in an area where the greatest amount of wear is likely to occur and an area in which there should be little or no wear. If this difference exceeds 0.1 mm (0.004 in) then the previously stated remedial action must be taken.

4   Establishment of the clearance between the cylinder bore and the piston can be made either by direct measurement of the cylinder bore and piston diameters and then by subtracting the latter figure from the former, or by actual measurement of the gap using a feeler gauge. In either case, if the clearance exceeds the maximum wear limit there is evidence that a new piston or a rebore and new piston is required. Seek the advice of a qualified dealer if there is any doubt about what action to take.

5   Check the surface of the cylinder bore for score marks or any other damage that may have resulted from an earlier engine seizure or displacement of the gudgeon pin. A rebore will be necessary to remove any deep indentations, irrespective of the amount of bore wear, otherwise a compression leak will occur.

6   Check the external cooling fins are not clogged with oil or road dirt, otherwise the engine will overheat.

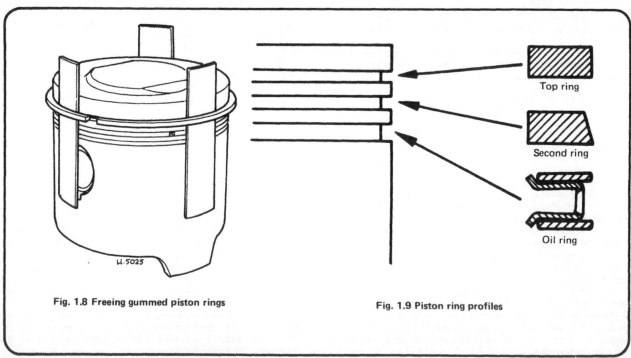

Fig. 1.8 Freeing gummed piston rings

Fig. 1.9 Piston ring profiles

25.4 Measure the amount of piston ring to groove clearance

27 Examination and renovation: valves, valve seats, valve guides and valve springs

1 Remove each valve in turn, using a valve spring compressor, and place the valves, springs, seats and collet halves in a suitable box or bag marked to denote inlet or exhaust as appropriate. Assemble the valve spring compressor in position on the cylinder head, and gradually tighten the threaded portion to place pressure on the upper spring seat. Do not exert undue force to compress the springs, the tool should be placed under slight load, and then tapped on the end to jar the collet halves free. Continue to compress the springs until the collet halves can be dislodged using a small screwdriver. Note that the valve springs exert considerable force, and care should be taken to avoid the compressed assembly flying apart. To this end, a small magnet is invaluable for retrieving the collet halves, being more practical than fingers.

2 After cleaning the valves to remove all traces of carbon, examine the heads for signs of pitting and burning. Examine also the valve seats in the cylinder head. The exhaust valve and its seat will probably require the most attention because this is the hotter running of the two. If the pitting is slight, the marks can be removed by grinding the seats and valves together, using fine valve grinding compound.

3 Valve grinding is a simple task, carried out as follows. Smear a trace of fine valve grinding compound (carborundum paste) on the seat face and apply a suction grinding tool to the head of the valve. With a semi-rotary motion, grind in the valve head to its seat. It is advisable to lift the valve occasionally, to distribute the grinding compound evenly. Repeat this operation until an unbroken ring of light grey matt finish is obtained on both valve and seat. This denotes the grinding operation is complete. Before passing to the next operation, make quite sure that all traces of the grinding compound have been removed from both the valve and its seat and that none has entered the valve guide. If this precaution is not observed, rapid wear will take place, due to the abrasive nature of the carborundum base.

4 When deeper pit marks are encountered, it will be necessary to use a valve refacing machine and also a set of three valve seat cutters to restore the seat to the original width and condition. This course of action should be resorted to only in an extreme case because there is risk of pocketing the valve and reducing performance. For the same reason, never resort to excessive grinding; reduced engine efficiency will result.

5 Because of the expense of purchasing the seat cutters and because of the accuracy with which cutting must be carried out, it is strongly recommended that the cylinder head be returned to an official Honda Service Agent who will not only be able to carry out the job to the degree of accuracy required

but will also be able to give advice as to the need to renew the valves should there be any doubt as to their condition.

6 Examine the condition of the valve collets and the groove on the valve in which they seat. If there is **any** sign of damage, new replacements should be fitted. If the collets work loose whilst the engine is running, a valve will drop in and cause extensive damage.

7 Measure the valve stems for wear, making reference to the tolerance values given in the Specifications Section of this Chapter.

8 Check the clearance between each valve stem and the guide in which it operates. The valve stem/guide clearance can be measured with the use of a dial gauge and a new valve. Place the new valve into the guide and measure the amount of shake with the dial gauge tip resting against the top of the stem. If the amount of wear is greater than the wear limit, the guide must be renewed.

9 To remove the worn guide, first remove any carbon deposits which may have built up on the guide end projecting into the port. Carbon deposits will impede the progress of the guide out of the cylinder head and this may also cause damage to the head.

10 A special valve guide removal tool is made by Honda. If this tool cannot be borrowed or hired from a Service Agent then it is possible to make up a shouldered drift with which to carry out the job. The smaller diameter of the drift should be close to that of the valve stem and the larger diameter slightly smaller than that of the valve guide.

11 With the cylinder head well supported on the work surface, drive out the guide from the valve port side. If the guide will not move from the cylinder head, do not use excessive force because this will damage the alloy casting. Place the cylinder head in an oven and heat it to approximately 100°C (212°F) before again attempting to drift the guide out of its location. To prevent distortion of the large alloy casting it is essential that the cylinder head is heated evenly. For this reason an oven **must** be used in preference to a blow torch or other methods of heating.

12 To fit the new valve guides, fit a new O-ring on the guide, place the cylinder head with the head to cylinder barrel mating face downwards on the work surface, oil the guide to cylinder head contact surfaces and drive the guide into its location in the head. Before carrying out this procedure ensure that the work surface is padded sufficiently to avoid damage to the cylinder head mating face. The guide bores in the cylinder head must be free from all contamination.

13 Do not use excessive force when driving the valve guides into their respective bores. Heating the cylinder head, as mentioned previously in this Section, will aid fitting. Use a soft-metal drift and hammer to fit the guides and inspect the guides for damage when fitted.

14 The valve guide will need to be reamed, when fitted, to the standard bore diameter quoted in the Specifications Section. A service tool is supplied by Honda for this purpose. Ensure the cylinder head is cleaned thoroughly on completion to remove any metal particles. Whenever a new guide has been fitted the valve seat must be re-cut to ensure centralisation of the seat relative to the guide axis. Following this, valve grinding should be carried out in the normal way.

15 Check the free length of each of the valve springs. The springs have reached their serviceable limit when they have compressed to the limit readings given in the Specifications Section of this Chapter.

16 Before reassembling the valve and valve springs by reversing the dismantling procedure, the cylinder head should be de-carbonised as detailed in the following Section. Ensure that all the springs are fitted with the close coils downward towards the cylinder head. Fit new oil seals to each valve stem and oil both the valve stem and guide bore prior to reassembly. Take special care not to damage this seal when inserting the valve into the head. As a final check after assembly, give the end of each valve stem a sharp tap with a hammer, to make sure the split collets have located correctly.

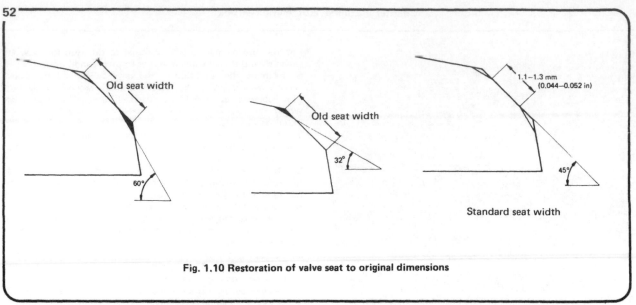

Fig. 1.10 Restoration of valve seat to original dimensions

27.16a Fit the valve spring seating washers ...

27.16b ... and insert the valve into the cylinder head

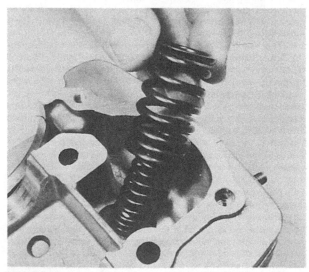

27.16c Ensure that all springs are fitted with the close coils downward towards the cylinder head ...

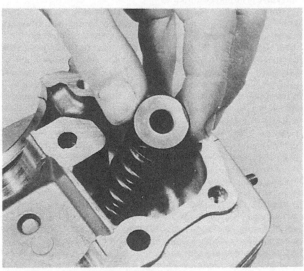

27.16d ... and fit the valve spring collar ...

27.16e ... followed by the collets

28 Examination and decarbonisation: cylinder head

1 Remove all traces of carbon from the combustion chamber and valve port surfaces by using either a hardwood or plastic scraper. Do not use anything that may score the light alloy surfaces, otherwise hot spots and leakages may occur. Finish by polishing the surfaces, using metal polish and a piece of cloth. A power operated polishing wheel is ideal for this purpose. Never use emery cloth.
2 It is best to have the valves fitted when decarbonising the combustion chamber as this will allow the valve heads to be polished and also prevent any possible damage to the valve seats.
3 With the valves removed, check to make sure the valve guide bores are free from carbon or any other foreign matter that may cause the valves to stick.
4 Make sure the cylinder head fins are not clogged with oil or road dirt, otherwise the engine will overheat. If necessary, use a degreasing agent and brush to clean between the fins. Check that no cracks are evident, especially in the vicinity of the holes through which the holding down studs and bolts pass, and near the sparking plug threads.
5 If leakage problems have been experienced between the cylinder head and cylinder barrel mating surfaces, the cylinder head should be checked for distortion by placing a straight edge across the mating surface and attempting to slide a 0.1 mm (0.004 in) feeler gauge between the straight edge and the mating surface. If this can be done then the cylinder head must be machined flat or a new head fitted. Most cases of cylinder head distortion can be traced to unequal tensioning of the cylinder head nuts and by tightening them in the incorrect sequence.
6 When refitting the valves in their guides, ensure that the stems are liberally coated in oil. The guide bores must be cleaned of all traces of metal polish, carbon deposits, etc, before the valves are fitted.

29 Examination and renovation: camshaft, camshaft sprockets and cam chain

1 The camshaft lobes should have a smooth surface and be entirely free from scuff marks or indentations. It is unlikely that severe wear will be encountered during the normal service life of the machine unless the lubrication system has failed, causing the case hardened surface to wear through. Details of cam height are given in the Specifications at the beginning of

this Chapter. If either cam is below the service limit given, then the camshaft must be renewed.
2 Thoroughly clean the camshaft so that all oilways and grooves are clear of any obstruction. After cleaning the component in petrol/paraffin, use an air supply to blow through the oilways.
3 On XR80, XL80 S and XL100 S models, clean the camshaft block and blow through the internal oilways to remove any obstruction.
4 Ensure the timing marks on the camshaft sprocket are clearly visible, because they are easily obscured by old oil. It will be necessary to refer to these marks during engine re-assembly.
5 Examine the camshaft chain sprockets for worn, broken or chipped teeth, an unusual occurrence that can often be attributed to the presence of foreign bodies or particles from some other broken engine component. Renewal of the camshaft driven sprocket, ie the one fitted to the camshaft, is straight-forward. The crankshaft-mounted drive sprocket, however, will require drawing from position using a suitable puller and, as noted in Section 22, on some models this can only be accomplished if the timing side main bearing is removed simultaneously. On all models the driven sprocket is a tight interference fit on the crankshaft, but the actual positioning of the sprocket can be infinite, by virtue of there being no keyway. To maintain the correct valve timing relationship the sprocket should be refitted with reference to a known datum line; the datum in this case being an extended centre line of the generator rotor locating keyway. This can be seen clearly in the accompanying illustrations. On XL125 S, XL185 S, XR185 and XR200 models the sprocket should be repositioned so that the centre line of the keyway passes through the centre line of any tooth. On XL80 S and XL100 S models the keyway centre line should pass through the dead centre between any two teeth. The correct positioning for the XR80 model is not specified; for this reason the position of the sprocket must be recorded accurately **before** the old sprocket is removed.
6 Examine the camshaft chain for excessive wear or cracked or broken rollers. An indication of wear is given by the extent to which the chain can be bent sideways; if a pronounced curve is evident, the chain should be renewed.
7 The chain tensioner, like the chain, does not normally give trouble because it is well lubricated. It is well, however, to check that the coating of the tensioner blade and guide blade has not worn through, and that the tension pushrod slides freely in its location.
8 On the XR80, XL80 S and XL100 S models, in order to renew the tensioner spring, the tensioner assembly must be removed from the cylinder barrel by unhooking the spring and slackening the adjusting bolt before withdrawing the assembly from its location.
9 On all other models, the tensioner arm and spring may be examined for signs of excessive wear whilst in situ and either part renewed by removing the single securing bolt.
10 On all models, in cases of doubt, err on the side of safety and renew components without question. The suspect item may be.taken to an official Honda Service Agent who will give his opinion as to whether renewal is necessary. Although a rare happening, a broken camshaft chain may cause extensive engine damage, with risk of the valves tangling with the piston.

30 Examination and renovation: rocker arms and spindles

1 It is unlikely that excessive wear will occur in either the rocker arms or the rocker shafts unless the flow of oil has been impeded or the machine has covered a very high mileage. A clicking noise from the rocker area is the usual symptom of wear in the rocker gear, which should not be confused with a somewhat similar noise caused by excessive valve clearances.
2 If any shake is present and the rocker arm is loose on its shaft, a new rocker arm and/or shaft should be fitted.

3 Check the tip of each rocker arm at the point where the arm makes contact with the cam. If signs of cracking, scuffing or break through in the case hardened surface are evident, fit a new replacement.

4 Check also the thread of the tappet adjusting screw, the thread of the rocker arm into which it fits and the thread of the locknut. The hardened end of the tappet adjuster must also be in good condition.

XR80, XL80 S and XL100 S models

5 To extract the rocker arm spindles from the camshaft holder, thread a bolt of the correct size into the threaded end of the spindle and pull the spindle out of its location in the holder. The rocker arms may then be removed for inspection.

XL125 S, XL185 S, XR185 and XR200 models

6 To extract the rocker arm spindles from the cylinder head cover, remove the plate retaining the spindles by unscrewing the crosshead screw. This will allow a 6 mm bolt to be inserted into the threaded end of each shaft. The shafts may then be drawn clear of the head cover and the rocker arms detached.

All models

7 Examination of the rocker arms and spindles should be carried out on a general basis as already stated. If excessive wear is suspected, refer to the Specifications at the beginning of this Chapter for the limits of wear and renew components as considered necessary.

8 Reassembly of the rocker components is carried out by reversing the dismantling procedure. Note that on the models referred to in paragraph 6 of this Section, the recess milled in each spindle must line up with the hole through which the cylinder securing studs pass when they are refitted. On the XL80 S, XR80 and XL100 S this applies only to the inlet rocker spindle.

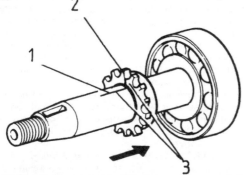

Fig. 1.11 Camchain sprocket installation marks on crankshaft — XL80 S and XL100 S

1 Crankshaft mark   3 Correct alignment positions
2 Sprocket mark

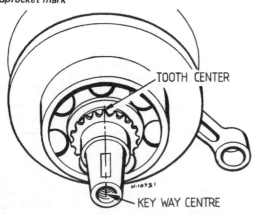

Fig. 1.12 Correct positioning of cam chain sprocket on crankshaft

28.5 Check the cylinder head mating surface for distortion

29.2 Thoroughly clean and inspect the camshaft components

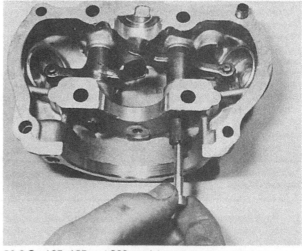

30.6 On 125, 185 and 200 models, extract the rocker arm spindles from the cylinder head ...

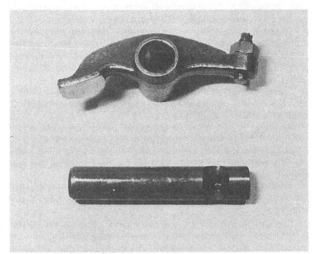

30.7 ... examine the rocker arms and spindles for wear ...

30.8 ... and refit the components in the cylinder head, locking the spindles in position with the plate

31 Examination and renovation: valve lifter assembly - XL185 S, XR185 and XR200 models

1 The cylinder head cover valve lifter assembly should not require any maintenance until the machine has been in service for a considerable period of time. If the lever shaft is suspected of having seized or the return spring is seen to have broken or weakened, the shaft may be removed from the cover by unscrewing the pin bolt located at the front right-hand upper corner of the cover. The spring may be unclipped from the lever before the lever is withdrawn.

2 Inspect the lever shaft for any signs of seizure in the cover bore. If necessary, clean the shaft and bore with a rag lightly moistened in paraffin and oil the same before reassembly. If the shaft or bore is found to be heavily scored due to grit finding its way through the seal, then it will be necessary to renew the seal. Failure of the seal is indicated by oil seeping from behind the lever with the assembly in situ. Removal of the seal is easily achieved by levering it from the cylinder cover recess using the flat of a screwdriver. Fit the new seal by pressing it into the cover recess squarely and in an identical position to the one removed.

3 Refitting of the valve lifter assembly is the reverse procedure to that for removal. Ensure that the pin bolt locates correctly in the lever shaft recess.

4 Removal of the cam follower assembly at the kickstart lever end is easily achieved after removal of the right-hand crankcase cover. Remove the return spring after having carefully noted its fitted position for reference when refitting. Remove the circlip retaining the shaft in position and withdraw the lever and shaft from the cover, allowing the cam follower to detach from the shaft.

5 Wear should not occur on these components until they have been in service for some considerable period of time. Inspect each individual item for signs of deterioration and wear, taking special note of the bearing surfaces on the cam follower lever and shaft. Service the shaft and oil seal as described in paragraph 2 of this Section. Refitting of the cam follower assembly is the reverse procedure to that for removal.

32 Examination and renovation: trochoid oil pump

The trochoid oil pump is removed as a sub-assembly as described earlier in this Chapter. Dismantling and renovation are covered in Chapter 2. The condition of the pump should

be checked as a matter of course if the engine is being overhauled, especially if signs of scuffing are evident on the various shafts and bushes.

33 Examination and renovation: clutch and primary drive

1 Give the plain and the inserted plates a wash with a petrol/paraffin mix and remove all traces of clutch insert debris. If this precaution is not taken, a gradual build-up of debris will occur and eventually affect clutch action.

2 After a considerable mileage has been covered, the bonded linings of the clutch friction plates will wear down to or beyond the specified wear limit, allowing the clutch to slip.

3 The degree of wear on the friction plates is measured across the faces of the friction material; the standard or new sizes being as given in the Specifications at the beginning of this Chapter. If the plates have worn to the service limit, they should be renewed even if clutch slip is not yet apparent. Check the plates for signs of warpage.

4 The plain plates should be free from scoring and signs of overheating, which will be apparent in the form of blueing. The plates should also be flat. If the plate warpage is more than that stated in Specifications, clutch judder or snatch may result.

5 Measure the free length of the clutch springs. If they have taken a permanent set to a length less than that given as a service limit in Specifications, they should be renewed. Always renew the springs as a set.

6 Check the condition of the thrust bearing assembly and pushrod, which are located in the clutch centre. Excessive play or wear will cause noise and erratic operation.

7 Check the condition of the slots in the outer surface of the clutch centre and the inner surfaces of the outer drum. In an extreme case, clutch chatter may have caused the tongues of the inserted plates to make indentations in the slots of the outer drum, or the tongues of the plain plates to indent the slots of the clutch centres. These indentations will trap the clutch plates as they are freed, and impair clutch action. If the damage is only slight the indentations can be removed by careful work with a file and the burrs removed from the tongues of the clutch plates in a similar fashion. More extensive damage will necessitate renewal of the parts concerned.

8 The clutch release mechanism is located within the right-hand crankcase cover and takes the form of a spindle with an integral or separate cam. A light return spring ensures that pressure is taken off the end of the pushrod once the handlebar lever has been released. No attention is normally required, other

than greasing prior to reassembly.

9 The primary drive is by means of a crankshaft-mounted gear driving the clutch by way of teeth on the outer drum. The two gears should be examined for signs of wear or chipped teeth, and replaced as necessary, preferably as a pair.

34 Examination and renovation: gearbox components

1 Examine each of the gear pinions to ensure that there are no chipped or broken teeth and that the dogs on the end of the pinions are not rounded. Gear pinions with these defects must be renewed; there is no satisfactory method of reclaiming them. If damage or wear warrants renewal of any gear pinions the assemblies may be stripped down, displacing the various shims and circlips as necessary.

2 The accompanying illustrations show how both clusters of the gearbox are assembled on their respective shafts. It is imperative that the gear clusters, including the thrust washers, are assembled in **exactly** the correct sequence, otherwise constant gear selection problems will occur. In order to eliminate the risk of misplacement, make rough sketches as the clusters

are dismantled. Also strip and rebuild as soon as possible to reduce any confusion which might occur at a later date.

3 Examine the selector forks carefully, ensuring that there is no scoring or wear where they engage in the gears, and that they are not bent. Damage and wear rarely occur in a gearbox which has been properly used and correctly lubricated, unless very high mileages have been covered.

4 Check the selector fork shaft for straightness by rolling it on a sheet of plate glass. A bent rod will cause difficulty in selecting gears.

5 The tracks in the selector drum, which co-ordinate the movement of the selector forks, should not show signs of undue wear. Check also that the drum stopper arm (detent arm) spring has not weakened, and that no play has developed in the gear selector linkages.

6 Refer to the Specifications at the beginning of this Chapter for the maximum wear allowed on the selector forks, shaft and drum.

7 Note the condition of the bush within the starter pinion. If this bush appears to be overworn or damaged in any way, the pinion should be returned to an official Honda Service Agent who will be able to decide whether or not the component should be renewed.

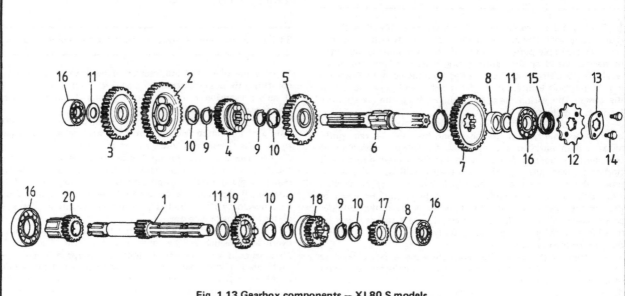

Fig. 1.13 Gearbox components — XL80 S models

1	Mainshaft	11	Thrust washer
2	Layshaft 1st gear pinion	12	Final drive sprocket
3	Kickstart idler gear	13	Locking washer
4	Layshaft 4th gear pinion	14	Bolt – 2 off
5	Layshaft 3rd gear pinion	15	Oil seal
6	Layshaft	16	Bearing
7	Layshaft 2nd gear pinion	17	Mainshaft 2nd gear pinion
8	Collar	18	Mainshaft 3rd gear pinion
9	Circlip	19	Mainshaft 4th gear pinion
10	Splined washer	20	Kickstart sleeve gear

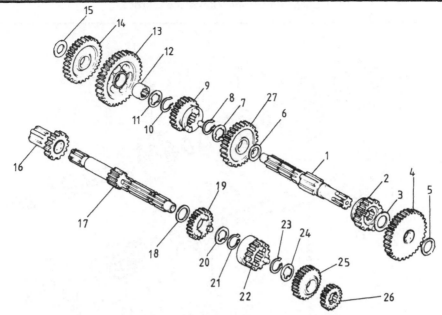

Fig. 1.14 Gearbox components — XR80 and XL100 S models

1	Layshaft	10	Circlip	19	Mainshaft 4th gear pinion
2	Layshaft 5th gear pinion	11	Splined washer	20	Splined washer
3	Thrust washer	12	Low gear collar — XL100 S only	21	Circlip
4	Layshaft 2nd gear pinion	13	Layshaft low gear pinion	22	Mainshaft 3rd gear pinion
5	Thrust washer	14	Kickstart idler gear	23	Circlip
6	Thrust washer — XL100 S only	15	Washer	24	Splined washer
7	Splined washer	16	Kickstart sleeve gear	25	Mainshaft 5th gear pinion
8	Circlip	17	Mainshaft	26	Mainshaft 2nd gear pinion
9	Layshaft 4th gear pinion	18	Thrust washer	27	Layshaft 3rd gear pinion

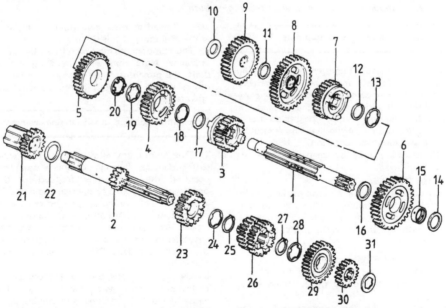

Fig. 1.15 Gearbox components — XL125, XR185 and XR200

1	Layshaft	12	Circlip	22	Thrust washer
2	Mainshaft	13	Splined washer	23	Mainshaft 5th gear pinion
3	Layshaft 6th gear pinion	14	Thrust washer	24	Splined washer
4	Layshaft 4th gear pinion	15	Bush	25	Circlip
5	Layshaft 3rd gear pinion	16	Thrust washer	26	Mainshaft 3rd and 4th gear pinion
6	Layshaft 2nd gear pinion	17	Circlip	27	Circlip
7	Layshaft 5th gear pinion	18	Splined washer	28	Splined washer
8	Layshaft low gear pinion	19	Splined washer	29	Mainshaft 6th gear pinion
9	Kickstart idler gear*	20	Lock washer	30	Mainshaft 2nd gear pinion
10	Thrust washer	21	Kickstart driven gear	31	Splined washer
11	Thrust washer				

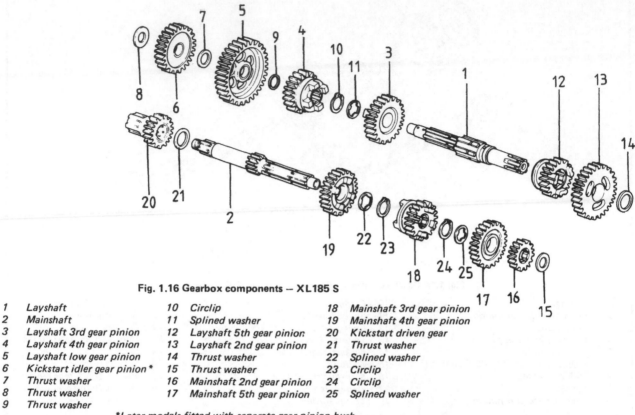

Fig. 1.16 Gearbox components — XL185 S

1	Layshaft	10	Circlip	18	Mainshaft 3rd gear pinion
2	Mainshaft	11	Splined washer	19	Mainshaft 4th gear pinion
3	Layshaft 3rd gear pinion	12	Layshaft 5th gear pinion	20	Kickstart driven gear
4	Layshaft 4th gear pinion	13	Layshaft 2nd gear pinion	21	Thrust washer
5	Layshaft low gear pinion	14	Thrust washer	22	Splined washer
6	Kickstart idler gear pinion *	15	Thrust washer	23	Circlip
7	Thrust washer	16	Mainshaft 2nd gear pinion	24	Circlip
8	Thrust washer	17	Mainshaft 5th gear pinion	25	Splined washer
9	Thrust washer				

Later models fitted with separate gear pinion bush

35 Examination and renovation: kickstart components

1 Check the condition of the kickstart components. If slipping has been encountered a worn ratchet and pawl will invariably be traced as the cause. Any other damage or wear to the components will be self-evident. If either the ratchet or pawl is found to be faulty, both components must be replaced as a pair. Examine the kickstart return spring, which should be renewed if there is any doubt about its condition.

2 The kickstart shaft assembly may be dismantled into its individual components as follows.

XR80, XL80 S and XL100 S models

3 Remove the circlip from the groove at the end of the spindle. This will release the ratchet spring which may then be removed from the spindle together with the drive ratchet. Remove the second circlip, together with the thrust washer, ratchet pinion and second thrust washer.

XL125 S, XL185 S, XR185 and XR200 models

4 Draw the collar off the kickstart lever end of the spindle, followed by the return spring and guide plate. Remove the ratchet pinion.

5 Remove the thrust washer from the opposite end of the spindle and remove the 18 mm circlip. Draw off the seat plate, ratchet spring, drive ratchet and second thrust washer. Finally, remove the second circlip from the spindle.

All models

6 In general, kickstart mechanisms are reliable devices and do not wear quickly unless a malfunction occurs or engine oil level is allowed to fall. Inspect each individual component for signs of damage or wear; looking for chipped or broken teeth on the ratchet/pinion assembly, broken or weakened springs and wear of the spindle/pinion/ratchet splines. Measure the pinion internal diameter and the kickstart spindle outer diameter at the point where the pinion slides. The maximum amount of wear allowed at these two points is given in the Specifications at the beginning of this Chapter.

7 The components may be refitted by reversing the dismantling sequence. Note that when refitting the drive ratchet to the shaft, the punch mark on each should be in alignment with the other. If this alignment is not correct, the kickstart will not operate correctly.

36 Examination and repair: engine casings and covers

1 The aluminium alloy casings and covers are unlikely to suffer damage through ordinary use. However, damage can occur if the machine is dropped, or if sudden mechanical breakages occur, such as the rear chain breaking.

2 Small cracks or holes may be repaired with an epoxy resin adhesive, such as Araldite, as a temporary expedient. Permanent repairs can only be effected by argon-arc welding, and a specialist in this process is in a position to advise on the viability of proposed repair. Often it may be cheaper to buy a new replacement.

3 Damaged threads can be economically reclaimed by using a diamond section wire insert, of the Helicoil type, which is easily fitted after drilling and re-tapping the affected thread. The process is quick and inexpensive, and does not require as much preparation and work as the older method of fitting brass, or similar inserts. Most motorcycle dealers and small engineering firms offer a service of this kind.

4 Sheared studs or screws can usually be removed with screw extractors, which consist of tapered, left-hand thread screws, of very hard steel. These are inserted by screwing anti-clockwise, into a pre-drilled hole in the stud, and usually succeed in dislodging the most stubborn stud or screw. If a problem arises which seems to be beyond your scope, it is worth consulting a professional engineering firm before condemning an otherwise sound casing. Many of these firms advertise regularly in the motorcycle papers.

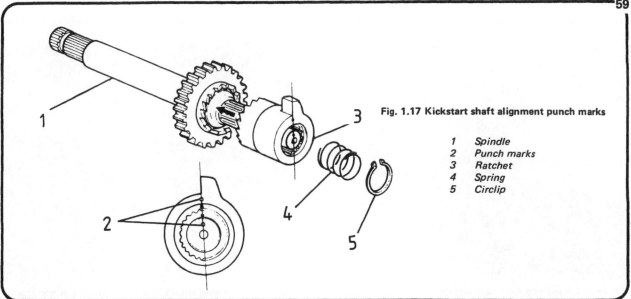

Fig. 1.17 Kickstart shaft alignment punch marks

1   Spindle
2   Punch marks
3   Ratchet
4   Spring
5   Circlip

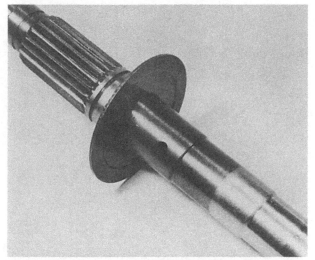

35.7a The kickstart mechanism fitted to the 125, 185 and 200 models may be assembled by sliding the guide plate onto the spindle ...

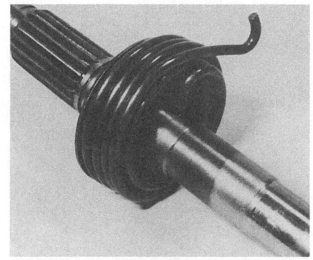

35.7b ... followed by the return spring ...

35.7c ... and the spring collar

35.7d From the opposite end of the spindle, fit the pinion ...

35.7e ... followed by the thrust washer and circlip

35.7f Fit the drive ratchet, ensuring that the punch marks align ...

35.7g ... followed by the spring and seat plate

35.7h Lock the assembly in position with the circlip and fit the end thrust washer

37 Engine reassembly: general

1 Before reassembly of the engine/gearbox unit is commenced, the various component parts should be cleaned thoroughly and placed on a sheet of clean paper, close to the working area.

2 Make sure all traces of old gaskets have been removed and that the mating surfaces are clean and undamaged. One of the best ways to remove old gasket cement is to apply a rag soaked in methylated spirit. This acts as a solvent and will ensure that the cement is removed without resorting to scraping and the consequent risk of damage. If a gasket becomes bonded to the surface through the effects of heat and age, a new sharp razor blade can be used to effect removal. Old gasket compound can also be removed using a soft brass wire brush of the type used for cleaning suede shoes. A considerable amount of scrubbing can take place without fear of damaging the mating surfaces.

3 Gather together all the necessary tools and have available an oil can filled with clean engine oil. Make sure that all new gaskets and oil seals are to hand, also all replacement parts required. Nothing is more frustrating than having to stop in the middle of a reassembly sequence because a vital gasket or replacement has been overlooked.

4 Make sure that the reassembly area is clean and that there is adequate working space. Many of the smaller bolts are easily sheared if over-tightened. Always use the correct size screwdriver bit for the crosshead screws and never an ordinary screwdriver or punch. If the existing screws show evidence of maltreatment in the past, it is advisable to renew them as a complete set.

5 If the purchase of a replacement set of screws is being contemplated, it is worthwhile considering a set of socket or Allen screws. These are invariably much more robust than the originals, and can be obtained in sets for most machines, in either black or nickel plated finishes. The manufacturers of these screw sets advertise regularly in the motorcycle press.

38 Engine reassembly: gear cluster and selector mechanism reassembly

1 Having examined and renewed the gearbox components as necessary, the clusters can be built up and assembled as a complete unit for installation in the engine/gearbox casings.

2 Select the appropriate model line drawing accompanying this text and study it carefully. Assemble both the mainshaft and layshaft components in the exact order shown ensuring that all thrust washers and circlips are correctly positioned.
3 Note that on the XL100 S models and other models fitted with a low gear collar, the oil hole in the collar must align with the corresponding hole in the layshaft.
4 The components must be inspected for any signs of contamination by dirt or grit during assembly and liberally coated in oil on the mating faces.

39 Reassembling the engine/gearbox unit: refitting the crankcase components

XR80, XL80 S and XL100 S models

1 Support the left-hand crankcase half on blocks placed on the work surface. Check that room is allowed underneath the component for the shafts to protrude when fitted.
2 Cupping the previously assembled gear clusters together in both hands with the gears correctly meshed, carefully lower them into the crankcase half. Ensure any end of shaft shims are not displaced when fitting the assembly.
3 Fit the gearchange drum, taking care to ensure the neutral switch rotor on the end of the drum is correctly aligned with the neutral switch.
4 Fit the selector forks in the previously noted positions, ensuring that the markings stamped on their sides face upwards. Fit the selector fork shaft and rotate the mainshaft by hand to see if the gears rotate freely. Fit the starter pinion, preceded by its thrust washer (where fitted), over the mainshaft end.
5 Fit the kickstart assembly into the crankcase so that the boss on the drive ratchet is set on the guide plate as shown in the accompanying illustration.
6 Feed the cam chain into the crankcase through its aperture in the crankcase mouth and spread the chain so that the sprocket on the crankshaft may be allowed to pass through it.

XL125 S, XL185 S, XR185 and XR200 models

7 The procedure for fitting the gearbox components into the crankcase is similar to that noted in the previous paragraphs in this Section with the following exceptions.
8 The kickstart assembly is fitted into the right-hand crankcase half, the end of the return spring being located over the abutment in the crankcase casting. Once fitted, the drive ratchet boss should be located in the groove in the crankcase casting by rotating the spindle clockwise (viewed from inside the casing).

The stop bolt and O-ring may then be fitted and tightened.
9 The cam chain need not now be fitted as the crankshaft sprocket is readily accessible after the crankshaft has been fitted.

All models

10 Before proceeding further, lubricate all the gearbox components with engine oil. Lubricate the crankshaft big-end and main bearings, also with engine oil, and lower the crankshaft assembly into position in the crankcase half. The main bearing bosses will not require to be heated in order that the bearings may be fitted, as they are fitted with steel inserts which means that a sliding fit is possible.
11 The cam chain may now either be pulled onto the crankshaft sprocket (80 and 100 cc models) or fitted over the sprocket and passed through its tunnel in the crankcase mouth.

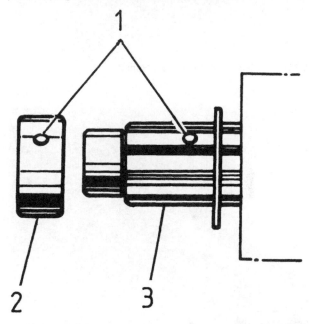

Fig. 1.18 Layshaft low gear collar alignment marks — XL100 S only

1   Alignment holes   2   Low gear collar   3   Layshaft

38.2a On the XL 125 S models, reassemble the mainshaft components by first fitting the 5th gear pinion, spline washer and retaining circlip ...

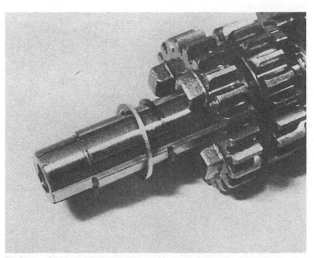

38.2b ... followed by the 3rd and 4th gear pinion, the retaining circlip and the spline washer ...

38.2c ... followed by the 6th gear pinion ...

38.2d ... and finally the 2nd gear pinion and special spline washer

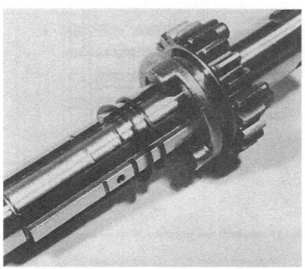

38.2e On the XL 125 S models, reassemble the layshaft components by first fitting the 6th gear pinion, retaining circlip and spline washer ...

38.2f ... followed by the 4th gear pinion, spline washer and lockwasher ...

38.2g ... followed by the 3rd gear pinion, spline washer and retaining circlip ...

38.2h ... 5th gear pinion ...

38.2i ... 1st pinion and thrust washer ...

38.2j ... the collar and starter idle gear ...

38.2k ... and the end thrust washer

38.2l From the opposite end of the layshaft, fit the thrust washer ...

38.2m ... followed by the collar and 2nd gear pinion ...

38.2n ... and finally, the end thrust washer

38.2p The assembled mainshaft and layshaft components, together with the starter pinion and thrust washer, are now ready to be fitted into the crankcase (XL 125 S)

39.7a Note the condition of the bush within the starter pinion

39.7b Lower the assembled gear clusters into the crankcase ...

39.7c ... followed by the gearchange drum ...

39.7d ... selector forks ...

39.7e ... selector fork shaft ...

39.7f ... and starter gear pinion and thrust washer (XL 125 S model shown)

39..8a ... On 125, 185 and 200 models, fit the kickstart assembly into the right-hand crankcase half ...

39.8b ... ensuring the end of the return spring is correctly located ...

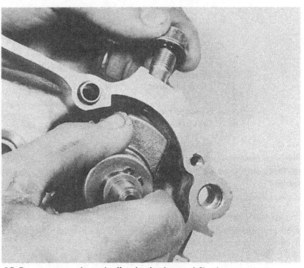

39.8c ... rotate the spindle clockwise and fit the stop bolt.

39.10 Fit the crankshaft assembly into the crankcase

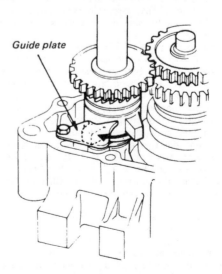

Fig. 1.19 Kickstart ratchet guide plate positioning —
80 and 100 models

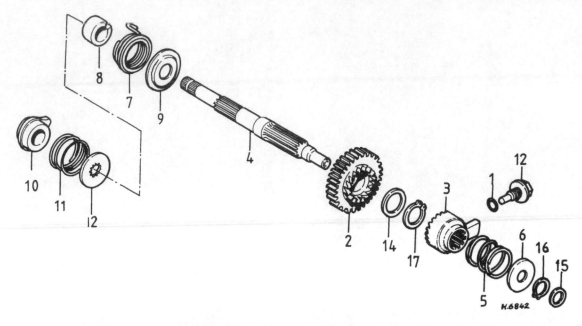

Fig. 1.20 Kickstart mechanism — 125, 185 and 200 models

1	O-ring	7	Return spring	13	Spring seat — not XL125
2	Ratchet pinion	8	Collar	14	Thrust washer
3	Drive ratchet	9	Guide plate	15	Thrust washer
4	Spindle	10	Decompressor cam — not XL125	16	Circlip
5	Ratchet spring	11	Spring — not XL125	17	Circlip
6	Seat plate	12	Stop bolt		

40 Reassembling the engine/gearbox unit: joining the crankcase halves

1 Ensure that the two dowels are located correctly in the left-hand crankcase half and fit a new gasket. The gasket may be held in position by a few light smears of gasket compound. Check that none of this compound is near any oil feed holes in the casting and these holes are in no way obscured by the gasket.

2 With the left-hand crankcase half still supported on the work surface, lower the right-hand crankcase half down firmly and squarely onto it. Take care to locate the right-hand main bearing correctly in its housing and the gearshafts in their respective bearings.

3 Note that on the XL125 S, XL185 S, XR185 and XR200 models, the shim on the kickstart shaft end will need to be retained by a small blob of grease so that it does not fall from the shaft as the right-hand crankcase half is inverted. Take care when inverting and fitting the right-hand crankcase half, to hold the kickstart assembly in position, otherwise the shaft will detach itself from the stop bolt under spring pressure.

4 Push the crankcase halves together with hand pressure, noting that the two dowels locate correctly in their recesses in the right-hand crankcase half. It may be necessary to give the right-hand crankcase half a few light taps with a soft-faced mallet before the jointing surfaces will mate up correctly. On no account must force be used when joining the crankcase halves.

5 If the crankcase halves will not align, then it is possible that one of the main bearings is not seating correctly. Alternatively, it may well be the kickstart mechanism that needs to be manoeuvred to enable the casing halves to locate correctly. To facilitate assembly of the casing halves, rotate the various shafts during the assembly operation.

6 On XR80, XL80 S and XL100 S models, refit and evenly

tighten the two crosshead crankcase retaining screws. Refit the kickstart return spring and spring collar over the exposed shaft.

7 On all other models, refit and tighten the ten hexagon-headed crankcase retaining screws, evenly and in a diagonal sequence. Fit and tighten also the single bolt located at the front of the crankcase mouth, not forgetting the clutch cable guide it retains.

8 The operation and free running of the crankshaft and gearbox assembly should be ascertained at this stage. Check the operation of the kickstart by temporarily refitting the lever. Any tight spots or resistance felt during operation of the assemblies must be investigated and rectified before further reassembly takes place.

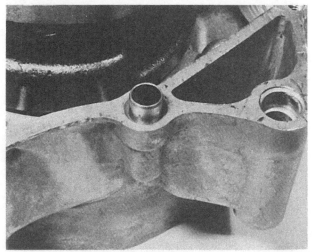

40.1 Insert the crankcase locating dowels ...

40.2 ... and fit the crankcase halves together

40.7 On 125, 185 and 200 models, do not forget to fit the clutch cable guide

41 Reassembling the engine/gearbox unit: refitting the gearchange mechanism, oil pump and primary drive pinion

XR80, XL80 S and XL100 S models
1 Refit the gearchange mechanism by first lightly greasing the oil seal through which the gearchange shaft passes and then carefully feeding the shaft into position, ensuring that the splines do not damage the seal.
2 Refit the four pins into the end of the gearchange drum. Place the index plate in position and fit and tighten the bolt retaining it to the drum end. Finally, refit the index arm and spring and retain it in position with the shouldered bolt.

XL125 S, XL185 S, XR185 and XR200 models
3 Refit the change drum stopper arm (detent lever) and spring in position in the casing and tighten the single retaining bolt. Check that the arm is free to pivot. Place the camplate locating pin in the end of the selector drum, followed by the camplate and its retaining bolt. Tighten the bolt.
4 The gearchange spindle can now be slid into place in its bore in the crankcase, and the operating claw engaged with the camplate. The return spring on the spindle should be fitted as shown in the accompanying photograph.

All models
5 Fit two new O-rings into their recesses in the oil pump seating face and place the complete pump assembly in position. Align the holes in the top cover and pinion and fit and tighten fully the two countersunk retaining screws.
6 Slide the crankshaft-mounted primary drive pinion into position on the splined mainshaft end, ensuring that its teeth mesh with that of the pump drive gear.
7 On XR80, XL80 S and XL100 S models, it is necessary to fit the collar before sliding the primary drive pinion onto the mainshaft. Secure the pinion by fitting the pressure relief valve assembly, lockwasher and retaining nut; tightening the nut to a torque of 3.5 - 4.5 kgf m (25 - 33 lbf ft). Prevent the crankshaft from rotating whilst tightening this nut by following the procedure given in paragraph 3 of Section 12 of this Chapter.

42 Reassembling the engine/gearbox unit: refitting the clutch assembly and centrifugal oil filter

1 Place the clutch outer drum in position on the gearbox mainshaft end, ensuring that its gearteeth mesh with those of the primary drive pinion. Refit the splined thrust washer.
2 Assemble the inner clutch assembly on a clean work surface. Start by fitting the clutch plates to the outer clutch centre half; friction plate first then a plain plate, building up the layers of plates and finishing with a friction plate. Finally, fit the inner clutch centre half to the outer half.
3 Lower the inner clutch assembly into the outer drum, having first aligned the friction plate outer tongues. Make sure that the assembly is held tightly together at all times during the fitting procedure. Some small amount of movement may have to be made to the friction plates so that each individual plate drops into the outer drum.
4 On the XR80, XL80 S and XL100 S models, secure the clutch centre and plates with the retaining circlip.
5 On all other models, the clutch centre is retained by a locknut and washer. Fit the washer with the OUTSIDE mark uppermost and fit and tighten the nut to a torque of 4.0 - 5.0 kgf m (29 - 36 lbf ft). The crankshaft should be prevented from rotating whilst tightening this nut by following the procedure given in paragraph 3 of Section 12 of this Chapter.
6 Fit the four clutch springs over the projecting threaded pillars. The thrust plate and fitted thrust bearing can now be placed in position, and the four retaining bolts fitted and tightened. Tighten the bolts down evenly in a diagonal sequence, in two or three stages. Fit the pushrod in the centre of the thrust bearing.
7 On the XL125 S, XL185 S, XR185 and XR200 models, place the inner half of the centrifugal oil filter over the crankshaft end. Fit the special washer with the OUTSIDE mark uppermost and fit and tighten the slotted securing nut to a torque of 4.0 - 5.0 kgf m (29 - 36 lbf ft), again preventing the crankshaft from rotating.
8 Prime the filter unit with a few squirts of clean engine oil from an oil can. Install a new gasket on the inner half mating surface, followed by the outer half of the filter unit. Tighten the three crosshead retaining screws evenly to avoid any risk of warpage. Note that there is a centre piece in the cap with a spring attached to it. This device acts as a pressure relief valve. It is retained to the outer half of the filter unit by an R-clip fitted to the inner face of the outer half.

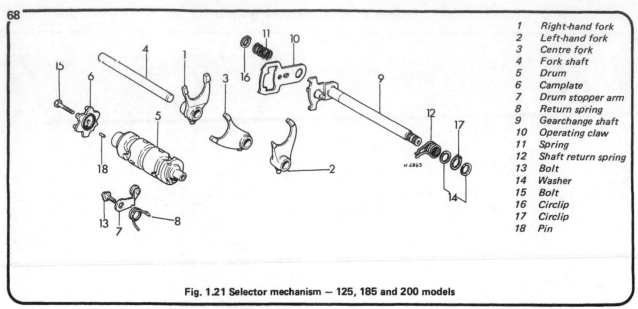

1 Right-hand fork
2 Left-hand fork
3 Centre fork
4 Fork shaft
5 Drum
6 Camplate
7 Drum stopper arm
8 Return spring
9 Gearchange shaft
10 Operating claw
11 Spring
12 Shaft return spring
13 Bolt
14 Washer
15 Bolt
16 Circlip
17 Circlip
18 Pin

Fig. 1.21 Selector mechanism — 125, 185 and 200 models

41.1 On 80 and 100 models, refit the gearchange shaft assembly ...

41.2 ... followed by the pins, index plate and index arm

41.3 On 125, 185 and 200 models, fit the detent lever, camplate pin and camplate ...

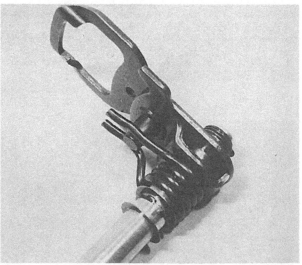

41.4a ... correctly locate the gearchange spindle return spring ...

41.4b ... and fit the gearchange spindle assembly

41.5 Fit new O-rings before fitting the oil pump

41.6 Slide the primary drive pinion onto the crankshaft end

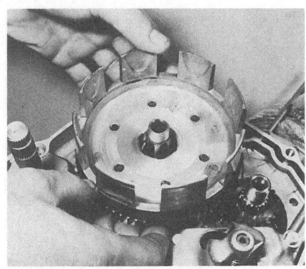
42.1a Place the clutch outer drum in position ...

42.1b ... followed by the splined thrust washer

42.2 Assemble the inner clutch components ...

42.3 ... and fit them into the outer drum ...

42.4 ... retaining them with the circlip (80 and 100) ...

42.5 ... or marked washer and locknut (125, 185 and 200)

42.6a Fit the clutch springs over the projecting pillars ...

42.6b ... followed by the thrust plate and bearing ...

42.6c ... and the clutch push rod

42.7a On 125, 185 and 200 models, fit the inner half of the centrifugal filter housing, followed by the marked washer ...

42.7b ... and the slotted securing nut

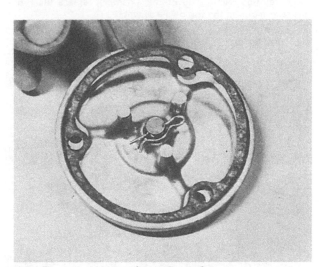

42.8a Fit a new gasket to the mating surface ...

42.8b ... and fit the filter cap

42.8c The pressure relief valve components

43 Reassembling the engine/gearbox unit: refitting the right-hand crankcase cover

1 On the XR185 and XR200 models, fit the spring seat, spring and kickstart cam for the decompressor over the kickstart spindle. Ensure that the punch mark on the cam is aligned with the corresponding mark on the kickstart spindle.

2 Make a final check to ensure that all components are correctly fitted and tightened and that the mating faces are clean. Check that the two locating dowels are correctly fitted in the right-hand crankcase half. Place a new gasket in position over the dowels and use light smears of jointing compound to hold it in position. Note that on the XR80, XL80 S and XL100 S models, the gauze filter must be correctly located in its slot at the bottom of the crankcase half before the gasket is fitted.

3 Ensure the clutch operating mechanism contained within the cover is correctly positioned so that it aligns with the clutch pushrod end. Lower the cover over the kickstart spindle. On the XL125 S, XL185 S, XR185 and XR200 models, take great care not to damage the cover oil seal with the splined end of the kickstart spindle. Light greasing of the splines will lessen the risk of damage to the oil seal.

4 Locate the cover on the two dowels and if necessary, tap it gently into position with a soft-faced mallet. If the casing will not seat properly, do not use excessive force, but raise it slightly and discover the cause of the trouble. A potential cause of difficulty is the kickstart shaft being pulled slightly out of alignment by the return spring. Check also that the gearchange shaft is located properly in the crankshaft casing by pushing in on the end of the shaft.

5 Refit and tighten the cover retaining screws, evenly and in a diagonal sequence to avoid distortion of the cover.

44 Reassembling the engine/gearbox unit: refitting the neutral indicator switch, cam chain and cam chain tensioner

1 Fit the neutral indicator switch into its location in the crankcase casting, preceded by a new O-ring and followed by its end spacer.

XR80, XL80 S and XL100 S models
2 Insert the cam chain tensioner and spring into its location in the cylinder barrel and position it so that it is at its full extension. Fit a new O-ring, followed by the tensioner adjusting screw and locknut. The cam chain would have been already fitted prior to fitting the crankshaft. Before finally tightening

the screw, check that the spring end is correctly hooked into the cylinder groove.

XL125 S, XL185 S, XR185 and XR200 models
3 With the engine unit supported on wooden blocks on the work surface with the left-hand side uppermost, fit the gauze crankcase filter, the spring and the large hexagon-headed cover. Ensure the O-ring seal on the cap/cover is in good condition; replace it with a new item if damage is apparent.

4 Fit the tensioner plunger and blade into their respective positions in the crankcase. Feed the cam chain down through its tunnel in the crankcase mouth and loop its end over the crankshaft sprocket. When the chain is correctly located over the sprocket, secure the upper end in position.

5 Position the cam chain tensioner arm and spring and secure it by fitting and tightening the single retaining bolt. Note that the lower end of the pushrod bears directly on the tensioner arm.

6 Fit the two angled bushes onto the pushrod through the aperture in the top of the crankcase. Screw in the adjuster bolt, having first checked the condition of its sealing O-ring, and then refit the sealing washer and the 6 mm blind bolt at the top of the assembly. When the cam chain tension has been set (see Section 53 of this Chapter), replace the rubber protective cap on the top of the assembly.

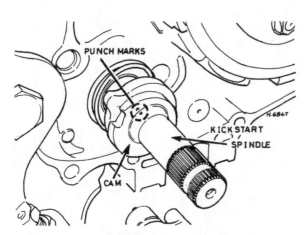

Fig. 1.22 Decompressor cam positioning marks

43.2 On 80 and 100 models, locate the gauze oil filter in its crankcase slot

43.3a The clutch operating mechanism (80 and 100)

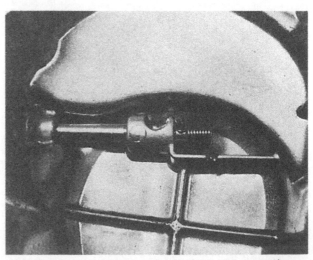

43.3b The clutch operating mechanism (125, 185 and 200)

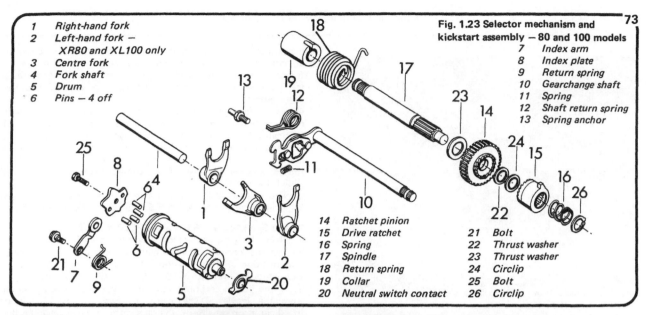

Fig. 1.23 Selector mechanism and kickstart assembly — 80 and 100 models

1 Right-hand fork
2 Left-hand fork — XR80 and XL100 only
3 Centre fork
4 Fork shaft
5 Drum
6 Pins — 4 off
7 Index arm
8 Index plate
9 Return spring
10 Gearchange shaft
11 Spring
12 Shaft return spring
13 Spring anchor
14 Ratchet pinion
15 Drive ratchet
16 Spring
17 Spindle
18 Return spring
19 Collar
20 Neutral switch contact
21 Bolt
22 Thrust washer
23 Thrust washer
24 Circlip
25 Bolt
26 Circlip

44.1 Fit the neutral indicator switch into the crankcase housing

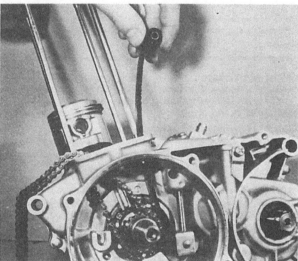

44.4 On 125, 185 and 200 models, fit the cam chain tensioner plunger and blade, cam chain ...

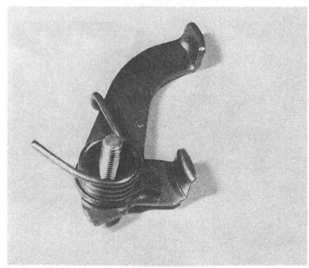

44.5 ... the tensioner arm and spring ...

44.6a ... followed by the two angled bushes ...

44.6b ... and the adjuster bolt assembly

45 Reassembling the engine/gearbox unit: refitting the flywheel generator

1 Refit the Woodruff key in the keyway provided in the tapered left-hand end of the crankshaft. Lock the crankshaft in position by following the procedure stated in paragraph 3 of Section 12 of this Chapter.

XR80, XL80 S and XL100 S models

2 Reconnect the neutral switch lead (where applicable). Inspect the condition of the O-ring on the stator plate and renew it if found to be in any way damaged. Fit the stator plate and secure it with the two retaining screws (three on the XL100 S model). Push the rubber grommet fitted over the electrical leads into the crankcase recess. Route the neutral switch lead so that it passes behind the clip situated midway between the switch itself and the rubber grommet; this will prevent it from becoming trapped between the crankcase and cover mating faces.

3 Check that the automatic advance mechanism is operating freely and the springs are in good condition. It is advisable to

check also whether the contact breaker points require attention at this stage, otherwise it will be necessary to withdraw the flywheel rotor again in order to gain access. Reference to Chapter 3 will show how the contact breaker points are renovated and adjusted.

4 Before fitting the generator rotor, ensure that the crankshaft/ rotor mating surfaces are clean and that the felt pad which lubricates the contact breaker cam is lightly lubricated.

5 Feed the rotor onto the crankshaft so that the keyway lines up with the Woodruff key. The rotor may have to be turned to clear the heel of the contact breaker before it will slide fully home.

6 The washer and rotor nut can now be fitted and the nut fully tightened to a torque of 6.5 - 7.5 kgf m (47 - 54 lbf ft).

XL125 S, XL185 S, XR185 and XR200 models

7 Place the generator rotor over the end of the crankshaft, line up the keyway with the Woodruff key and push the rotor into position. Fit the central retaining bolt and plain washer and tighten to a torque of 4.0 - 5.0 kgf m (29 - 36 lbf ft).

8 If the stator assembly has, for any reason, been removed from the inside of the generator cover, it should now be fitted in position. Refit the generator cover, checking that its O-ring is not displaced whilst doing so, and retain it with its four hexagon-headed screws. Tighten the screws evenly and in a diagonal sequence to avoid distortion to the cover. Do not fit the inspection caps to the cover at this stage.

9 Route the neutral switch lead under the clip cast into the back of the cover and connect it to the switch. Push the generator leads between the projecting flanges of the casting, as shown in the accompanying photograph, and secure it in position with the retaining plate and crosshead screw.

46 Reassembling the engine/gearbox unit: refitting the final drive sprocket

1 Place the final drive sprocket in position on the splined shaft end, followed by its retaining plate, which is similarly splined. When the plate is aligned with the groove in the shaft, it can be rotated sufficiently to allow the securing bolts to be fitted and tightened.

2 If the rear chain was removed together with the sprocket, the fitting of the sprocket should be left until the engine is fitted in the frame.

45.3 Check the condition of the contact breaker points before refitting the rotor (80 and 100)

45.7a On 125, 185 and 200 models, fit the generator rotor over the crankshaft end ...

45.7b ... and secure it with the retaining bolt and plain washer

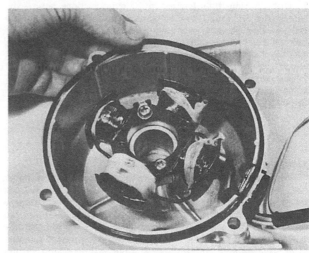

45.8 Check the generator cover O-ring is not displaced

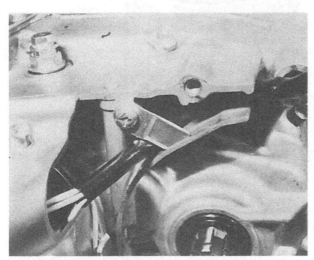

45.9a ... fit the cover and route the generator leads as shown ...

45.9b ... before reconnecting the neutral indicator switch

47 Reassembling the engine/gearbox unit: refitting the piston, cylinder barrel and cylinder head

1 Place the crankcase unit upright on the workbench, taking care that the camshaft drive chain does not slide back into the crankcase.

2 Raise the connecting rod to its highest point and with the crankcase mouth packed with clean rag to prevent any component parts falling into the crankcase, proceed to fit the piston to the connecting rod.

3 Position the piston on the connecting rod so that the IN mark on the piston crown faces the rear of the engine. Slide the gudgeon pin into position. The pin should be a light sliding fit but if it proves to be tight, warm the piston in hot water to expand the metal around the gudgeon pin bosses. Use new circlips to retain the gudgeon pin, and double check to ensure that each is correctly located in the piston boss groove. If a circlip works loose, it will cause serious engine damage. The circlips should be fitted so that the gap between the circlip ends is well away from the piston cutout underneath the gudgeon pin hole. Finally, check that the piston rings have not been disturbed from the positions quoted in Section 25 of this Chapter.

4 Trim the crankcase gasket material protruding up from the crankcase mouth mating surface. Fit the two dowels, one to each left-hand cylinder head and barrel retaining stud, and carefully lower the gasket over the studs, cam chain and tensioner blade. There is no need to use any jointing compound with this gasket.

5 Oil the cylinder bore and piston rings with engine oil. Position two blocks of wood across the crankcase mouth, one either side of the connecting rod, and carefully lower the piston onto the blocks. This will provide positive support to the piston whilst easing the rings into the bore. Obtain a length of wire and attach it to the cam chain at the point furthest from the crankshaft sprocket.

6 Place the cylinder barrel in position over the studs and pass the length of wire up through the tunnel in the barrel. Ensure that tension is kept on the cam chain at all times to prevent it falling off the crankshaft sprocket.

7 Carefully lower the barrel down over the studs and cam chain tensioner blade until it is positioned with the sleeve just above the piston crown. Guide the piston crown into the bore and push in on each side of the piston rings so that they slide into the bore. There is a generous lead in on the base of the bore which will aid this operation.

8   With the rings fully located in the cylinder bore, remove the two blocks of wood from underneath the piston and remove also the rag packed into the crankcase mouth. Lower the cylinder barrel down onto the two locating dowels. If necessary, tap the barrel gently with a soft-faced mallet to locate it properly on the dowels.

9   The cam chain guide may now be inserted into its location in the cylinder barrel. Ensure the two spigots on the guide fit correctly into the corresponding slots in the barrel casting.

XR80, XL80 S and XL100 S models

10   Whilst pushing down on the top of the cam chain tensioner rod, slacken and retighten the tensioner adjusting screw. Lock the screw in position by tightening the locknut. This will place the tensioner blades in the retracted position and, therefore, aid insertion of the camshaft and fitting of the sprocket. Fit the two locating dowels, one over the forward right-hand cylinder head retaining stud and one over the rear left-hand stud. Fit the new O-ring over the rear right-hand stud and carefully lower the new gasket over the studs, cam chain and cam chain tensioner and guide blades.

11   Again tensioning the cam chain by pulling up on the attached length of wire, feed the chain through the tunnel in the cylinder head whilst lowering the head onto the cylinder barrel. Ensure the upper ends of the tensioner and guide blades do not foul the edges of the tunnel as the head is positioned.

12   Gently tap the cylinder head with a soft-faced mallet to locate it properly on the two locating dowels and fit finger-tight the single cylinder head retaining bolt, located on the left-hand side of the head.

13   Refit the fine adjuster for the cam chain, ensuring it locates properly with the end of the cam chain tensioner. Fit a new O-ring, followed by the set plate and bolt (finger-tight). Align the punch mark on the adjuster with the set plate as shown in the accompanying illustration, and tighten the bolt.

XL125 S, XL185 S, XR185 and XR200 models

14   Fit a dowel pin over each of the left-hand cylinder head retaining studs. Fit a dowel pin with a new O-ring over the rear right-hand retaining stud. Carefully lower a new gasket, over the four retaining studs and cam chain tensioner and guide blades, onto the clean mating surface. The cam chain should be held up by the length of wire and threaded through the tunnel section of the gasket as the gasket is lowered.

15   Lower the cylinder head into position, whilst feeding the cam chain up through the tunnel. When the chain emerges, secure it with a length of stiff wire or by passing a rod through the chain loop. If the cylinder head proves to be a tight fit over the locating dowels, lightly tap the head with a soft-faced mallet to seat it correctly.

16   With the cylinder head seated correctly, align the hole in the top of the cam tensioner blade with the hole in the left-hand side of the cylinder head. Insert the retaining pivot bolt and its sealing washer, ensuring that the bolt has located correctly with the hole in the top of the blade before the bolt is tightened.

17   Insert the one long 6 mm bolt through the hole in the left-hand side of the cylinder head; just below the aperture where the camshaft will emerge. Tighten the bolt finger-tight.

18   Remove the rubber cap from the head of the cam chain tensioner adjusting bolt, remove the inner 6 mm bolt and slacken the adjuster bolt. Push the tensioner plunger rod down by inserting a screwdriver through the hole in the centre of the adjuster bolt. When the rod reaches the end of its travel, tighten the adjuster bolt and refit the 6 mm bolt and rubber cap. This will ensure that the maximum amount of cam chain length is available when fitting the camshaft sprocket.

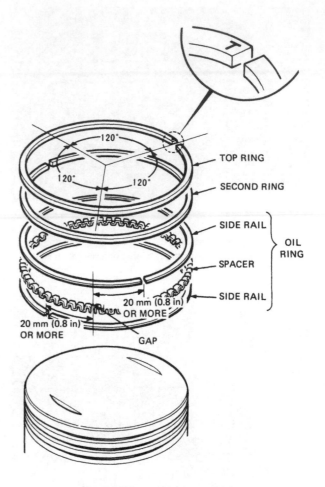

Fig. 1.24 Piston ring gap positions

47.3a Note the piston crown is marked for reference when refitting

47.3b The gudgeon pin should be a light sliding fit ...

47.3c ... and the circlips fitted with the gap well away from the piston cutout

47.4a Fit the two dowels over the cylinder barrel retaining studs ...

47.4b ... followed by the gasket

47.7 Carefully lower the cylinder barrel over the piston

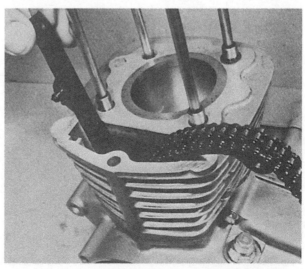

47.9 Locate the cam chain guide in the cylinder barrel tunnel

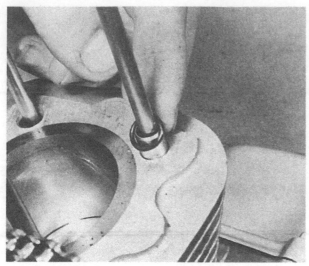

47.14a On 125, 185 and 200 models, fit the O-ring over the rear right-hand dowel ...

47.14b ... before fitting the head gasket ...

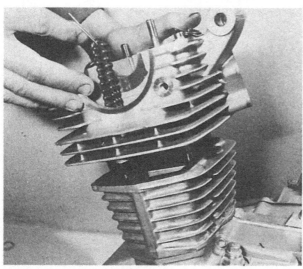

47.15 ... followed by the cylinder head

47.16 Insert the cam tensioner blade pivot bolt and sealing washer

47.18 Push the tensioner plunger rod down to ensure the maximum amount of cam chain length is available

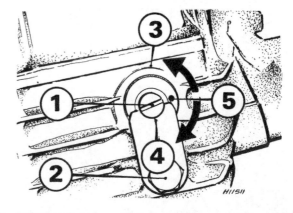

Fig. 1.25 Cam chain fine adjustment tensioner — XR80 and XL100 S

1   Adjusting screw
2   Set plate
3   Position where chain is excessively slack
4   Position where chain is excessively tight
5   Punch mark

48 Reassembling the engine/gearbox unit: refitting the overhead camshaft and timing the valves

XR80, XL80 S and XL100 S models

1 Check the camshaft for cleanliness and coat the journals with molybdenum disulphide grease. Fit the camshaft into position in the cylinder head.

2 Fit the two dowels over the retaining studs; one on the forward right-hand stud and one on the rear left-hand stud. Fit the camshaft holder block over the camshaft. The rocker assemblies should be already fitted in the block and the valve adjusting screws loosened. Fit the four camshaft holder retaining nuts and tighten them evenly and in a diagonal sequence to a torque of 1.8 - 2.0 kgf m (13 - 14 lbf ft). Do not forget to fit the four thick plain washers, one under each nut.

XL125 S, XL185 S, XR185 and XR200 models

3 Before replacing the camshaft in the cylinder head, the cylinder head cover must be refitted. If the rocker assemblies are already replaced in the head cover, then the complete assembly can be fitted to the cylinder head. Ensure that the valve adjusting screws are loosened before fitting the cover.

4 Fit the camshaft bush into its recess in the cylinder head. Ensure its locating spigot fits correctly into the cutout in the recess. Fit the two dowel pins into their recesses in the cylinder head mating surface. Before fitting the cylinder head cover, coat the camshaft bearing surfaces inside the cylinder head and cover with molybdenum disulphide grease.

5 Because there is no gasket fitted between the cylinder head and head cover, apply a **thin** coat of gasket cement to ensure an oiltight joint. Before lowering the head cover down, ensure the small rubber 'bung', vital to the flow of oil around the head components, is refitted in its recess in the mating surface of the head. Also, make sure the cam chain is pulled up through the circular aperture formed when the head cover is mated to the cylinder head.

6 Position the head cover on the cylinder head mating surface ensuring it seats correctly. Fit the four copper sealing washers and the domed nuts over the protruding ends of the holding down studs. Fit the four 6 mm socket bolts to the 'corners' of the head cover.

7 Tighten the cylinder head nuts in the sequence given in the accompanying illustration, a little at a time to obviate the risk of cylinder head distortion. Final tightening should be accomplished with a torque wrench. Torque the domed nuts down to a setting of 1.8 - 2.0 kgf m (13 - 14 lbf ft). Tighten the four 6 mm socket bolts down after torque loading the domed nuts.

8 Before fitting the camshaft, check it for cleanliness, fit the large thrust washer and smear the journals with molybdenum disulphide grease.

9 If the cam chain is retained temporarily by a wire hook,

use the wire to pull the chain up and over the end of the camshaft being inserted. Alternatively, displace the metal rod being used to retain the cam chain and, using your fingers, open out the chain and slide the camshaft through the opening into the head. The camshaft must be fitted with the cam sprocket mounting holes at 12 o'clock and 6 o'clock.

All models

10 Rotate the crankshaft anti-clockwise so that the T mark on the generator rotor is exactly in line with the index mark on the left-hand crankcase/cover. In this position the piston is exactly at top dead centre (TDC). Make sure the cam chain does not become slack and slip off the lower sprocket or slide off the camshaft and fall down into the chain tunnel. The cam chain sprocket may now be fitted to the end of the camshaft as follows.

XR80, XL80 S and XL100 S models

11 Rotate the camshaft until the sprocket fixing holes are in the 3 and 9 o'clock positions (parallel with the cylinder head surface). The counterbored hole should be on the intake side.

12 Clean the cam chain sprocket to find the tooth marked with an O. Thread the chain round the sprocket such that when the chain is tight against the front guide the O mark lines up with a cast line on top of the camshaft block.

13 Fit the sprocket onto the camshaft and retain it with the two bolts. Tighten the bolts to a torque loading of 1.0 - 1.4 kgf m (7 - 10 lbf ft). Take great care not to drop the bolts into the cylinder head casting during the fitting process.

XL125 S, XL185 S, XR185 and XR200 models

14 Slide the upper cam chain sprocket over the end of the camshaft. The sprocket must be positioned so that the O mark on the sprocket, aligns with the index mark on the cylinder head cover, with the front run of the chain tight. As the sprocket is being fitted to the camshaft, the cam chain should be looped over the sprocket.

15 Install the two sprocket retaining bolts and fully tighten them. A further check on the correct positioning of the sprocket and camshaft is that the two sprocket retaining bolts should be above and below the end of the camshaft. The sprocket retaining bolts should be tightened to a torque setting of 0.8 - 1.2 kgf m 6 - 9 lbf ft). Finally, adjust the cam chain tension by removing the rubber cover from the tensioner and slackening, then tightening, the housing nut. This will serve to take up any excess play in the chain and will prevent any danger of the chain jumping the sprockets once the engine is started. When the engine is running, the chain must be properly adjusted as described in Section 53.

All models

16 The valve timing procedure is now completed but should be checked for accuracy before proceeding further.

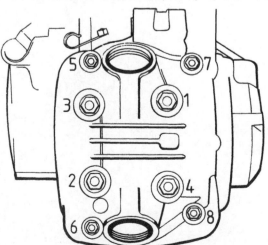

Fig. 1.26 Cylinder head cover
tightening sequence — 125, 185 and 200 models

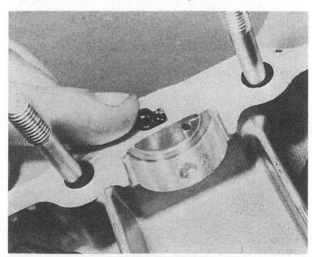

48.4a On 125, 185 and 200 models, ensure the small rubber
bung is correctly located ...

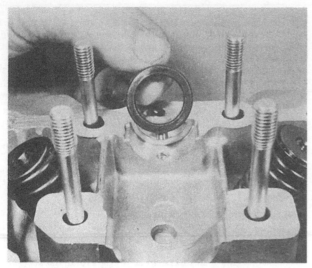

48.4b ... before fitting the camshaft bush into its recess

48.5 A thin coat of gasket cement on the mating surfaces will ensure an oiltight joint

48.7 Do not omit to tighten the fifth cylinder head retaining bolt

48.9 On 125, 185 and 200 models, fit the camshaft with the sprocket mounting holes in the 12 and 6 o'clock positions ...

48.14 ... before aligning the camshaft sprocket ...

48.15 ... and fitting and torque loading the sprocket securing bolts

49 Reassembling the engine/gearbox unit: refitting the CDI pulse generator to the cylinder head and timing the ignition - XL125 S, XL185 S, XR185 and XR200 models

1 The CDI unit should have already been inspected, cleaned and the defective parts renewed, prior to storing it ready for reassembly and refitting to the cylinder head. Lay out the component parts of the unit on a clean, dry work surface and proceed to fit them as follows.

2 Fit the pulser base over the end of the camshaft, taking care not to damage the seal. Fit and tighten evenly and in a diagonal sequence the two hexagon-headed retaining bolts. Push the dowel pin into its location in the camshaft end.

3 Before fitting the pulser rotor/automatic advance unit, ensure the punch mark on the rotor is aligned with the index mark on the automatic advance unit plate. Fit the pulser rotor assembly over the camshaft end, making sure that the groove in the boss of the automatic advance unit plate is in line with the protruding end of the dowel pin. Push the unit home over the pin and secure it with the central retaining bolt and plain washer. Torque load the bolt to 0.8 - 1.2 kgf m (6 - 9 lbf ft).

4 The pulser generator/baseplate assembly may now be fitted by installing the assembly in the housing and retaining it in position with the two crosshead screws. The screws should only be tightened finger-tight so that the plate may be rotated to align the previously made mark on the plate with that on the unit housing.

5 At this point, rotate the crankshaft anti-clockwise so that the F mark on the generator rotor aligns exactly with the index pointer on the crankcase cover. The timing mark on the pulser rotor should now align exactly with the timing mark on the pulser generator. If the marks are seen to be out of alignment, the generator baseplate will have to be rotated so that the marks align. Tighten the two crosshead screws to hold the baseplate in position.

6 The CDI unit cover may now be refitted to the cylinder head. Check that the rubber grommet fitted over the output wires is correctly located in the housing recess before placing the cover in position and securing it with the two retaining screws.

7 Further details of ignition timing and servicing for the CDI system are given in Chapter 3. Check that the timing is correct soon after starting the engine for the first time.

49.2 On 125, 185 and 200 models, with the dowel pin located in the camshaft end, fit the pulser base ...

49.3a ... followed by the pulser rotor assembly ...

49.3b ... the rotor retaining bolt and washer ...

49.4 ... and the pulser generator/baseplate assembly

49.6 Fit the CDI unit cover on completion of the timing procedure

50 Reassembling the engine/gearbox unit: setting the ignition timing - XR80, XL80 S and XL100 S models

1   The static timing on these models is set by changing the gap between the contact breaker points. With the generator cover removed, the points may be viewed through one of the slots in the generator rotor. Set the points gap to 0.3 - 0.4 mm (0.012 - 0.016 in) with the points in the fully open position.

2   Check the condition of the contact breaker points, as described in Chapter 3. Obtain an ordinary torch battery and bulb and three lengths of wire. Connect one end of a wire to the positive (+) terminal of the battery and one end of another wire to the negative (—) terminal of the battery. The negative lead should be earthed to a point on the engine casing. Ensure the casing point is clean and that the wire is positively connected, a crocodile clip fastened to the wire is ideal. Take the free end of the positive lead and connect it to the bulb. The third length of wire may now be connected between the bulb and the black output lead from the generator; as the final connection is made, the bulb should light with the points closed.

3   Rotate the generator rotor anti-clockwise and align the F mark on the rotor with the index mark on the crankcase. If the timing is correct, the light should dim as both marks align. If the ignition timing proves to be incorrect, loosen the contact breaker locking screw and adjust the points so that the light dims at the correct point. Tighten the locking screw and rotate the rotor until the points are fully open. Recheck the gap between the points. If the gap is within the specified range of 0.3 - 0.4 mm (0.012 - 0.016 in) all is well. If, however, the gap is outside the range it is evident that the contact breakers have worn to a point where correct ignition timing and correct contact breaker clearance cannot be maintained simultaneously. If this is the case the contact breaker assembly must be renewed as described in Chapter 3.

51 Reassembling the engine/gearbox unit: resetting the valve clearances

1   On XL125 S, XL185 S, XR185 and XR200 models, access to the valve adjusters is obtained by removing the two circular threaded covers in the cylinder head cover. Removal of each cover offers access to one set of adjusters (one for the inlet valve and one for the exhaust valve). On all other models the adjusters are accessible until the cylinder head cover is refitted.

2   Rotate the crankshaft anti-clockwise until the T mark on the generator rotor aligns exactly with the index mark on the

casing. The engine should now be on top dead centre (TDC) on the compression stroke with both valves closed. To ensure this is so, inspect the valve mechanisms; if they are rocking when the T mark is lined up, the engine is in the wrong position. Turn it over again until the T mark lines up and the valves do not move when the rotor is moved slightly.

3   Using a feeler gauge, check the clearance between each rocker arm adjuster and the end of the valve stem with which it makes contact. The inlet and exhaust valves should both have a clearance identical with that given in the Specifications at the beginning of this Chapter.

4   If the clearances are not correct, adjustment is made by slackening the locknut on the end of the rocker arm concerned and by turning the adjuster until the correct setting is obtained. Hold the adjuster still and tighten the locknut. Then recheck the setting before passing to the next valve.

5   Valve clearances should always be adjusted with the engine COLD, otherwise a false reading will be obtained. Badly adjusted tappets normally give a pronounced clicking noise from the vicinity of the cylinder head.

6   On the XL125 S, XL185 S, XR185 and XR200 models, the circular threaded covers may be refitted. Prior to fitting the covers, check the condition of the O-ring fitted to each, and renew if necessary.

7   The XR80, XL80 S and XL100 S models may now have the cylinder head cover fitted. Prior to fitting the cover, check the condition of the cover gasket and of the two rubber washers through which the special bolts pass. Renew these components if found to be damaged or perished. Position the cover on the cylinder head and fit and tighten evenly the two special retaining bolts.

51.4 Adjust the valve clearances by turning the adjuster scew (125, 185 and 200)

52 Refitting the engine/gearbox unit in the frame

1   It is worth checking at this stage that nothing has been omitted during the rebuilding sequences. It is better to discover any left-over components at this stage rather than just before the rebuilt engine is to be started.

2   Installation is, generally speaking, a reversal of the removal sequence. With the help of an assistant, position the engine in the frame and fit the engine plates and mounting bolts. Do not tighten any bolts until all the plates and bolts have been fitted. On the XR80, XL80 S and XL100 S models, all the mounting bolt nuts should be torque loaded to 1.8 - 2.3 kgf m (13 - 16 lbf ft), whereas on the XL125 S, XL185 S, XR185 and XR200 models, the 8 mm bolt nuts should be torque loaded to 2.7 - 3.3 kgf m (20 - 24 lbf ft) and the 10 mm bolt nuts to 3.0 -

4.0 kgf m (22 - 29 lbf ft).

3 Reconnect the generator output leads at the connector block. Reconnect also the CDI unit leads (where fitted) and the neutral indicator switch. Ensure that all connections are clean and free from moisture before reconnecting.

4 Reconnect the clutch, tachometer and decompressor cables (where fitted), ensuring that all the control cables are correctly routed and not passed through tight curves or between components where they may become jammed or damaged. Do not allow cables to rest directly on hot engine castings. Adjust the clutch cable as described in Section 54 of this Chapter.

5 Before refitting the carburettor and exhaust system, check that all O-rings and gaskets have been renewed and are located correctly in their respective positions. Refit the throttle valve assembly which will have been left attached to the throttle cable, and screw down the mixing chamber top. Check the throttle for correct operation.

6 If the rear chain was not separated, mesh the gearbox sprocket with the chain and fit the sprocket onto the splined output shaft. Secure the sprocket with the plate and two bolts. If the chain was separated when it was detached, loop the chain round the engine sprocket and reconnect the two ends with the master link. Ensure that the spring link is correctly fitted with the closed end facing the normal direction of chain travel. The sprocket guard plate may now be fitted (the

generator cover on the 80 and 100 cc models) and secured with the retaining screws. Tighten these screws evenly and in a diagonal sequence to avoid distortion to the cover.

7 Refit the gearchange lever and its retaining pinch bolt. Similarly, refit the kickstart lever, and its single retaining pinch bolt. Check the operating angle of both levers before tightening the pinch bolts.

8 On the machines fitted with a battery, reconnect the battery leads, observing the polarity markings. Ensure that the connections are clean and free from corrosion; to protect the connections from corrosion, smear them with a liberal coating of petroleum jelly. Do not use ordinary grease.

9 After the fuel tank has been refitted, reconnect the fuel pipe to the carburettor. Turn the tap to the 'On' position and carefully check both ends of the pipe for any signs of fuel leakage. On no account should fuel be allowed to come into contact with hot engine castings; if this is allowed to happen fire may result causing serious personal injury.

10 Before refitting the sparking plug, ensure that the electrodes are clean and correctly gapped. Route the HT lead so that it cannot chafe on any frame or engine component parts.

11 Finally, check around the engine to ensure that all components have been fitted correctly and are functioning properly. Check all electrical, vent and breather pipe connections and recheck all bolts, nuts and screws for security.

52.2 Do not omit to fit the rear upper engine mounting bolt spacer (125, 185 and 200)

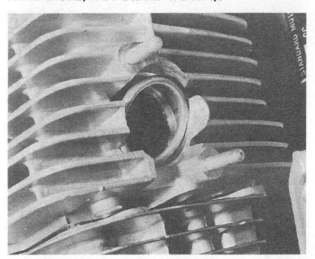

52.5a Fit the exhaust pipe gasket ..

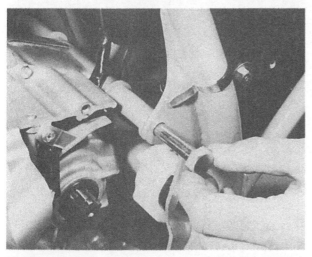

52.5b ... followed by the split collar ...

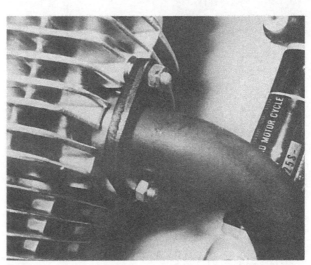

52.5c ... and secure the pipe in position with the two retaining nuts (125, 185 and 200)

52.5d The HT lead guide is fitted under the inlet stub right-hand securing screw head (125, 185 and 200)

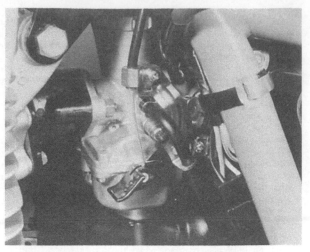

52.5e Ensure the hand-operated choke cable is correctly secured to the carburettor attachment points

52.6a Secure the gearbox sprocket with the locking plate and two bolts

52.6b Ensure the chain spring link is facing the direction of travel

53 Resetting the cam chain tension

1   With the type of cam chain tensioning device fitted to these models, the tension on the chain can only be correctly set with the engine running.

XR80, XL80 S and XL100 S models

2   Set the cam chain fine adjuster as described in paragraph 13 of Section 47 of this Chapter. Moving to the main adjuster situated on the rear of the barrel, slacken its locknut and the central adjuster screw; adjustment will now be made automatically. Tighten the adjuster screw and its locknut.

3   Make fine adjustment by starting the engine, slackening the fine adjuster set plate bolt, and turning the adjuster to achieve the correct setting (see Fig. 1.25).

XL125 S, XL185 S, XR185 and XR200 models

4   To adjust the cam chain tension on these models, start the engine and allow it to run at a fast tick-over. Prise the rubber boot from the tensioner pushrod housing and slacken the housing by applying a spanner to the large hexagon. The tension will adjust automatically. Retighten the housing. This will cause the two angled bushes to lock the tensioner pushrod in position, so maintaining the pre-set tension until adjustment is required again. Refit the rubber boot to the housing.

52.11 Do not omit to fit the sump guard forward securing bolt lockwasher (XL 125 S)

All models

5   Adjustment of the cam chain can, of course, take place only after installation of the engine is complete. **Do not** in any circumstances forget to carry out this operation soon after the engine is first run.

6   It should be noted that if the cam chain is too slack, the valve timing will be incorrect and poor engine performance will result. A slack cam chain will produce a chattering noise. A cam chain that is too tight, however, will produce a whining noise and place undue strain not only on itself but on the crankshaft and camshaft sprockets and bearings. Noting the sound produced from the area of the cam chain when the engine is running, will indicate whether the chain is too slack or too tight. The remedy in the form of adjustment is the same in both cases.

54 Final adjustments and preparation

1   Check the free play in the clutch cable and, if necessary, reset it to give 10 - 20 mm (0.4 - 0.8 in) of free play at the handlebar lever end. The adjustment procedure is as follows.

XR80, XL80 S and XL100 S models

2   Minor adjustments to the clutch cable free play may be made by means of the adjuster and locknut at the handlebar lever housing. When making major adjustments, however, both the handlebar mounted adjuster and cable adjuster must be turned in all the way. With this done, loosen the adjusting screw locknut, situated at the base of the clutch cover projection just forward of the kickstart spindle. Turn the adjusting screw anti-clockwise until resistance is felt. Turn the screw clockwise 1/8 to ¼ of a turn. Hold the screw in position with the screwdriver and tighten the locknut. The cable adjuster may now be turned out until there is 25 mm (1.0 in) of free play at the handlebar lever end and the locknut then tightened. The handlebar mounted adjuster may now be turned to obtain the specified amount of free play at the handlebar lever end and the locknut tightened.

XL125 S, XL185 S, XR185 and XR200 models

3   Major adjustments to clutch cable free play on these models is made by means of the adjuster passing through the bracket located at the base of the cylinder barrel. Adjustment is achieved simply by loosening the locknut, turning the adjuster the required amount and then retightening the locknut. The same principle should be applied to the handlebar mounted adjuster when fine adjustment is required.

All models

4   If the rear brake setting has been disturbed, when removing the right-hand crankcase cover with the engine fitted in the frame, reset the pedal height by following the procedure given in Chapter 5 of this Manual.

5   The throttle cable should be adjusted to give 2 - 6 mm (1/8 - ¼ in) of free play at the handlebar twistgrip. Major adjustments may be made by means of the adjuster located underneath the rubber cap in the centre of the removable mixing chamber top. Fine adjustment may be carried out by loosening the locknut and turning the adjuster at the twistgrip end. Tighten this locknut securely on completion of adjustment.

6   Ensure that the engine oil drain plug (where fitted) has been refitted and tightened to the specified torque loading. The drain plug sealing washer must be in good condition. Fill the crankcase with the recommended oil to the level marked on the dipstick. The dipstick should be rested on top of the crankcase hole into which it screws when determining the oil level.

XL185 S, XR185 and XR200 models

7   Ensure that the engine decompressor mechanism fitted to these models is adjusted correctly. Excessive free play at the

53.2 On 80 and 100 models, set the cam chain fine adjuster ...

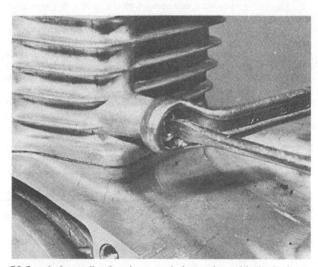

53.3 ... before adjusting the cam chain tension with the locknut and screw

54.2 On 80 and 100 models, make major clutch adjustments here

cylinder head lever will cause the starting procedure to be made more difficult, whereas insufficient free play may cause erratic idling and burning of the valves.

8 Adjustment of this mechanism should be carried out only after the exhaust valve clearance has been correctly adjusted. Align the T mark on the generator rotor with the index mark on the crankcase cover upper inspection hole, ensuring that the piston is on the compression stroke. Measure the amount of free play at the end of the cylinder head lever. This should be 1 - 2 mm (0.04 - 0.08 in). If necessary, adjust the amount of free play by loosening the locknut securing the cable adjuster and rotating the adjuster the necessary number of turns to give the correct amount of free play. Tighten the locknut.

55 Starting and running the rebuilt engine

1 Open the petrol tap to allow fuel to flow to the carburettor, close the carburettor choke and start the engine. Raise the choke as soon as the engine will run evenly and keep it running at a low speed for a few moments to permit the oil to circulate through the lubrication system.
2 Make the necessary adjustment to the cam chain tensioning device as described in Section 53 of this Chapter.
3 Bear in mind that the engine parts should be liberally coated with oil during assembly, so the engine will tend to smoke heavily for a few minutes until the excess oil is burnt away. Do not despair if the engine will not fire up at first, as it is quite likely that the excess oil will foul the sparking plug, necessitating its removal and cleaning. When the engine does start, listen carefully for any unusual noises, and if present, establish, and if necessary rectify, the cause. Check around the engine for any signs of leaking gaskets, or oil leaks.
4 Make sure each gear engages correctly and that all controls function effectively, particularly the brakes. This is an essential last check before taking the machine on the road.

56 Taking the rebuilt machine on the road

1 Any rebuilt machine will need time to settle down, even if parts have been replaced in their original order. For this reason it is highly advisable to treat the machine gently for the first few miles to ensure oil has circulated throughout the lubrication system and that new parts fitted have begun to bed down.
2 Even greater care is necessary if the engine has been rebored or if a new crankshaft has been fitted. In the case of a rebore, the engine will have to be run in again, as if the machine were new. This means greater use of the gearbox and a restraining hand on the throttle until at least 500 miles have been covered. There is no point in keeping to any set speed limit; the main requirement is to keep a light loading on the engine and to gradually work up performance until the 500 mile mark is reached. These recommendations can be lessened to an extent when only a new crankshaft is fitted. Experience is the best guide since it is easy to tell when an engine is running freely.
3 If at any time a lubrication failure is suspected, stop the engine immediately and investigate the cause. If an engine is run without oil, even for a short period, irreparable engine damage is inevitable.
4 When the engine has cooled down completely after the initial run, recheck the various settings, especially the valve clearances. During the run most of the engine components will have settled down into their normal working locations.

57 Fault diagnosis: engine

Symptom	Cause	Remedy
Engine will not start XR80, XL80 S and XL100 S models	Defective sparking plug	Remove the plug and lay it on cylinder head. Check whether sparking occurs when ignition is switched on and engine rotated.
	Dirty or closed contact breaker points	Check condition of points and whether gap is correct.
	Faulty or disconnected condenser	Check whether points arc when separated. Replace condenser if evidence of arcing.
XL125 S, XL185 S, XR185 and XR200 models	Faulty CDI unit	Check unit for electrical resistance. Renew if faulty.
	Faulty CDI source coil or pulser	Check circuits. Renew components if faulty.
Engine runs unevenly	Ignition and/or fuel system fault	Check each system independently, as though engine will not start.
	Blowing cylinder head gasket	Leak should be evident from oil leakage where gas escapes.
	Incorrect ignition timing	Check accuracy and if necessary reset.
Lack of power	Fault in fuel system or incorrect ignition timing	See above.
Heavy oil consumption	Cylinder barrel in need of rebore	Check for bore wear, rebore and fit oversize piston if required.
	Damaged oil seals	Check engine for oil leaks.
Excessive mechanical noise	Worn cylinder barrel (piston slap)	Rebore and fit oversize piston.
	Worn camshaft drive chain (rattle)	Adjust tensioner or replace chain.
	Worn big end bearing (knock)	Fit replacement crankshaft assembly.
	Worn main bearings (rumble)	Fit new journal bearings and seals.
Engine overheats and fades	Lubrication failure	Stop engine and check whether internal parts are receiving oil. Check oil level.

58 Fault diagnosis: clutch

Symptom	Cause	Remedy
Engine speed increases as shown by tachometer but machine does not respond	Clutch slip	Check clutch adjustment for free play at handlebar lever. Check thickness of inserted plates.
Difficulty in engaging gears. Gear changes jerky and machine creeps forward when clutch is withdrawn. Difficulty in selecting neutral	Clutch drag	Check clutch adjustment for too much free play. Check clutch drums for indentations in slots and clutch plates for burrs on tongues. Dress with file if damage not too great.
Clutch operation stiff	Damaged, trapped or frayed control cable	Check cable and replace if necessary. Make sure cable is lubricated and has no sharp bends.

59 Fault diagnosis: gearbox

Symptom	Cause	Remedy
Difficulty in engaging gears	Selector forks bent Gear clusters not assembled correctly	Replace. Check gear cluster arrangement and position of thrust washers.
Machine jumps out of gear	Worn dogs on ends of gear pinions Stopper arms not seating correctly	Replace worn pinions. Remove right-hand crankcase cover and check stopper arm action.
Gearchange lever does not return to original position	Broken return spring	Replace spring.
Kickstarter slips	Ratchet assembly worn	Part crankcase and replace all worn parts.
Kickstarter does not return when engine is turned over or started	Broken or poorly tensioned return spring	Replace spring or re-tension.

Chapter 2 Fuel system and lubrication

For information relating to 1981 — 1987 models, see Chapter 7

Contents

Specifications

	XR80	XR185	XR200
Fuel tank			
Overall capacity	3.6 litre (0.8 Imp gal, 0.95 US gal)	7.0 litre (1.54 Imp gal, 1.85 US gal)	7.0 litre (1.54 Imp gal, 1.85 US gal)
Reserve capacity	0.8 litre (1.40 Imp pint, 1.70 US pint)	1.5 litre (2.64 Imp pint, 3.17 US pint)	1.5 litre (2.64 Imp pint, 3.17 US pint)
Carburettor			
Make	Keihin	Keihin	Keihin
ID number	PC10B-A (US 1979,UK) PC10C-A (US 1980)	PD31A-A	PD32A-A/B
Venturi diameter	20 mm	24 mm	26 mm
Main jet	95	115	115
Pilot screw setting (turns out from fully in)	1¾	2¼	1¾
Jet needle position	4th groove from top	4th groove from top	3rd groove from top
Float level	21.5 mm (0.85 in)	12.5 mm (0.50 in)	12.5 mm (0.50 in)
Idling speed	1500 ± 100 rpm	1300 rpm	1300 ± 100 rpm
Air cleaner			
Element type	Oiled, polyurethane foam		
Lubrication			
System type	Forced pressure, wet sump		
Oil capacity	0.9 litre (1.58 Imp pint, 1.90 US pint)	1.1 litre (1.94 Imp pint, 2.32 US pint)	1.1 litre (1.94 Imp pint, 2.32 US pint)
Oil pump			
Type	Trochoid	Trochoid	Trochoid
Outer rotor to housing clearance	0.15 mm (0.006 in)	0.30 - 0.36 mm (0.012 - 0.014 in)	0.30 - 0.36 mm (0.012 - 0.014 in)
Service limit	0.20 mm (0.008 in)	0.40 mm (0.016 in)	0.40 mm (0.016 in)
Inner rotor to outer rotor tip clearance	0.15 mm (0.006 in)	0.15 mm (0.006 in)	0.15 mm (0.006 in)
Service limit	0.20 mm (0.008 in)	0.20 mm (0.008 in)	0.20 mm (0.008 in)
Rotor end clearance	—	0.15 - 0.20 mm (0.006 - 0.008 in)	0.15 - 0.20 mm (0.006 - 0.008 in)
Service limit	—	0.25 mm (0.010 in)	0.25 mm (0.010 in)

Fuel tank	XL80 S	XL100 S	XL125 S	XL185 S
Overall capacity	3.6 litre (0.79 Imp gal, 0.95 US gal)	7.0 litre (1.54 Imp gal, 1.85 US gal)	7.0 litre (1.54 Imp gal, 1.85 US gal)	7.0 litre (1.54 Imp gal, 1.85 US gal)
Reserve capacity	0.8 litre (1.40 Imp pint, 1.70 US pint)	1.5 litre (2.64 Imp pint, 3.17 US pint)	1.5 litre (2.64 Imp pint, 3.17 US pint)	1.5 litre (2.64 Imp pint, 3.17 US pint)
Carburettor				
Make	Keihin	Keihin	Keihin	Keihin
Type	PF65A-A	PD20A-A (US 1979) PD30A-A (US 1980) PD22A-A (XL100S-Z)	PD14A-A (US 1979) PD13A-A/B (US 1980) PD21A-A (XL125S-Z) PD21A-A/B (XL125S-A)	PD14B-A (US 1979) PD13B-A/B (US 1980) PD15B-A (XL185S-Z) PD15D-A (XL185S-A)
Venturi diameter	20 mm	22 mm	22 mm	22 mm
Main jet	72	102	98	102
Pilot screw setting (turns out from fully in)	2¾	1¾	1¾	1½
Jet needle position	—	—	4th groove from top	4th groove from top
Float level	16.6 mm (0.65 in)	12.5 mm (0.49 in)	12.5 mm (0.49 in)	12.5 mm (0.49 in)
Idling speed	1500 ± 100 rpm	1300 ± 100 rpm	1300 ± 100 rpm	1300 rpm
Air cleaner				
Element type...	Oiled, polyurethane foam			
Lubrication				
System type	Forced pressure, wet sump			
Oil capacity	0.9 litre (1.58 Imp pint, 1.90 US pint)	1.0 litre (1.76 Imp pint, 2.11 US pint)	1.1 litre (1.94 Imp pint, 2.32 US pint)	1.1 litre (1.94 Imp pint, 2.32 US pint)

Oil pump	XL80 S	All other models
Type	Trochoid	Trochoid
Outer rotor to housing clearance	0.15 - 0.20 mm (0.006 - 0.008 in)	0.15 - 0.21 mm (0.006 - 0.008 in)
Service limit	0.25 mm (0.0098 in)	0.40 mm (0.016 in)
Inner rotor to outer rotor tip clearance ...	0.15 mm (0.006 in)	0.15 mm (0.006 in)
Service limit	0.20 mm (0.008 in)	0.20 mm (0.008 in)
Rotor end clearance	0.02 - 0.07 mm (0.0008 - 0.0028 in)	0.15 - 0.20 mm (0.006 - 0.008 in)
Service limit	0.25 mm (0.010 in)	0.25 mm (0.010 in)

1 General description

The fuel system comprises a petrol tank from which petrol is fed by gravity to the carburettor float chamber. It is controlled by a petrol tap with a built-in filter. The tap has three positions: 'Off', 'On', 'Reserve', the latter providing a reserve supply of petrol when the main supply has run out. For cold starting the carburettor has a choke (manually operated) which is operated at the rider's discretion. The machine should run on 'Choke' for the least amount of time possible.

A large capacity air cleaner, with a detachable oil-soaked plastic foam element is mounted on one side of the machine, within a moulded plastic box. It is attached directly to the carburettor intake by a short rubber hose.

The exhaust system fitted to the XR80 and XL80 S models takes the form of an integral exhaust pipe and silencer, whereas the system fitted to all other models comprises a detachable front pipe section and silencer assembly. Both types of exhaust system are carried on the right-hand side of the machine.

The lubrication system is of the pressure fed type, supplying oil to almost every part of the machine.

On the XL125 S, XL185 S, XR185 and XR200 models, there is a centrifugal filter mounted directly on the end of the crankshaft. Centrifugal force caused by the rotation of the engine throws the heavier impurities outwards where they stick to the walls, allowing only the clean, lighter oil through.

The centre of the filter rotor cap incorporates a pressure relief valve. This valve is contained in the centre of the special primary drive pinion retaining nut on models that do not have this type of filter fitted.

All the models covered in this Manual have the same type of oil pump fitted which operates as follows.

Oil is picked up by the oil pump and pressure fed through the right-hand crankcase where it is diverted into two routes. In one direction it goes through a passage in the right-hand crankcase cover and then through the oil filter to the crankshaft. The other direction takes the oil through a passage via a cylinder head stud to the rocker arms. On 125, 185 and 200 models the transmission also receives oil under pressure.

Besides the centrifugal filter already mentioned, the engine is protected by a second filter. The second, separate, filter, is of the metal gauze type.

On XR80, XL80 S and XL100 S models, this gauze screen takes the form of a flat plate which is located in a slot cast in the base of the right-hand crankcase. All other models have a cylindrical screen which is located behind a threaded circular cover in the left-hand crankcase cover.

2 Petrol tank: removal and refitting

1 Before the petrol tank is removed, the dualseat must be detached by unscrewing the two bolts which retain the seat base to the frame. This will give access to the single tank retaining bolt at the rear of the tank. Remove this bolt and the large diameter washer beneath it from the rear of the tank. Turn the fuel tap lever to the 'Off' position and ease the fuel feed pipe and clip from the tap stub. Release the breather pipe and clip (where fitted) from the tank cap stub. Raise the rear end of the tank and pull it backwards, moving it gently

from side to side in order to free the guide channels from the rubber buffers on each side of the steering head.

2   Store the tank in a safe place whilst it is removed from the machine, well away from any naked lights or flames. It will otherwise represent a considerable fire or explosion hazard. Check that the tap is not leaking and that it cannot be accidentally knocked into the 'On' position. It is well worth taking simple precautions to protect the paint finish of the tank whilst in storage. Placing the tank on a soft protected surface and covering it with a protective cloth or mat may well avoid damage being caused to the finish by dirt, grit, dropped tools, etc.

3   To refit the tank, reverse the procedure adopted for its removal. Move it from side to side before it is fully home, so that the rubber buffers engage with the guide channels correctly. If difficulty is encountered in engaging the front of the tank with the rubber buffers, apply a small amount of petrol to the buffers to ease location. Always carry out a leak check on the fuel pipe connections after fitting the tank and turning the tap lever to the 'On' position. Any leaks found must be cured; as well as wasting fuel, any petrol dropping onto hot engine castings may well result in a fire or explosion occurring.

3   Petrol tap: removal, examination and renewal

1   Before the petrol tap can be removed, it is first necessary to drain the tank. This is easily accomplished by removing the feed pipe from the carburettor float chamber and allowing the contents of the tank to drain into a clean receptacle, with the tap turned to the 'Reserve' position. Alternatively, the tank can be removed and placed on one side, so that the fuel level is below the tap outlet. Care must be taken not to damage the tank paintwork whilst doing this.

2   The tap unit is retained to the threaded boss on the tank by means of a gland nut. Undo this nut, remove the clip and fuel pipe from the fuel tap stub and withdraw the filter stack from the tank.

3   If fuel leakage is found around the tap lever pivot, it will be necessary to renew the tap as a complete unit. Refitting the tap is a reversal of the removal procedure.

4   Check that the feed pipe from the tap to the carburettor is in good condition and that the push-on joints are a good fit, irrespective of the retaining wire clips. If particles of rubber are found in the filter, replace the pipe, since this is an indication that the internal bore is breaking up.

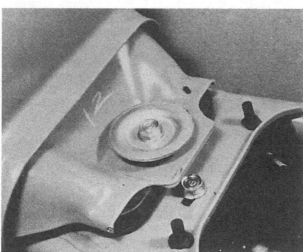

2.1a The petrol tank is retained by a single bolt at the rear ...

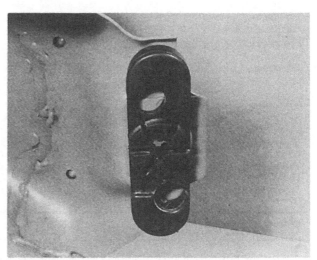

2.1b ... which passes through the rubber mounting

4   Petrol feed pipe: examination

The petrol feed pipe is made from thin walled synthetic rubber and is of the push-on type. It is only necessary to replace the pipe if it becomes hard or splits. It is unlikely that the retaining clip should need replacing due to fatigue as the main seal between the pipe and union is effected by an interference fit.

5   Carburettor: removal and refitting

1   On all but the XR80 and XL80 S models, it is advisable to improve access to the carburettor by removing the sidepanels, dualseat and petrol tank. This will make the removal process far easier and lessen the risk of damage to any paintwork. On the XR80 and XL80 S models, it is necessary only to remove the right-hand sidepanel in order to gain access to the air cleaner cover securing nuts.

2   If tank removal is necessary, follow the procedure stated in Section 2 of this Chapter. The carburettor top may be unscrewed at this stage but withdrawal of the throttle valve assembly is best left until the carburettor is detached from the

3.2 Withdraw the petrol tap and filter stack from the tank

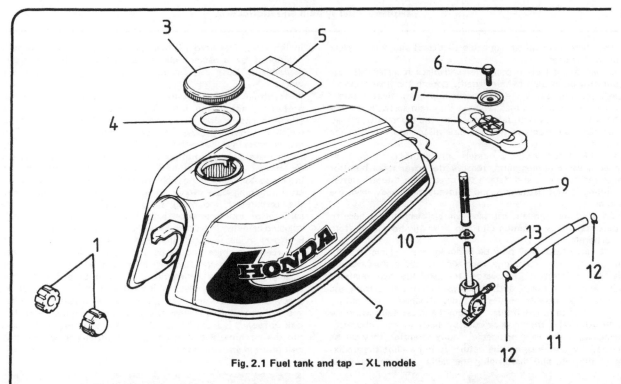

Fig. 2.1 Fuel tank and tap — XL models

1	Rubber buffers	6	Bolt	10	Washer
2	Fuel tank	7	Washer	11	Fuel delivery pipe
3	Filler cap	8	Rear tank cushion	12	Clip — 2 off
4	Cap gasket	9	Filter stack	13	Tap body
5	Caution label				

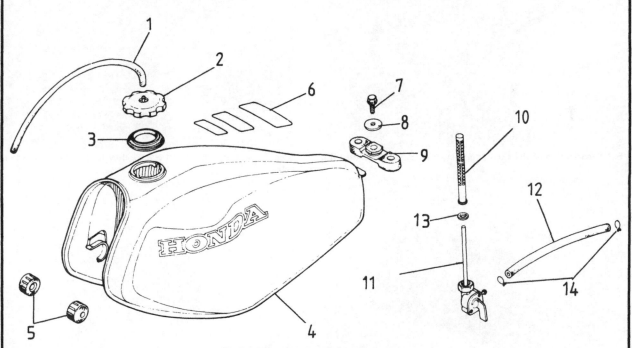

Fig 2.2 Fuel tank and tap — XR models

1	Breather pipe	6	Caution labels	11	Tap body
2	Filler cap	7	Bolt	12	Fuel delivery pipe
3	Gasket	8	Washer	13	Washer
4	Fuel tank	9	Rear tank cushion	14	Clip — 2 off
5	Rubber buffers	10	Filter stack		

engine. Disconnect the choke cable (if fitted) and place it clear of the carburettor.

3   Obtain a small clean container and place it underneath the carburettor drain pipe (where fitted). Loosen the float chamber drain screw and allow the fuel contained in the float chamber to drain into the container. Tighten the drain screw. Note that if the fuel tank has been left in position, the tap must be turned to the 'Off' position and the fuel pipe disconnected before the carburettor is drained.

4   On XR80 and XL80 S models only, remove the three air cleaner cover retaining nuts, remove the cover to carburettor mouth retaining clip and detach the cover from the carburettor by pulling it first backwards and then outwards away from the machine.

5   On all other models, displace the air cleaner case hose to carburettor intake retaining clip and prise the hose away from the carburettor.

6   The carburettor may now be removed from the inlet stub by removing the two nuts, one each side of the stub. Carefully ease the carburettor backwards, to clear the two retaining studs, and then down and sideways to remove it from the machine. Take care to ensure all pipes attached to the carburettor are free to follow it during removal and carefully withdraw the throttle valve from the carburettor body as the carburettor is removed. If the valve or needle require attention, they can be detached by compressing the return spring against the underside of the top, and disengaging the cable end from its recess

in the valve. The needle is held by a spring clip, which is itself positioned by a second clip inside the valve. It is normally advisable to leave this assembly undisturbed unless obviously worn.

7   Refitting of the carburettor is a reversal of the removal procedure, noting the following points. Great care must be taken to ensure that all connections are free from air leaks, which will have an adverse effect on engine performance. This means checking the carburettor body to intake stub O-ring for damage or deterioration and renewing it if necessary; the air cleaner case to carburettor mouth hose should also be inspected for signs of possible air leakage. Tighten the carburettor retaining nuts evenly to avoid any risk of distortion to the carburettor mounting flange. Finally, check that the carburettor top is screwed fully home and the throttle cable adjusted as stated in paragraph 5 of Section 54 of Chapter 1.

8   Note: it is essential that before the engine is run, all disturbed fuel pipe and carburettor connections are checked for leaks. Switch the fuel tap lever to the 'On' position and closely inspect the connections, allowing time for the fuel to fill the float chamber. Whilst carrying out this inspection, check that all fuel and vent pipes have been correctly routed and are not chafing against frame or engine components. If a leak is found, it must be cured before attempting to start and run the machine. Petrol dripping onto hot engine castings may well result in an explosion or fire.

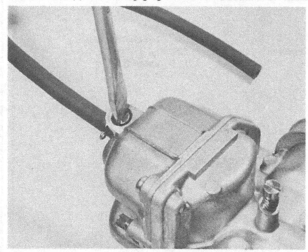

5.3 The location of the float chamber drain screw (all except 80)

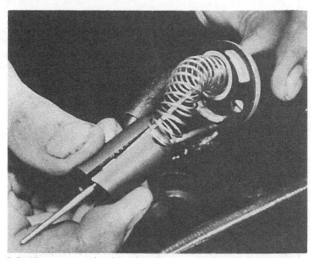

5.6a If necessary, the throttle valve may be detached from the cable

5.6b The V-spring ...

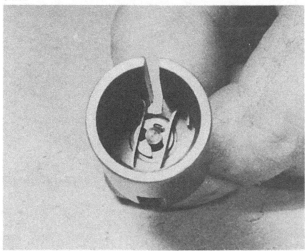

5.6c ... locates within the valve and retains the needle spring clip

6 Carburettor: dismantling, examination and reassembly

1 Before dismantling the carburettor, cover an area of the work surface with clean paper or rag. This will not only prevent any components that are placed upon it from becoming contaminated with dirt, moisture or grit but, by making them more visible, will also prevent the many small components removed from the carburettor body from becoming lost.

2 Detach the float chamber by inverting the carburettor and removing the three screws and spring washers that retain it to the main body (two on the XR80 and XL80 S models). There is a sealing gasket around the edge of the float chamber, which will remain either with the float chamber or the main body of the carburettor.

3 Pull out the pivot pin from the twin float assembly and lift the floats away. The float needle can now be displaced from its seating and should be put aside in a safe place for examination at a later stage. It is very small and easily lost if care is not taken to store it in a safe place.

4 Remove the plastic anti-surge baffle (where fitted) from the main jet and unscrew the main jet from the central column between the two floats. The needle jet holder is immediately below the main jet, and may also be unscrewed. Invert the carburettor and allow the needle jet to fall out of its recess.

5 On XR80 models, remove the slow running jet, situated next to the needle jet holder. On all other models, the jet is in a similar position but fixed.

6 On XR80 models, the pilot adjustment screw is situated on the side of the carburettor, being the lower, rearmost of the two screws with the carburettor in the installed position. On all other models, the screw is situated just in front of the float chamber. When removing the pilot screw, count the number of turns needed to detach it from the carburettor body and note the information for reference when refitting. Care should be taken to retain the spring after the screw has been removed. On 1980 XL100 S (US) models the pilot screw is fitted with a limiter cap to conform to EPA regulations. The limiter cap allows rotation of the screw only within fixed limits (about 300° of rotation) and thus prevents unjudicious adjustment of the screw which might increase mixture strength and thus exceed the EPA regulations. A pilot screw fitted with a limiter cap can be removed only after removal of the float bowl. On reassembly it should be fitted in exactly the position in which it was found. If a new screw is to be fitted the mixture strength should be set accurately by following the procedure given in paragraph 3 of the following Section. Having done this a new limiter cap should be fitted so that the stop prevents rotation of the cap in an anti-clockwise direction and thus prevents a richer mixture setting being provided. On XL80 S models the

pilot screw is covered by a tight-fitting plug drifted into a recess in the carburettor body. Because of this, pilot screw removal or adjustment is not possible subsequent to factory installation.

7 Remove the throttle stop screw, located on the side of the carburettor just above the float chamber, taking care to count and note the number of turns needed for removal and to retain the spring after the screw has been removed.

8 Check that the floats are in good order and not punctured. It is not possible to effect a permanent repair. In consequence, a new replacement should always be fitted if damage is found.

9 The float needle seating will wear after lengthy service and should be closely examined with a magnifying glass. Wear usually takes the form of a ridge or groove, which will cause the float needle to seat imperfectly. In extreme cases the seat will wear, and because it is not removable the complete carburettor body must be renewed.

10 The hand-operated cold starting choke should not require attention. Wear is unlikely to occur unless the machine has covered a very high mileage, in which case the whole carburettor will then require renewal.

11 Although the carburettor top has been unscrewed, no further dismantling should prove necessary nor is advisable. It is unlikely that the position of the needle will need to be changed or the slide renewed, except after a long period of service in the latter case.

12 Check that all mating surfaces on the carburettor body are flat by using a straight-edge laid across the mating surface. Ensure all O-rings and sealing gaskets are renewed when reassembling and refitting the carburettor and that, where applicable, they are correctly seated in their retaining grooves. The springs on the pilot jet and throttle stop screws should be carefully inspected for signs of corrosion and fatigue and renewed if necessary.

13 Before the carburettor is reassembled, by reversing the dismantling procedure, it should be cleaned out thoroughly using compressed air. Avoid using a piece of rag since there is always risk of particles of lint obstructing the internal passageways or the jet orifices.

14 Never use a piece of wire or any pointed metal object to clear a blocked jet. It is only too easy to enlarge the jet under these circumstances, and increase the rate of petrol consumption. If compressed air is not available, a blast of air from a tyre pump will usually suffice. As a last resort, a fine **nylon** bristle may be used.

15 Do not use excessive force when reassembling a carburettor because it is easy to shear a jet or some of the smaller screws. Furthermore, the carburettor is cast in a zinc based alloy which itself does not have a high tensile strength. If any of the castings are damaged during reassembly, they will almost certainly have to be renewed.

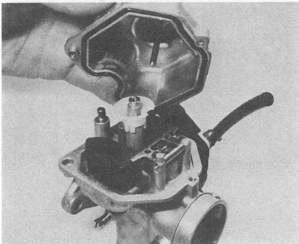

6.2 Detach the carburettor float chamber ...

6.3a ... pull out the pivot pin to free the float assembly ...

6.3b ... and lift the float assembly clear

6.3c The float needle is very small and easily lost

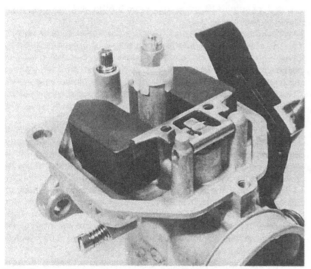

6.4a Remove the plastic anti-surge baffle (where fitted) ...

6.4b ... and remove the main jet ...

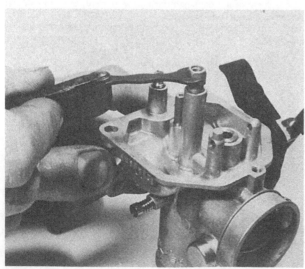

6.4c ... and needle jet holder

6.6a The location of: 1 The throttle stop screw, 2 The pilot adjustment screw (80)

6.6b The location of the pilot adjustment screw ...

6.7 ... and throttle stop screw (all models except 80)

6.10 The choke assembly should not require attention

6.12a Check that all mating surfaces are flat ...

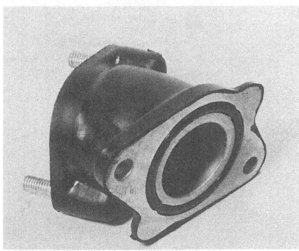

6.12b ... and the O-rings are correctly seated

7 Carburettor: adjustment

1 If flooding of the carburettor or excessive weakness have been experienced it is wise to start operations by checking the float level, which will involve detaching the carburettor, if not already removed, inverting it and removing the float chamber bowl.

2 Place the carburettor on a flat level surface with its air intake mouth uppermost and with the float valve just closed. Check that the distance between the mating surface of the carburettor body and the furthest vertical surface of the float is as stated in the Specifications at the beginning of this Chapter. If necessary, adjust the setting by carefully bending the float arm.

3 Refit the carburettor, and with the engine running at normal operating temperature, turn the pilot jet screw inwards until the engine misfires or decreases in speed. Note the position of the screw, then turn it outwards until similar symptoms are observed. The screw should be set exactly between these two positions. This position is normally close to the number of turns quoted in the Specifications from the fully closed position.

4 If the idling speed is now too low or too high, adjust the idling speed (throttle stop) screw until the engine will tick-over at the recommended idling speed.

5 Note that these adjustments should always be made with the engine at normal operating temperature and with the air cleaner connected, otherwise a false setting will be obtained.

8 Carburettor: settings

1 Some of the carburettor settings, such as the sizes of the needle jet, main jet, and needle position etc are predetermined by the manufacturer. Under normal circumstances, it is unlikely that these settings will require modification, even though there is provision made. If a change appears necessary, it can often be attributed to a developing engine fault.

2 As an approximate guide, the pilot jet setting controls engine speed up to 1/8 throttle. The throttle slide cutaway controls engine speed from 1/8 to ¾ throttle. The size of the main jet is responsible for engine speed at the final ¾ to full throttle. It should be added however, that these are only guide-lines. There is no clearly defined demarcation line due to a certain amount of overlap that occurs between the carburettor components involved.

3 Always err slightly on the side of a rich mixture, since a weak mixture will cause the engine to overheat. Reference to Chapter 3 will show how the condition of the sparking plug can be interpreted with some experience as a reliable guide to carburettor mixture strength.

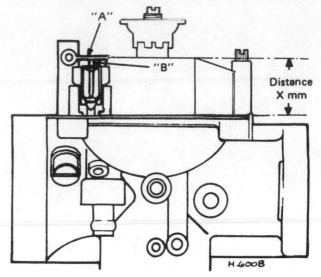

Fig. 2.3 Float height adjustment

A   Float tongue
B   Float valve
X   Float height

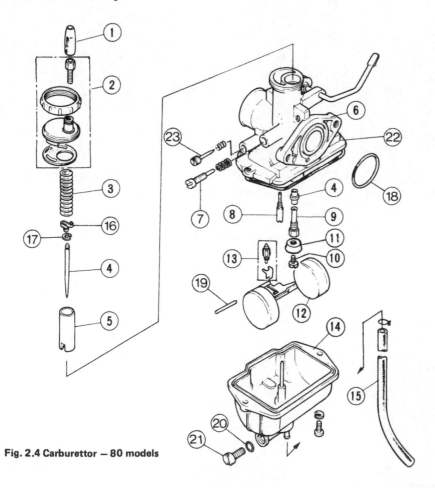

Fig. 2.4 Carburettor — 80 models

1   Rubber cover
2   Carburettor top assembly
3   Spring
4   Jet needle
5   Throttle valve
6   Carburettor body
7   Pilot jet screw
8   Slow running jet *
9   Needle jet holder
10   Main jet
11   Anti-surge baffle
12   Float
13   Float needle and holder
14   Float chamber
15   Drain pipe
16   Spring clip
17   E-clip
18   O-ring
19   Pivot pin
20   Sealing washer
21   Drain screw
22   Float chamber gasket
23   Throttle stop screw

*XR80 only

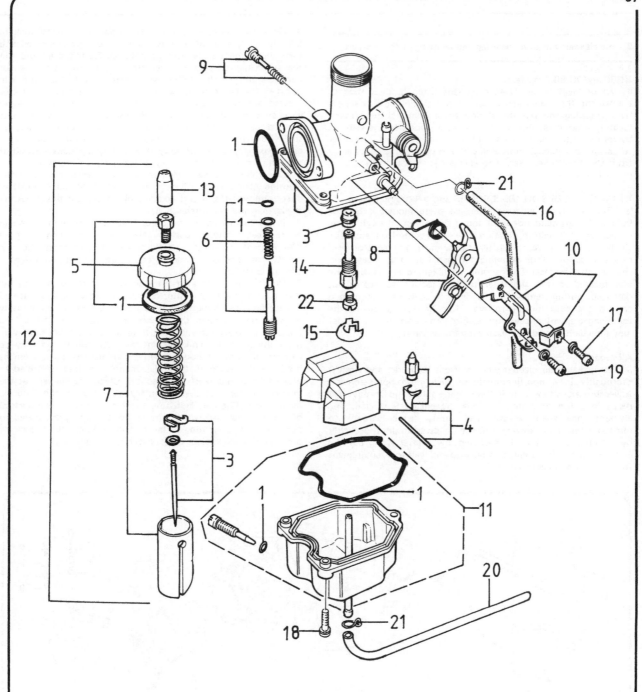

Fig. 2.5 Carburettor — XR185 (others similar)

1	Seal set	9	Throttle stop screw	16	Vent hose
2	Float needle assembly	10	Cable anchor	17	Screw
3	Jet needle assembly	11	Float bowl assembly	18	Screw — 3 off
4	Float/pivot	12	Throttle assembly	19	Screw
5	Carburettor top	13	Rubber cap	20	Drain hose
6	Pilot screw	14	Jet holder	21	Clip — 2 off
7	Throttle slide/spring	15	Main jet	22	Main jet
8	Choke arm				

8   Air cleaner: removal, cleaning and refitting of the element

XR80 and XL80 S models

1   To remove the air cleaner element fitted to these models, expose the three air cleaner case cover retaining nuts or screws by unclipping the right-hand side panel from its three frame locating grommets. Access may now be gained to the element by removing the three nuts and plain washers, releasing the cover to carburettor mouth retaining clip and pulling the cover away from the air cleaner case. The element is attached to the back of the cover.

XL100 S, XL125 S, XL185 S, XR185 and XR200 models

2   The air cleaner element fitted to these models is located behind the left-hand sidepanel and is fitted around a cylindrical holder which is held in position by a single central wingnut. The sidepanel may be removed by unscrewing the three crosshead retaining screws, thus allowing the panel to drop away from the air cleaner case. Care should be taken to ensure that the seal located in the sidepanel to case joining surface is not damaged during removal. It should be correctly positioned around the case mouth to avoid loss or damage during the following element cleaning procedure. The element and holder may now be removed by unscrewing the wingnut.

All models

3   Remove the air cleaner element from its holder and wash it thoroughly in a non-flammable or high flash point solvent. Never use petrol or a low flash point solvent as it may well result in a fire or explosion once the engine is started. The element may then be squeezed gently (never wrung) to remove most of the solvent and then allowed to dry thoroughly by standing it in a well ventilated area. Never wring out the element, as the foam will be damaged and lead to a requirement for premature replacement.

4   Once the element is properly dried, allow it to soak in SAE 80 - 90 gear oil. Gently squeeze out any excess oil and slide the element back over its housing. Refit the assembly in the air cleaner case using a reversal of the procedure used for removal, noting the following points. Check that the seal between the cover or sidepanel and air cleaner case is correctly located and undamaged. Where the element holder is marked with an arrow and the word TOP, ensure that the arrow points upward before tightening the retaining nut. Ensure that, on the XR80 and XL80 S machines, the carburettor to air cleaner cover connection is correctly made. Any air leaks allowed at this connection will affect the performance of the engine.

5   Clean the air cleaner element at the intervals quoted in the Routine Maintenance Chapter at the front of this Manual. In dusty atmospheres, it will be advisable to increase the frequency of cleaning and re-impregnating.

6   If the foam element becomes torn or perforated, it should be replaced without question. Never run the engine without the element connected to the carburettor because the carburettor is specially jetted to compensate for the addition of this component.

7   On models where a primary air filter is fitted beneath the seat, gain access to the filter housing by removing the seat and if necessary, the tank to fully expose the head of the housing securing bolt. Detach the rubber cowl from the rear of the housing, unscrew the retaining bolt and lift the housing away from the machine. The element with its holder can now be withdrawn from the housing and the element removed from its holder and cleaned in a similar manner to that described in paragraphs 3 and 4 of this Section. Clean the element at the intervals stated for the secondary air filter in the Routine Maintenance Chapter. Refitting the filter assembly is a reversal of the removal procedure; ensure no air leaks are allowed to occur around the joining surfaces of the rubber cowl and renew the cowl if it shows signs of deterioration or splitting.

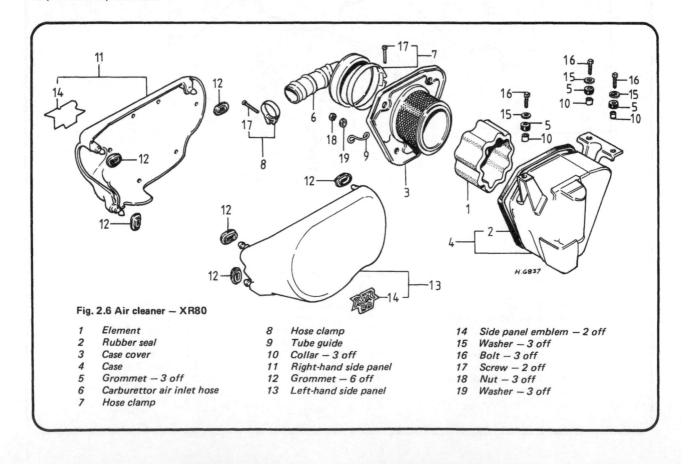

Fig. 2.6 Air cleaner — XR80

1	Element	8	Hose clamp	14	Side panel emblem — 2 off
2	Rubber seal	9	Tube guide	15	Washer — 3 off
3	Case cover	10	Collar — 3 off	16	Bolt — 3 off
4	Case	11	Right-hand side panel	17	Screw — 2 off
5	Grommet — 3 off	12	Grommet — 6 off	18	Nut — 3 off
6	Carburettor air inlet hose	13	Left-hand side panel	19	Washer — 3 off
7	Hose clamp				

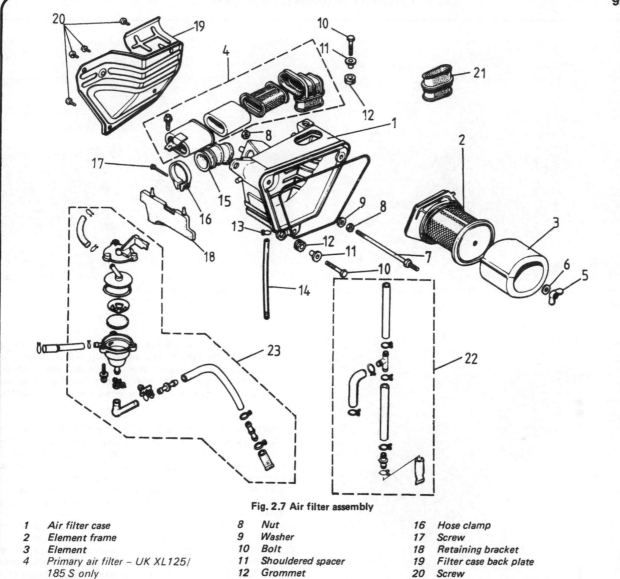

Fig. 2.7 Air filter assembly

1	Air filter case	8	Nut	16	Hose clamp
2	Element frame	9	Washer	17	Screw
3	Element	10	Bolt	18	Retaining bracket
4	Primary air filter – UK XL125/	11	Shouldered spacer	19	Filter case back plate
	185 S only	12	Grommet	20	Screw
5	Wingnut	13	Pipe clip	21	Hose
6	Washer	14	Pipe	22	Breather assembly – standard type
7	Stud	15	Carburettor air inlet hose	23	Breather assembly – EPA type

9.4a Check that the element holder is correctly positioned ...

9.4b ... and the cover seal correctly located

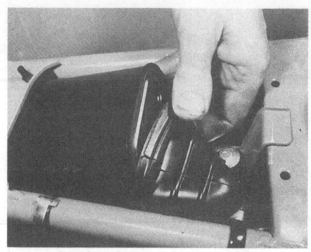

9.6a To remove the primary air filter housing, detach the rubber cowl ...

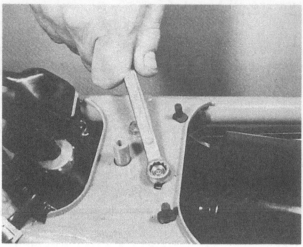

9.6b ... and unscrew the securing bolt

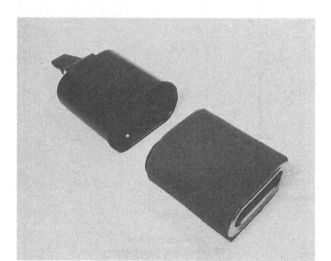

9.6c The filter element and holder may be withdrawn from the housing

---

10 Exhaust system: general

---

1   The exhaust system on a four-stroke motorcycle will require very little attention, as, unlike two-stroke machines, it is not prone to the accumulation of carbon. The only points requiring attention are the general condition of the system, including mountings and protective finish, and ensuring that the system is kept airtight, particularly at the exhaust port.

2   Air leaks at the exhaust port, ie the joint between the exhaust pipe and the cylinder head, will cause mysterious backfiring when the machine is on overrun, as air will be drawn in causing residual gases to be ignited in the exhaust pipe. To this end, make sure that the composite sealing ring is renewed each time the system is removed.

3   Cleaning of the exhaust system spark arrester and silencer baffles varies between the different types of machines and should be carried out as follows.

XR80 and XL80 S models

4   The spark arrester fitted to the exhaust system on these machines is detachable and may be removed for cleaning by unscrewing the three retaining screws at the silencer end. Clean any carbon deposits off the spark arrester with a wire brush and clean out the exhaust system by first removing the drain plug located in the exhaust pipe just under the right-hand sidepanel. The engine should now be started and revved approximately twenty times to clear the exhaust of any carbon build up.

XL100 S, XL125 S, XL185 S, XR185 and XR200 models

5   On these machines the spark arrester is a permanently fixed item. The system is cleaned of any carbon build up by removing the plate located underneath the rear of the silencer. An impact driver may be necessary to free the two crosshead retaining screws. If this tool is used, the silencer must be well supported to avoid any strain being placed upon the exhaust system mounting points.

6   With the plate and gasket removed, start the engine and allow it to idle. Place a thick pad of rag over the end of the silencer, thereby creating back pressure in the system, and rev the engine approximately twenty times to blow any carbon out of the silencer plate hole. Before refitting the plate, check the condition of the screws and gasket and renew if thought necessary.

All models

7   When carrying out the above listed cleaning procedures it is strongly advised that the following safety precautions be observed. The machine should be run only in a well ventilated area. The area should be free of any combustible materials, due to there being an increased fire hazard once the spark arrester is removed. Always wear some form of eye protection to protect against dislodged particles of carbon. Wait until the exhaust system is cool to the touch before removing or refitting any components; this system becomes very hot during engine operation.

8   There is no means of detaching the baffles from the silencer to facilitate their cleaning; usually the silencer will have reached the point of visible deterioration long before there is any chance of obstruction through carbon build up.

9   The silencer, as its name implies, effectively reduces the exhaust noise to an acceptable level without having any adverse effects on engine performance. Do not tamper with or remove any baffles from within the silencer. Although a much louder exhaust note may give the impression of greater speed, this is rarely the case in practice. Tampering with the standard system, designed to give optimum performance with a low noise level, will upset the balance and cause reduced performance, even though the changed exhaust note creates the illusion of speed.

11 Exhaust system: removal and refitting - XL100 S, XL125 S, XL185 S, XR185 and XR200 models

1 The exhaust system fitted to these models is a two-piece unit, the front section of exhaust pipe being attached to the silencer assembly by a clamp and Allen bolt. Removal and refitting of the front section of the system is described in Chapter 1, Sections 5 and 52 respectively.

2 To remove the silencer assembly, proceed as follows. Unscrew the two seat retaining bolts and remove the seat. Note that the right-hand seat retaining bolt also passes through the rear silencer mounting bracket. Remove the right-hand sidepanel by pulling it free of its three frame mounting grommets. Remove the battery case (where fitted) upper securing bolt and move the case rearwards to allow access to the front section of the silencer assembly. Note the level of the acid in the battery whilst moving the case and do not allow it to clear the tops of the battery plates. The maximum amount the case should be moved from the vertical is 45°.

3 In practice, it was found that in order to allow easy withdrawal of the silencer assembly from the rear of the machine, removal of the rear wheel was necessary. Removal of the rear wheel should therefore be carried out in accordance with the instructions given in Chapter 5.

4 Remove the rear section of the rear mudguard by unscrewing the two retaining bolts, located one either side of the mudguard, and the single dome nut located just forward of the rear light assembly. Unclip the electrical connections, noting each connection for reference when refitting. The forward section of the rear mudguard may now be detached from its location points on the frame and removed from the machine after the electrical loom has been unclipped from the mudguard.

5 Unscrew and remove the two forward silencer to frame attachment bolts and withdraw the silencer assembly rearwards from the machine.

6 Refitting the exhaust system is essentially a reversal of the removal procedure. Once fitted, the system should be checked for any signs of exhaust gas leakage at the pipe connections and the seal or gasket renewed if necessary.

12 Engine lubrication

1 The engine oil, which is common also to the gearbox, is contained within a crankcase compartment formed by the right-hand crankcase cover. Note that checking the oil level must be done before the engine has been run for any time, because the oil would otherwise be somewhat dispersed around the engine. Also note that to check correctly the oil level, the dipstick/cap should be rested on the edge of the filler orifice and not screwed home.

2 The lubrication works on the wet sump principle. The trochoid type oil pump picks up oil from the crankcase compartment, which passes under pressure through an oil filter gauze before it is delivered to the engine components such as the overhead camshaft and rocker gear, crankshaft, connecting rod and main bearings and to the gearbox shafts. Apart from the filter already mentioned, the XL125 S, XL185 S, XR185 and XR200 models have a centrifugal filter attached to the right-hand end of the crankshaft. This filter removes impurities from the oil, so that oil that has drained back into the crankcase can be recirculated. A few of the engine and gearbox components are lubricated on the splash principle, namely those that are subjected to less stress. See also the Section at the beginning of this Chapter for a description of the passage of the lubricant around the engine.

3 The need to change the engine oil at the prescribed regular intervals and to clean out the two filters at the recommended intervals, is of paramount importance to the long-term reliability of these machines. If the machine is used only for short journeys, especially in the lower temperature range, it is prudent to make the oil changes more frequent. Low running temperatures encourage condensation within the engine and if the lubricating oil becomes contaminated with water, it will emulsify, leading to severe corrosion of many of the working parts.

4 The oil pump is unlikely to give trouble unless particles of metal find their way into the rotor assembly and cause scoring. This will usually only occur if some engine component has broken up during service, and underlines the necessity to strip completely and clean the engine if any such unfortunate incident occurs. Impurities in the oil will normally be trapped by one or other of the filters, before they can cause any harm.

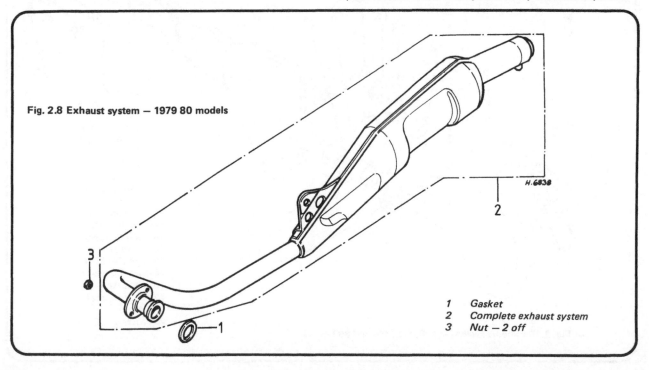

Fig. 2.8 Exhaust system — 1979 80 models

H.6838

1 Gasket
2 Complete exhaust system
3 Nut — 2 off

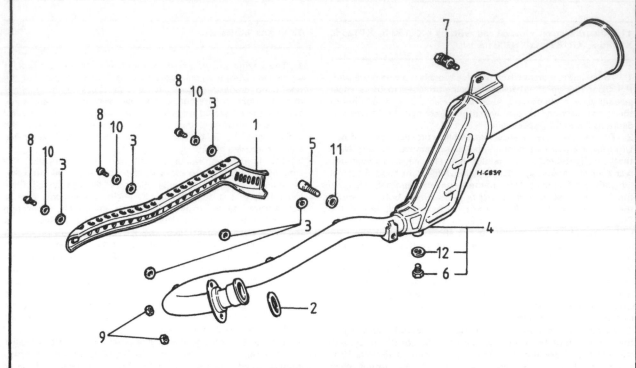

Fig. 2.9 Exhaust system — 1980 80 models

1	Heat shield	5	Allen screw	9	Nut — 2 off
2	Gasket	6	Drain bolt	10	Washer — 3 off
3	Packing — 6 off	7	Bolt and washer	11	Washer
4	Exhaust system	8	Screw — 3 off	12	Washer

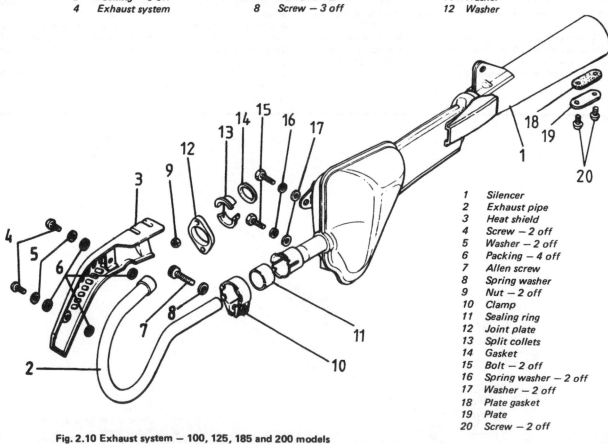

1 Silencer
2 Exhaust pipe
3 Heat shield
4 Screw — 2 off
5 Washer — 2 off
6 Packing — 4 off
7 Allen screw
8 Spring washer
9 Nut — 2 off
10 Clamp
11 Sealing ring
12 Joint plate
13 Split collets
14 Gasket
15 Bolt — 2 off
16 Spring washer — 2 off
17 Washer — 2 off
18 Plate gasket
19 Plate
20 Screw — 2 off

Fig. 2.10 Exhaust system — 100, 125, 185 and 200 models

11.2 Remove the battery case upper securing bolt (XL models)

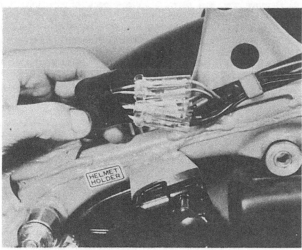

11.4a Unclip the electrical connections ...

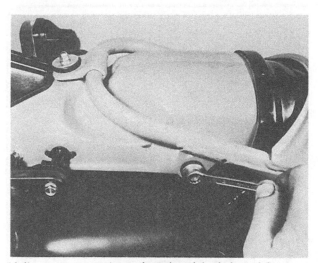

11.4b ... and remove the mudguard retaining bolts and dome nut

11.5a Remove the forward silencer to frame attachment bolts ...

11.5b ... and withdraw the silencer assembly rearwards

13 Oil pump: removal, examination, renovation and refitting

1   There is no necessity to strip the engine in order to gain access to the oil pump, although it should be stressed that this component should not be removed and dismantled unless really essential. It will give good service during the normal service life of the machine without attention and is likely to give trouble only if metallic particles or other foreign bodies contaminate the oil and score the pump rotors.

2   To remove the pump without dismantling the engine, drain off the engine oil by removing the crankcase drain plug and sealing washer, then remove the right-hand crankcase cover by following the procedure given in Sections 11 or 12 of Chapter 1.

3   The driving gear housing will be seen to have two holes in its outer face. Turn the engine over until the holes in the driving gear are in line, then pass a screwdriver through to release the two countersunk retaining screws. The pump can now be lifted away. If the two countersunk crosshead screws are particularly tight, an impact driver may have to be used. Take great care in this case, that no damage is inflicted on the pump drive gear or housing etc.

4   Remove the end plate from the back of the unit; it is retained by two further countersunk crosshead screws. Note the locating pip and the corresponding recess in the rear of the pump body which must align during reassembly. Shake out the inner and outer rotors. The driving gear housing can now be separated, if desired; it is held together by the two hexagon-headed bolts.

5   It should be noted that there is a worm gear which forms part of the pump spindle. On machines fitted with a tachometer, this gear engages with a drive mechanism fitted to the inside of the right-hand crankcase cover, thereby providing a drive for the tachometer via the cable. Care must be taken not to lose any thrust washer fitted to the outer end of the pump spindle during dismantling, and to refit it correctly upon reassembly.

6   Examine each component for signs of scuffing and wear. Note especially the condition of the rotors and pump body. If these are at all worn, the pump must be renewed as a complete unit.

7   Refit the rotors and measure the clearance between the outer rotor and the pump body using feeler gauges. If this clearance is found to be more than that given in the Specifications, the pump must be renewed. Check the clearance

between any two rotor peaks of the inner and outer rotors in a similar manner. If this clearance exceeds that given in the Specifications the pump must be renewed. Finally, check the end clearance of the two rotors by laying a straight-edge across the new gasket fitted to the pump body to end plate mating surface and measuring the clearance between the straight-edge and the side face of the rotors with a feeler gauge. If this clearance exceeds 0.25 mm (0.010 in) then again the pump must be renewed.

8   Reassembly and refitting the oil pump is a reversal of the removal and dismantling procedures, noting the following points. Lubricate each component with clean engine oil, and make sure that a new gasket is fitted. Note that the end plate pip and the depression in the pump body must align. Note also that there is a flat on the inner end of the tachometer drive gear/pump spindle. There is a similar flat in the centre of the drive gear pinion and in the centre of the inner rotor. With the pump spindle inserted correctly, all three flats will align. Refit the unit to the engine using new O-rings.

9   Refit the right-hand crankcase cover, following the procedure given in Section 43 of Chapter 1. Refit and tighten the crankcase drain plug (with a new sealing washer) and refill the oil compartment with the correct quantity and type of oil.

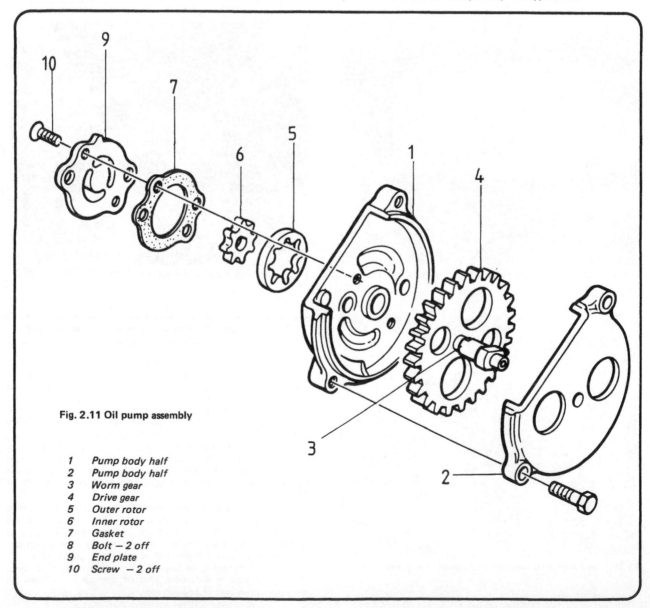

Fig. 2.11 Oil pump assembly

1   Pump body half
2   Pump body half
3   Worm gear
4   Drive gear
5   Outer rotor
6   Inner rotor
7   Gasket
8   Bolt — 2 off
9   End plate
10 Screw — 2 off

13.7a Refit the oil pump rotors ...

13.7b ... and measure the clearance between the outer rotor and pump body ...

13.7c ... and the clearance between the rotor peaks of the inner and outer rotors

13.8a Fit a new gasket ...

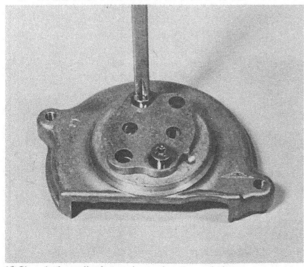

13.8b ... before aligning and securing the end plate

13.8c Fit new O-rings ...

13.8d ... and secure the pump body in position

13.8e Align the flat on the tachometer drive gear/pump spindle end with that of the inner pump rotor ...

13.8f ... and refit the drive gear and housing

14 Oil filters: location and cleaning

1   The type of engine used in the XL125 S, XL185 S, XR185 and XR200 is fitted with two types of oil filter; a gauze element located in the lower part of the left-hand crankcase behind a large hexagon-headed plug, and a crankshaft mounted centrifugal oil filter.

2   The engine used in the XR80, XL80 S and XL100 S models has only a gauze filter fitted. This takes the form of a rectangular plate which is located in a slot cast in the base of the right-hand crankcase. It is necessary to remove the right-hand crankcase cover to gain access to this filter.

3   Removal and cleaning of these components is covered in the Routine Maintenance Chapter at the beginning of this Manual.

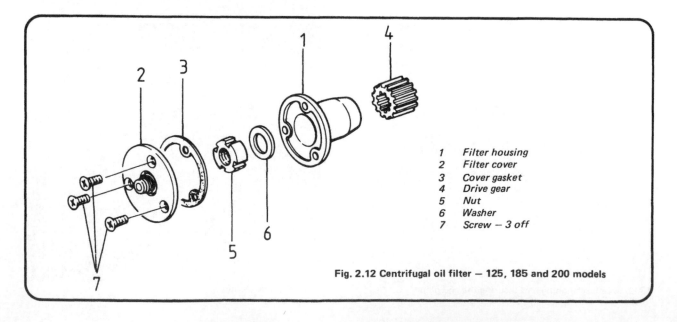

1   Filter housing
2   Filter cover
3   Cover gasket
4   Drive gear
5   Nut
6   Washer
7   Screw — 3 off

Fig. 2.12 Centrifugal oil filter — 125, 185 and 200 models

15 Fault diagnosis: fuel system and lubrication

Symptom	Cause	Remedy
Excessive fuel consumption	Air cleaner choked or restricted	Clean or renew.
	Fuel leaking from carburettor. Float sticking	Check all unions and gaskets. Float needle seat needs cleaning.
	Badly worn or distorted carburettor	Replace.
	Jet needle setting too high	Adjust as figure given in Specifications.
	Main jet too large or loose	Fit correct jet or tighten if necessary.
	Carburettor flooding	Check float valve and replace if worn. Check float height.
Idling speed too high	Throttle stop screw in too far	Adjust screw. Tighten top.
	Carburettor top loose	
	Throttle cable sticking	Disconnect and lubricate or replace.
Engine dies after running for a short while	Blocked air hole in filler cap	Clean.
	Dirt or water in carburettor	Remove and clean out.
General lack of performance	Weak mixture; float needle stuck in seat	Remove float chamber or float and clean.
	Air leak at carburettor joint	Check joint to eliminate leakage, and fit new O-ring.
	Blocked air filter	Clean and relubricate element.
Engines does not respond to throttle	Throttle cable sticking	See above.
	Petrol octane rating incorrect	Use correct grade (star rating) petrol.
Engine runs hot and is noisy	Lubrication failure	Stop engine immediately and investigate cause. Slacken cylinder head nut to check oil circulation. Do not restart until cause is found and rectified.

Chapter 3 Ignition system

For information relating to 1981 — 1987 models, see Chapter 7

Contents

Specifications

		XR80	XR185 and XR200
Flywheel generator			
Output	...	0.090 kw @ 5000 rpm	0.045 kw @ 5000 rpm
Ignition timing			
Static (F mark aligned)	...	15° BTDC	—
Dynamic:			
Initial	...	—	10° ± 2° BTDC @ 1950 ± 150 rpm
Full advance	...	—	30° ± 2° BTDC @ 3150 ± 150 rpm
Contact breaker gap	...	0.3 - 0.4 mm (0.012 - 0.016 in)	—
Condenser capacity	...	0.22 - 0.27 microfarad	—
Sparking plug			
Make ...	...	NGK/ND	NGK/ND
Type:			
UK...	...	CR7HS/U22FSR-L	DR8ES-L/X24ESR-U
US...	...	C7HS/U22FS	D8EA/X24ES-U
Gap ...	...	0.6 - 0.7 mm (0.024 - 0.028 in)	0.6 - 0.7 mm (0.024 - 0.028 in)
Ignition coil			
Primary winding resistance ...	...	—	0.2 - 0.8 ohm
Secondary winding resistance ...	...	—	8.0 - 15.0 K ohm

		XL80 S	XL100 S	XL125 S and XL185 S
Flywheel generator				
Output	...	0.058 Kw @ 5000 rpm	—	—
Ignition timing				
Static (F mark aligned)	...	15° BTDC	—	—
Dynamic:				
Initial	...	15° ± 2° BTDC @ 1800 ± 100 rpm	10° ± 2° BTDC @ 1800 ± 150 rpm	10° ± 2° BTDC @ 1950 ± 150 rpm
Full advance	...	28.5° - 31.5° BTDC @ 3250 rpm	28° ± 1.5° BTDC @ 4000 ± 200 rpm	34° ± 2° BTDC @ 3350 ± 150 rpm

	XL80 S	XL100 S	XL125 S and XL185 S
Contact breaker gap	0.3 - 0.4 mm (0.012 - 0.016 in)	0.3 - 0.4 mm (0.012 - 0.016 in)	—
Condenser capacity	0.27 - 0.33 microfarad	0.25 microfarad	—
Sparking plug			
Make	NGK/ND	NGK/ND	NGK/ND
Type:			
UK...	—	CR7HS/U22FSR-L	DR8ES-L/X24ESR-U
US	C7HS/U22FS	C7HS/U22FS	D8EA/X24ES-U
Gap	0.6 - 0.7 mm (0.024 - 0.028 in)	0.6 - 0.7 mm (0.024 - 0.028 in)	0.6 - 0.7 mm (0.024 - 0.028 in)
Ignition coil			
Primary winding resistance	—	1.6 ohm	0.2 - 0.8 ohm
Secondary winding resistance	—	8.0 K ohm	8.0 - 15.0 K ohm

1 General description

The models covered in this Manual are fitted with an ignition system either of the conventional contact breaker type or of the CDI (Capacitor discharge ignition) type, depending on the model involved. In both cases ignition source power is provided by a flywheel generator driven from the crankshaft left-hand end.

The contact breaker system, which is fitted to the XL80 S, XR80 and XL100 S models, functions as follows. As the flywheel generator rotor rotates, alternating current (ac) is generated in the ignition source coil mounted on the generator stator. Because the contact breaker is closed the power runs to earth. When the contact breaker points open the current is transferred to the primary windings of the ignition coil. In doing this a high voltage is produced in the ignition secondary windings (by means of mutual induction) which is fed to the sparking plug via the HT lead. As the energy flows to earth across the sparking plug electrodes a spark is produced and the combustible gases in the cylinder ignited. A capacitor (condenser) is fitted into the system to prevent arcing across the points; this helps reduce erosion due to burning.

The CDI system fitted to the remaining models in the range works on a principle similar to that described above except that the ignition firing point is timed by a pulser generator incorporated in the system. The pulser rotor is mounted on the end of the camshaft and the pick-up coil on the left-hand side of the cylinder head. As the flywheel generator rotor rotates an alternating current is produced in the ignition source coil. This is passed to the CDI unit, mounted above the engine, where it is converted to dc and stored in a capacitor. When the pulser rotor passes the pick-up unit a signal voltage is produced which activates an electronic switch in the CDI unit, allowing the capacitor to discharge through the ignition coil primary windings. This pulse of energy induces a high voltage in the coil secondary windings which is passed via the HT lead to the sparking plug.

By eliminating the mechanical contact breaker, frequent attention to the ignition system, and the inevitable alterations in ignition timing due to wear of moving parts, is eliminated.

2 Contact breaker: adjustment - XR80, XL80 S and XL100 S models

1 To gain access to the contact breaker assembly, it is first necessary to remove the left-hand crankcase cover by unscrewing the five retaining screws. The contact breaker points may then be viewed through one of the slots in the flywheel generator rotor.

2 Remove the sparking plug and rotate the generator rotor slowly until the points can be viewed in their fully-open position through one of the rotor slots. Examine the contact faces. If they are dirty, burnt or pitted, they should be removed for renewal as described in Section 4 of this Chapter. If the points are in sound condition, proceed with adjustment as follows.

3 Contact breaker adjustment is carried out by loosening the single contact breaker locking screw slightly and moving the contact points nearer or further apart as the case may be, by levering with a screwdriver in the plate indentation provided. Turn in the appropriate direction until the gap is within the range 0.3 to 0.4 mm (0.012 to 0.016 in) and then retighten the locking screw. It is imperative that the points are open FULLY whilst this adjustment is made, or a false reading will result. Recheck the setting when the screw has been tightened; it is not unknown for the setting to move slightly during the re-tightening operation.

4 If adjustment has proved necessary, it is likely that the ignition timing will have been affected, and this should be checked. See Section 6 of this Chapter for the relevant procedure.

5 Refit the crankcase cover by fitting and tightening the five retaining screws, evenly and in a diagonal sequence. Ensure the cover gasket is undamaged and correctly located before doing so. Refit the sparking plug.

3 Ignition timing: checking and resetting - XR80, XL80 S and XL100 S models

1 In order to check the accuracy of the ignition timing, it is first necessary to remove the flywheel generator rotor cover by unscrewing the five retaining screws. It will be observed that the rotor is inscribed with two lines marked 'T' and 'F' and that there is a small pointer cast in the crankcase mating surface just below the centre line of the cylinder barrel. In addition, there are two further parallel lines inscribed on the rotor. These indicate the point of full advance when they align with the fixed pointer.

2 If the ignition timing is correct, the 'F' line on the rotor will coincide exactly with the fixed pointer when the contact breaker points are just beginning to separate. The moment at which the contact points separate may be determined as follows.

3 Obtain an ordinary torch battery and bulb and three lengths of wire. Connect one end of a wire to the positive (+) terminal of the battery and one end of another wire to the negative (−) terminal of the battery. The negative lead may now be earthed to a point on the engine casing. Ensure the earth point on the casing is clean and that the wire is positively connected, a crocodile clip fastened to the wire is ideal. Take the free end

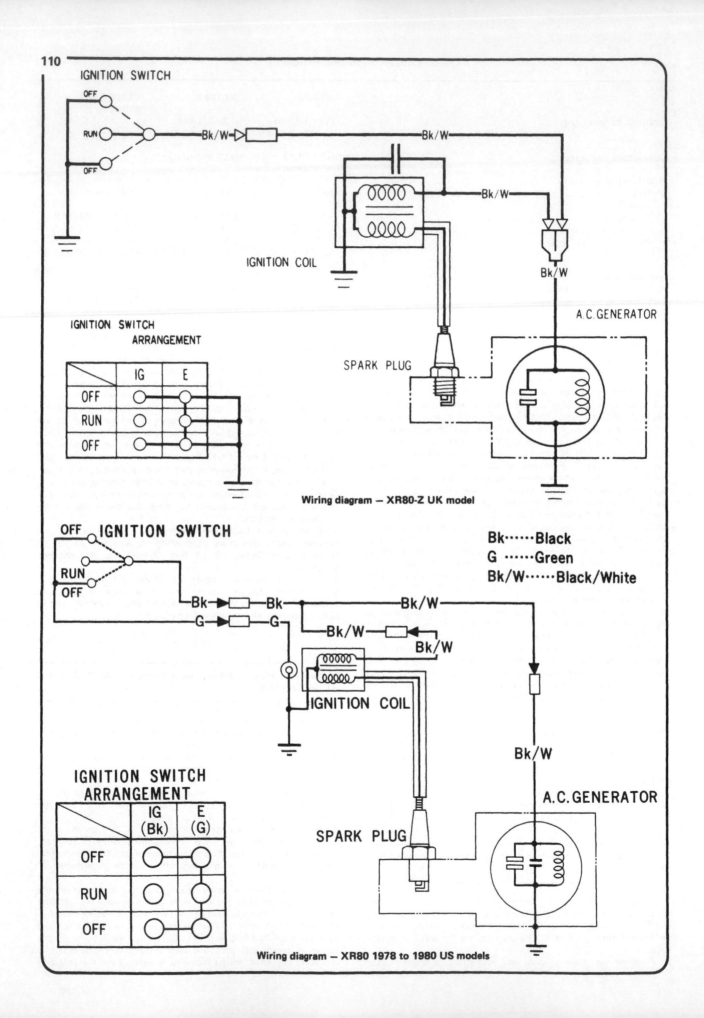

IGNITION SWITCH

OFF
RUN
OFF

Bk/W

IGNITION COIL

Bk/W

Bk/W

SPARK PLUG

A.C. GENERATOR

IGNITION SWITCH ARRANGEMENT

	IG	E
OFF	○	○
RUN	○	○
OFF	○	○

Wiring diagram — XR80-Z UK model

IGNITION SWITCH

OFF
RUN
OFF

Bk·······Black
G ·······Green
Bk/W······Black/White

Bk   Bk   Bk/W
G   G
Bk/W   Bk/W

IGNITION COIL

Bk/W

A.C. GENERATOR

SPARK PLUG

IGNITION SWITCH ARRANGEMENT

	IG (Bk)	E (G)
OFF	○	○
RUN	○	○
OFF	○	○

Wiring diagram — XR80 1978 to 1980 US models

of the positive lead and connect it to the bulb. The third length of wire may now be connected between the bulb and the black output lead from the generator. As the final connection is made, the bulb should light with the points closed.

4 Rotate the generator rotor anti-clockwise until the 'F' line on the rotor aligns with the fixed pointer. If the timing is correct the light should dim as both marks align thus indicating that the points have begun to separate.

5 If the ignition timing proves to be incorrect, adjust the contact breaker gap as described in paragraph 3, Section 2 of this Chapter so that the light dims at the correct point. If, after adjusting the contact breaker points to achieve the correct ignition timing, the gap is not within the recommended limits of 0.3 - 0.4 mm (0.012 - 0.016 in) the points should be renewed as a complete set.

6 It cannot be overstressed that optimum performance depends on the accuracy with which the ignition timing is set. Even a small error can cause a marked reduction in performance and the possibility of engine damage as the result of overheating.

7 An alternative method of checking the ignition timing can be adopted, whilst the engine is running, using a stroboscopic lamp. When the light from the lamp is aimed at the timing marks it has the effect of 'freezing' the moving marks on the rotor in one position, and thus the accuracy of the timing can be seen. An additional advantage is also provided in that the correct and smooth functioning of the automatic advance-retard unit can be determined.

8 The stroboscopic lamp should be connected up as recommended by the lamp's manufacturer and the engine started. Up to about 1800 rpm the F mark should be in line with the fixed index mark. As the speed is increased, the two parallel advance marks should be seen to move closer to the fixed index point. Note the rpm reading at which the initial movement of the marks occurs and compare it with that reading given in the Specifications at the beginning of this Chapter for the particular type of machine. Note also the rpm reading at which point full advance is reached (ie the parallel advance marks on the rotor directly in line with the fixed index point) and compare it with that figure given in the Specifications. Alterations to the ignition timing should be made in the normal way as described in the first paragraphs of this Section.

9 If it is found that the static ignition timing is correct but that full advance cannot be reached, or that progression to full advance is erratic, there is some indication that the mechanical automatic timing unit is not functioning correctly. Refer to Section 11 of this Chapter for details relating to examination of this component.

4 Contact breaker points: removal, renovation and refitting - XR80, XL80 S and XL100 S models

1 If the contact breaker points are burned, pitted or badly worn, they should be removed for dressing. If it is necessary to remove a substantial amount of material before the faces can be restored, new replacements should be fitted. Only a limited amount of material can be removed because of the effect on maintaining the correct contact breaker gap and ignition timing accuracy.

2 It is necessary first to withdraw the flywheel generator rotor before access to the points can be gained. Refer to Section 9 of Chapter 1 for details of rotor removal. The fixed contact is removed by withdrawing the screw which holds the assembly to the stator plate of the generator. The moving contact is detached by releasing the circlip from the end of the pivot pin and by freeing the leaf return spring from its attachment point.

3 The points should be dressed with an oilstone or fine emery cloth. Keep them absolutely square during the dressing

operation, otherwise they will make angular contact when they are refitted and will burn away rapidly as a result.

4 Fit the contacts by reversing the dismantling procedure. Take particular care to fit any insulating washers in their correct sequence, otherwise the points will be isolated electrically and the ignition system will not function. Lightly grease the pivot pin before the moving contact is fitted and check that there is no oil or grease on the surfaces of the points.

5 Refit the flywheel rotor after greasing the internal contact breaker cam. It is also advisable to add a few drops of light oil to the lubricating wick which rubs on the contact breaker cam, if the wick has a dry appearance.

6 Re-adjust the contact breaker gap after the flywheel rotor has been locked in position and the centre retaining nut tightened to the recommended torque wrench setting of 6.5 - 7.5 kgf m (47 - 54 lbf ft).

7 On completion of refitting the contact breaker gap should be adjusted and the ignition timing checked as described in Sections 2 and 3 of this Chapter.

5 Condenser: removal, testing and refitting - XR80, XL80 S and XL100 S models

1 A condenser is included in the contact breaker circuitry to prevent arcing across the contact breaker points as they separate. The condenser is connected in parallel with the points and if a fault develops, ignition failure is liable to occur.

2 If the engine proves difficult to start, or misfiring occurs, it is possible that the condenser is at fault. To check, separate the contact breaker points by hand when the ignition is switched on. If a spark occurs across the points and they have a blackened and burnt appearance, the condenser can be regarded as unserviceable.

3 It is not possible to check the condenser without the appropriate equipment. In view of the low cost involved, it is preferable to fit a new replacement and observe the effect on engine performance. The condenser may be removed and the new item fitted as follows.

XR80 and XL80 S models

4 On XL80S models the condenser is mounted on the generator stator plate, so that it will be necessary to remove the rotor to gain access to it, as described in Section 9 of Chapter 1. Disconnect the wires using a soldering iron, withdraw the single retaining screw and remove the condenser. On reassembly, take care to ensure that the condenser is correctly installed and that the wires are soldered securely to the centre terminal only.

5 On XR80 models the condenser is mounted on the ignition HT coil and is not designed to be separated from it. If condenser failure is suspected, remove the complete coil/condenser unit and take it to a Honda Service Agent or a competent auto-electrician who has the necessary equipment to determine whether the condenser is at fault. If the condenser is found to be at fault the only answer is to renew the complete ignition coil assembly as separate condensers are not available as Honda replacement parts. It may, however, be possible for an experienced auto-electrician to devise some means of fitting a suitable condenser from another model.

XL100 S models

6 The condenser fitted to this model is mounted on a frame attachment just above and to the rear of the crankcase. Removal is achieved simply by disconnecting the single wire at the bullet connection and removing the condenser to frame securing bolt. Fitting the unit is a reversal of the removal procedure. Before fitting the unit, check that the frame fitting point is clean and free from corrosion, because this point provides the earth connection of the condenser. It follows that the bolt should also be clean and fully tightened.

6 Ignition timing: checking and resetting - XL125 S, XL185 S, XR185 and XR200 models

1 The above mentioned machines utilise an ignition system where the normal contact breaker unit is dispensed with; the ignition firing point being controlled entirely by electrical means. Because no mechanical components are used, and therefore wear is eliminated, the ignition timing should remain accurate almost indefinitely during normal usage.

2 Static timing of the system is covered in Section 49 of Chapter 1. If any components have been disturbed in the pulser generator unit then the static timing procedure must be carried out before proceeding further.

3 With the upper inspection cap removed from the flywheel generator cover, connect the strobe lamp into the system as recommended by the manufacturer of the lamp. Start the engine and aim the lamp at the index pointer and the rotor periphery. At the engine idle speed of 1300 rpm, the 'F' mark on the generator rotor should be aligned with the index mark on the rotor cover. As the engine speed is allowed to increase, the 'F' mark will be seen to move away from the index mark. Note the rpm reading at which this initial movement occurs and compare it with that reading given in the Specifications at the beginning of this Chapter for the particular type of machine. Note also the rpm reading at which point full advance is reached (ie the parallel advance marks on the rotor directly in line with the fixed index point) and compare it with that figure given in Specifications. Any alterations to the ignition timing should be made by following the previously mentioned static timing procedure.

4 If it is found that the static ignition timing is correct but that full advance cannot be reached, or that progression to full advance is erratic, there is some indication that the mechanical automatic timing unit is not functioning correctly. Refer to Section 11 of this Chapter for details relating to examination of this component.

7 Pulser generator: testing - XL125 S, XL185 S, XR185 and XR200 models

1 No dismantling of the pulser generator unit is necessary before testing can be carried out. Removal of the fuel tank is, however, necessary so that access may be gained to the generator electrical connections. Refer to Section 2 of the previous Chapter for the tank removal procedure.

2 Disconnect the two generator leads at their bullet connections and, using a multimeter, measure the resistance between the two wires. The resistance should be between 20 and 60 ohms. If an incorrect reading is obtained, then the unit may be removed as described in Section 7 of Chapter 1 and a new unit fitted. If the unit components are disturbed, then the ignition timing must be checked as described in Section 6 of this Chapter.

6.3 Timing marks may be found on the flywheel generator rotor (125, 185 and 200)

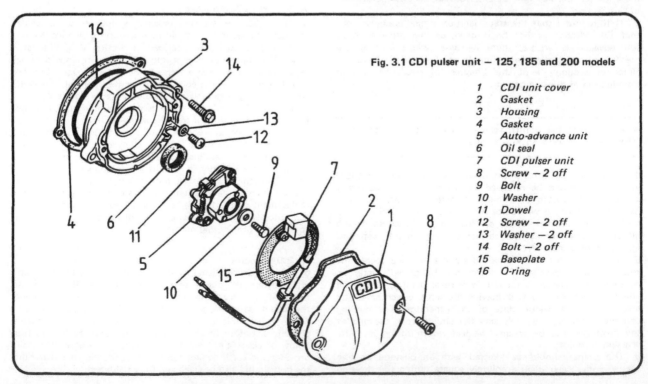

Fig. 3.1 CDI pulser unit — 125, 185 and 200 models

1   CDI unit cover
2   Gasket
3   Housing
4   Gasket
5   Auto-advance unit
6   Oil seal
7   CDI pulser unit
8   Screw — 2 off
9   Bolt
10 Washer
11 Dowel
12 Screw — 2 off
13 Washer — 2 off
14 Bolt — 2 off
15 Baseplate
16 O-ring

8 CDI unit: removal, testing and fitting - XL125 S, XL185 S, XR185 and XR200 models

1 The fully transistorised CDI unit is mounted underneath the fuel tank on the left-hand side of the frame top tube. It follows, therefore, that the tank must be removed to gain access to the unit. Refer to Section 2 of the previous Chapter for the tank removal procedure.

2 With the tank removed, disconnect the electrical leads to the unit at the snap connections, noting each connection in turn for reference when fitting the unit. Unscrew and remove the two unit to frame retaining nuts, noting the lead attached underneath the forward nut, and pull the unit off the two studs.

3 The unit is sealed for life and if any malfunction occurs, it must be renewed. Because no test data is available for the CDI unit, and because in any event testing requires very specialised equipment, malfunction of this component should be determined initially by a process of elimination.

4 If the flywheel generator is found to be serviceable after testing in accordance with Section 9 of this Chapter, and the ignition coil and pulser generator are in good order (see Sections 10 and 7) it may be assumed that the CDI unit is faulty. Before consigning the unit to the rubbish bin, it is advised that the machine is returned to a Honda Service Agent for further tests to be carried out.

5 Fitting the new or original unit is a reversal of the removal procedure. Ensure that all electrical connections are clean and correctly made and the unit correctly positioned on the frame attachment points before fitting and tightening the retaining nuts.

9 Flywheel generator: checking the output

1 The flywheel generator is instrumental in creating the power in the ignition system, and any failure or malfunction will affect the operation of the ignition system. If the machine will not start and there is no evidence of a spark at the sparking plug, a check should first be made to ensure there is no fault at either the contact breaker assembly (XR80, XL80 S and XL100 S models) or the pulser generator or CDI unit (XL125 S, XL185 S, XR185 and XR200 models). A test should also be made on the ignition coil itself. Refer to Sections 4, 7 and 8 of this Chapter for test procedures on these components.

2 If the above checks have failed to show any faults, the output from the generator itself should be suspected. Reference should be made, therefore, to Chapter 6, Section 2 for output checking procedure by basic test methods, but before doing this, a thorough check of the circuit wiring should be made to ensure the wiring connections are not badly corroded or contaminated by dirt or moisture. Check the wiring itself for signs of chafing against the frame or engine components or any indication of a break in the wiring.

10 Ignition coil: checking

1 The ignition coil is a sealed unit and designed to give long service without need for attention. The coil is located beneath the petrol tank, and is mounted on the right-hand side of the frame top tube. It follows, therefore, that the tank must be removed to gain access to the coil. Refer to Section 2 of the previous Chapter for the tank removal procedure.

2 If a weak spark and difficult starting causes the performance of the coil to be suspect, it should be tested as follows. Using a multimeter, connected as shown in the accompanying illustrations, test the resistance of the primary and secondary coils and compare the resistance shown on the meter with that given in the Specifications at the beginning of this Chapter.

3 Should these tests fail to produce the expected result, the coil should be taken to a Honda Service Agent or auto-electrician for a more thorough check. If the coil is found to be faulty, it must be replaced; it is not possible to effect a satisfactory repair.

4 On XR80, XL80 S and XL100 S models, a defective condenser in the contact breaker circuit can give the illusion of a defective coil and for this reason it is advisable to investigate the condition of the condenser before condemning the ignition coil. Refer to Section 5 of this Chapter for the appropriate details.

11 Auto-advance unit: removal, examination and refitting

1 Fixed ignition timing is of little advantage as the engine speed increases, and it is necessary to incorporate a method of advancing the timing by centrifugal means. A balance weight assembly located within the flywheel generator rotor (XR80, XL80 S and XL100 S models) or behind the pulser generator rotor (XL125 S, XL185 S, XR185 and XR200 models) provides this method of advancing the timing.

2 To gain access to the auto-advance unit fitted within the flywheel generator rotor it is necessary to remove the rotor from the crankshaft end. Full details for rotor removal are given in Section 9 of Chapter 1. Where the auto-advance unit is located behind the pulser generator rotor, it is necessary to remove the rotor by following the procedure given in Section 7 of Chapter 1.

3 The unit comprises spring loaded balance weights, which move outwards against the spring tension as centrifugal force increases. The balance weights must move freely on their pivots and be rust-free. The tension springs must also be in good condition. Keep the pivots lubricated and make sure the balance weights move easily, without binding. Most problems arise as a result of condensation within the engine, which causes the unit to rust and balance weight movement to be restricted.

4 If any malfunction or breakage has occurred, renew the complete unit, but if it appears to be in good condition, lightly oil it before refitting. Refitting procedures for both the flywheel generator rotor and the pulser generator rotor are given in Chapter 1, Sections 45 and 49 respectively.

8.1 The CDI unit is mounted underneath the fuel tank

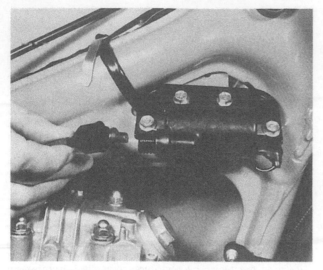

10.1 The ignition coil is mounted underneath the fuel tank

11.1 On the 80 and 100 models, the auto-advance unit is located within the flywheel generator rotor

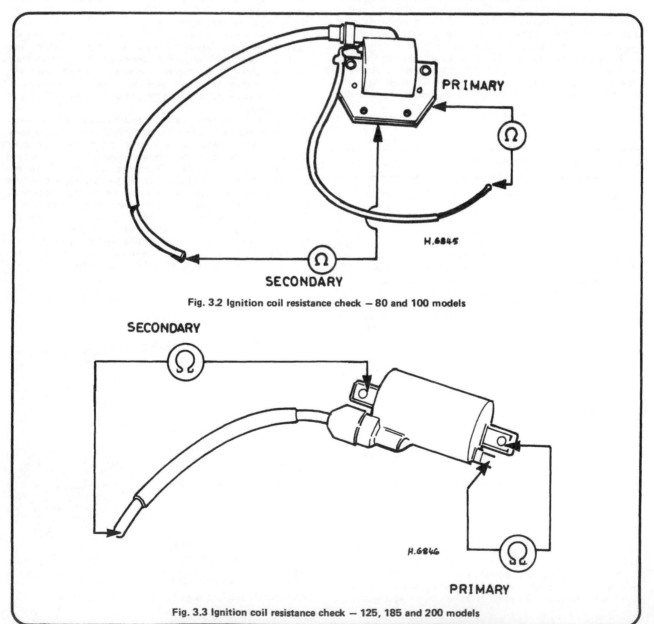

PRIMARY

SECONDARY

H.6845

Fig. 3.2 Ignition coil resistance check — 80 and 100 models

SECONDARY

PRIMARY

H.6846

Fig. 3.3 Ignition coil resistance check — 125, 185 and 200 models

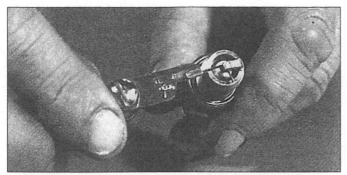

Electrode gap check - use a wire type gauge for best results

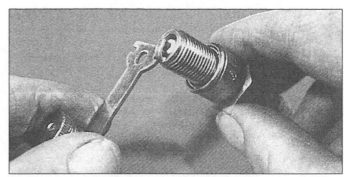

Electrode gap adjustment - bend the side electrode using the correct tool

Normal condition - A brown, tan or grey firing end indicates that the engine is in good condition and that the plug type is correct

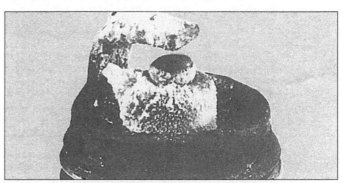

Ash deposits - Light brown deposits encrusted on the electrodes and insulator, leading to misfire and hesitation. Caused by excessive amounts of oil in the combustion chamber or poor quality fuel/oil

Carbon fouling - Dry, black sooty deposits leading to misfire and weak spark. Caused by an over-rich fuel/air mixture, faulty choke operation or blocked air filter

Oil fouling - Wet oily deposits leading to misfire and weak spark. Caused by oil leakage past piston rings or valve guides (4-stroke engine), or excess lubricant (2-stroke engine)

Overheating - A blistered white insulator and glazed electrodes. Caused by ignition system fault, incorrect fuel, or cooling system fault

Worn plug - Worn electrodes will cause poor starting in damp or cold weather and will also waste fuel

12 High tension (sparking plug) lead: examination

1   Erratic running faults and problems with the engine suddenly cutting out in wet weather can often be attributed to leakage from the high tension lead and sparking plug cap. If this fault is present, it will often be possible to see tiny sparks around the lead and cap at night. One cause of this problem is the accumulation of mud and road grime around the lead, and the first thing to check is that the lead and cap are clean. It is often possible to cure the problem by cleaning the components and sealing them with an aerosol ignition sealer, which will leave an insulating coating on both components.

2   Water dispersant sprays are also highly recommended where the system has become swamped with water. Both these products are easily obtainable at most garages and accessory shops. Occasionally, the suppressor cap or the lead itself may break down internally. If this is suspected, the components should be renewed.

3   Where the HT lead is permanently attached to the ignition coil, it is recommended that the renewal of the HT lead is entrusted to an auto-electrician who will have the expertise to solder on a new lead without damaging the coil windings.

13 Sparking plug: checking and resetting the gap

1   The type of sparking plug fitted to each model covered in this Manual must comply with that listed in the Specifications at the beginning of this Chapter. If in any doubt as to the type of sparking plug that should be fitted to a particular type of machine, consult an official Honda Service Agent who will be able to recommend the correct sparking plug grade for the conditions in which the machine is to operate. The type of plug

recommended by the manufacturer gives the best all round service.

2   Check the gap between the plug electrodes at the service interval recommended in the Routine Maintenance Chapter at the beginning of this Manual. To reset the gap, bend the outer electrode to bring it closer to the centre electrode and check that a 0.6 - 0.7 mm (0.024 - 0.028 in) feeler gauge can be inserted. Never bend the central electrode or the insulator will crack, causing engine damage if the particles fall in whilst the engine is running.

3   With some experience, the condition of the sparking plug electrodes and insulator can be used as a reliable guide to engine operating conditions.

4   Always carry a spare sparking plug of the recommended grade. In the rare event of plug failure, it will enable the engine to be restarted.

5   Beware of overtightening the sparking plug, otherwise there is risk of stripping the threads from the aluminium alloy cylinder head. The plug should be sufficiently tight to seat firmly on its sealing washer, and no more. Use a spanner which is a good fit to prevent the spanner from slipping and breaking the insulator.

6   If the threads in the cylinder head strip as a result of over-tightening the sparking plug, it is possible to reclaim the head by the use of a Helicoil thread insert. This is a cheap and convenient method of replacing the threads; most motorcycle dealers operate a service of this nature at an economic price.

7   Make sure the plug insulating cap is a good fit and has a rubber seal. They should also be kept clean to prevent tracking. The cap contains the suppressor that eliminates both radio and TV interference.

8   Before fitting a sparking plug in the cylinder head, coat the threads sparingly with a graphited grease to aid future removal.

14   Fault diagnosis: ignition system

Symptom	Cause	Remedy
Engine will not start	Faulty sparking plug	Adjust gap or renew plug.
	Defective ignition coil	Renew.
XR80, XL80 S and XL100 S models	Dirty contact breaker points	Clean or renew and check gap.
	Poor points contact	Check for arcing. Renew condenser.
XL125 S, XL185 S, XR185 and XR200 models	Faulty CDI unit	Renew.
	Faulty pulser generator	Renew after testing.
Engine starts but runs erratically	Plug gap incorrect or has whiskered	Adjust, clean or renew.
	Incorrect ignition timing	Check timing.
	Defective ignition coil	Renew after testing.
	Low output from flywheel generator	Test and renew if necessary.
	HT lead insulation breaking down	Check and renew.
XR80, XL80 S and XL100 S models	Weak points arm return spring	Renew.
XL125 S, XL185 S, XR185 and XR200 models	Faulty pulser generator	Test and renew.
	Faulty CDI unit	Renew.
Engine misfires	Fouled sparking plug	Clean or replace plug.
	Poor spark due to generator failure	Check output from generator.
XR80, XL80 S and XL100 S models	Faulty condenser in ignition circuit	Replace condenser and retest.
Engine lacks power and overheats	Retarded ignition timing	Check timing. Check whether auto-advance mechanism has jammed.
Engine 'fades' when under load	Pre-ignition	Check grade of plugs fitted; use recommended grades only.

Chapter 4 Frame and forks

For information relating to 1981 — 1987 models, see Chapter 7

Contents

Specifications

	XR80	XR185	XR200
Frame	Spine type, engine used as stressed member		

Front forks

	XR80	XR185	XR200
Type	Oil damped, telescopic		
Travel	126 mm (5.0 in)	216 mm (8.5 in)	216 mm (8.5 in)
Spring free length	—	568.5 mm (22.23 in)	524.5 mm (20.65 in) - main spring 42.7 mm (1.68 in) - top spring
Service limit	—	560.2 mm (22.06 in)	504.1 mm (19.8 in) - main spring 38.4 mm (1.51 in) - top spring
Piston OD	30.950 - 30.975 mm (1.2185 - 1.2195 in)	—	—
Service limit	30.90 mm (1.2165 in)	—	—
Lower leg ID	31.0 - 31.039 mm (1.2205 - 1.2220 in)		
Service limit	31.18 mm (1.2276 in)	—	—
Stanchion runout service limit	—	0.2 mm (0.008 in)	0.2 mm (0.008 in)
Oil capacity	118 cc (4.15 Imp fl oz, 3.98 US fl oz)	170 cc (5.98 Imp fl oz, 5.75 US fl oz)	170 cc (5.98 Imp fl oz, 5.75 US fl oz)
Oil type	Automatic transmission fluid (ATF)		

Rear suspension

	XR80	XR185 and XR200
Type	Swinging arm fork, controlled by two suspension units	
Travel	84.6 mm (3.3 in)	191 mm (7.5 in)
Swinging arm type	Welded tubular steel	Welded tubular steel
Bush to collar clearance	—	0.032 - 0.111 mm (0.0013 - 0.0044 in)
Service limit	—	0.32 mm (0.013 in)
Bush ID	—	18.0 - 18.052 mm (0.7087 - 0.7107 in)
Service limit	—	18.20 mm (0.717 in)
Collar OD	—	17.941 - 17.968 mm (0.7063 - 0.7074 in)
Service limit	—	17.88 mm (0.704 in)
Rear suspension units	Coil spring and hydraulic damper	Coil spring and gas/oil damper
Spring free length	188.7 mm (7.43 in)	346.5 mm (13.64 in)
Service limit	183.0 mm (7.09 in)	339.6 mm (13.37 in)

	XL80 S	XL100 S	XL125 S and XL185 S
Frame	Spline type, engine used as stressed member		

Front forks

	XL80 S	XL100 S	XL125 S and XL185 S
Type	Oil damped, telescopic	Oil damped, telescopic	Oil damped, telescopic
Travel	126 mm (5.0 in)	170 mm (6.7 in)	200 mm (7.9 in)
Spring free length	397.7 mm (15.657 in)	556.7 mm (21.92 in)	568.5 mm (22.38 in)
Service limit	370.0 mm (14.5669 in)	545.5 mm (21.48 in)	557.1 mm (21.93 in)
Piston OD	30.927 - 30.960 mm (1.2176 - 1.2189 in)	—	—
Service limit	30.90 mm (1.2165 in)	—	—
Lower leg ID	31.025 - 31.064 mm (1.2215 - 1.2230 in)	—	—
Service limit	31.1 mm (1.2244 in)	—	—
Stanchion runout service limit	—	0.2 mm (0.01 in)	0.2 mm (0.01 in)
Oil capacity	112 cc (4.00 Imp fl oz, 3.85 US fl oz)	140 cc (4.93 Imp fl oz, 4.73 US fl oz)	155 cc (5.46 Imp fl oz, 5.24 US fl oz)
Oil type	Automatic transmission fluid (ATF)		

Rear suspension

	XL80 S	XL100 S	XL125 S and XL185 S
Type	Swinging arm fork, controlled by two suspension units		
Travel	112 mm (4.4 in)	130 mm (5.1 in)	165 mm (6.5 in)
Swinging arm type	Welded tubular steel	Welded tubular steel	Welded tubular steel
Bush to collar clearance	—	0.032 - 0.111 mm (0.0013 - 0.0044 in)	0.032 - 0.111 mm (0.0013 - 0.0044 in)
Service limit	—	0.32 mm (0.013 in)	0.64 mm (0.026 in)
Bush ID	—	18.0 - 18.052 mm (0.7087 - 0.7107 in)	18.0 - 18.052 mm (0.7087 - 0.7107 in)
Service limit	—	18.20 mm (0.717 in)	18.20 mm (0.717 in)
Collar OD	—	17.941 - 17.968 mm (0.7063 - 0.7074 in)	17.941 - 17.968 mm (0.7063 - 0.7074 in)
Service limit	—	17.88 mm (0.704 in)	17.88 mm (0.704 in)
Rear suspension units	Coil spring and hydraulic damper	Coil spring and hydraulic damper	Coil spring and gas/oil damper
Spring free length	226.0 mm (8.898 in)	223.3 mm (8.79 in) - upper spring 70.6 mm (2.78 in) - lower spring	302.1 mm (11.89 in)
Service limit	214.7 mm (8.45 in)	219 mm (8.6 in) - upper spring 67 mm (2.6 in) - lower spring	296.1 mm (11.66 in)

Torque wrench settings
kgf m (lbf ft)

	XR80 and XL80 S	XL100 S, XL125 S, XL185 S, XR185 and XR200
Steering stem nut	6.0 - 9.0 (43 - 65)	6.0 - 9.0 (43 - 65)
Stanchion cap bolt	—	1.5 - 3.0 (11 - 22)
Top yoke pinch bolt nut	—	0.9 - 1.3 (7 - 9)
Bottom yoke pinch bolt or nut	2.0 - 3.0 (14 - 22)	2.0 - 2.5 (14 - 18)
Lower leg socket-headed bolt	—	0.8 - 1.2 (6 - 9)
Handlebar clamp bolts	0.8 - 1.2 (5.8 - 8.7)	2.0 - 2.5 (14 - 18)
Rear suspension unit bolts or nuts	3.0 - 4.0 (22 - 29)	3.0 - 4.0 (22 - 29)
Swinging arm pivot bolt nut	3.0 - 4.0 (22 - 29)	4.0 - 5.0 (29 - 36)
Kickstart lever clamp bolt	0.8 - 1.2 (6 - 9)	0.8 - 1.2 (6 - 9)

1 General description

The model types covered in this Manual utilise two variants of a spine frame, this being a composite structure of tubular construction with a monocoque spine. The engine unit is bolted between the front downtube and main frame structure to form a structural part of the frame.

The front forks are of the conventional telescopic type, having internal, oil-filled dampers. The fork springs are contained within the fork stanchions and each fork leg can be detached from the machine as a complete unit, without dismantling the steering head assembly.

Rear suspension is of the swinging arm type, using oil or oil and gas filled suspension units to provide the necessary damping action. The units are adjustable so that the spring ratings can be effectively changed within certain limits to match the load carried.

2 Front forks: removal - general

1 It is unlikely that the forks will require removal from the frame unless the fork seals are leaking or accident damage has been sustained. In the event that the latter has occurred, it should be noted that the frame may also have become bent and, whilst this may not be obvious when checked visually, could prove to be potentially dangerous.

2 If attention to the fork legs only is required, it is unnecessary to detach the complete assembly, the legs being easily removed individually. Refer to Section 4 before proceeding with dismantling.

3 If, on the other hand, the headstock bearings are in need of attention, the forks complete with bottom yoke must be removed.

4 Before any dismantling work can be undertaken, the machine should be placed on blocks positioned underneath the sumpguard and footrest assembly and blocked securely so that the front wheel is held off the ground. Detach the speedometer drive cable at the wheel, by unscrewing the crosshead screw which retains it, and pulling the cable clear.

5 Remove the front brake cable from the brake backplate attachment points by screwing both the handlebar lever and brake plate cable adjusters fully in, detaching the cable inner from the handlebar lever and detaching the cable inner from the brake backplate lever. The cable and adjuster may now be detached from the brake backplate and secured to a point on the frame so that it is clear of the front wheel. Take care to retain the large return spring as the cable is removed.

6 Remove the wheel spindle nut and split pin, and withdraw the spindle with the aid of a tommy bar. The wheel can now be lowered clear of the forks and put to one side.

3 Front forks: removing the fork assembly from the frame

1 Prior to removing the fork assembly from the frame, it will be necessary to remove the handlebars, complete with controls, the mudguard and the headlamp unit. Start by covering the petrol tank with an old blanket, or similar, to protect the paintwork from damage.

2 Slacken and remove the four handlebar clamp retaining bolts. Lift off the top halves of the clamps, and rest the handlebar assembly across the top of the tank. Note the punch marks on the top halves of the clamps and on the handlebar itself. The position of these marks should be noted for reference when refitting.

3 Remove the headlamp unit from its shell, by releasing the two screws, which secure the rim. The headlamp can be removed completely by disconnecting the colour coded leads inside the shell. Alternatively, the unit can be left in place, the two mounting bolts released, and the leads threaded around the stanchions as the forks are removed. Note that whilst this will save disconnecting and then reconnecting the headlamp wiring, care must be taken to avoid damage to the unit during the fork dismantling sequence. It is in any case, a simple task to unplug and reconnect any of the wires, because they are all colour-coded for straight-forward reconnection.

4 On XR models fitted with a combined headlamp/number plate unit, the complete unit may be removed by unscrewing the single, central lower retaining bolt and the two upper yoke bracket retaining bolts. Pull the unit forward to expose the headlamp wiring connections and disconnect the colour coded leads.

5 Remove the speedometer by first disconnecting the drive cable from the base of the instrument and then unscrewing the two retaining nuts. The instrument may now be lifted clear of the upper yoke assembly, detaching the bulb holders from the base as this is done. Repeat this procedure with the tachometer head (if fitted). Where there is an indicator lamp/ignition switch unit mounted next to the speedometer, disconnect the electrical leads to the unit.

6 On machines fitted with a combined speedometer, tachometer, indicator lamp and ignition switch unit, the unit may be detached from its mounting by disconnecting the two instrument drive cables from the base of the unit and unscrewing the two cap nuts retaining the unit to the upper fork bracket.

7 On machines fitted with a number plate only (XR models), remove the plate and its brackets by unscrewing the two retaining bolts, taking care to retain the plain and spring washers in their correct order.

8 Remove the horn (if fitted) by unscrewing its single retaining bolt. Alternatively, disconnect the leads and leave the unit in position.

9 On machines where the choke control is mounted on the upper yoke bracket, disconnect the control from the bracket by loosening the large cable retaining cap nut.

10 The headlamp pivot/instrument retainer assembly (where fitted) may now be removed from the steering head assembly by unscrewing its retaining bolts and pulling the unit up from the lower yoke to clear the wire spigots. Note the location of the two bracket to fork tube rubber buffers for reference when refitting.

11 Remove the mudguard by unscrewing the three bolts that retain it to the underside of the fork lower yoke. The fork top yoke may now be removed as follows.

12 Slacken the two retaining bolts which pass through the top yoke. On the XR80 and XL80 S models, these bolts should be removed and the O-ring fitted to each bolt inspected for damage or deterioration. Unscrew the large chromium plated nut which retains the top yoke to the steering head and remove the large plain washer located underneath it. The upper yoke can now be lifted off. If it proves stubborn, it may be tapped free using a soft-faced mallet. If the headlamp has been left in place, this should be positioned clear of the forks, as should any trailing cables.

13 Using a C-spanner of the correct size, slacken off the lock ring which retains the steering head column. Have to hand two small bins or jars in which the head race ball bearings can be kept safely. The balls from the lower race will drop free as the cup and cone part, and should be caught. Make sure none are left clinging to the race. Support the fork assembly while the lock ring and cone are removed, and the balls from the upper race are removed and placed in the second container. The lower yoke can now be removed complete with the fork legs.

3.6a Disconnect the drive cables from the base of the instrument unit ...

3.6b ... and remove the unit by unscrewing the two securing nuts

3.9 Disconnect the choke control by loosening the large securing nut

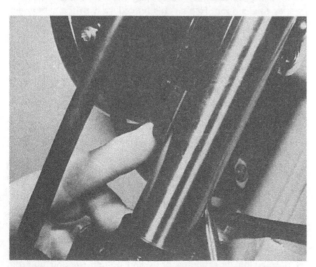

3.10 Note the location of the two headlamp bracket rubber buffers

3.12 Remove the two stanchion to upper yoke retaining bolts, noting the O-ring (80)

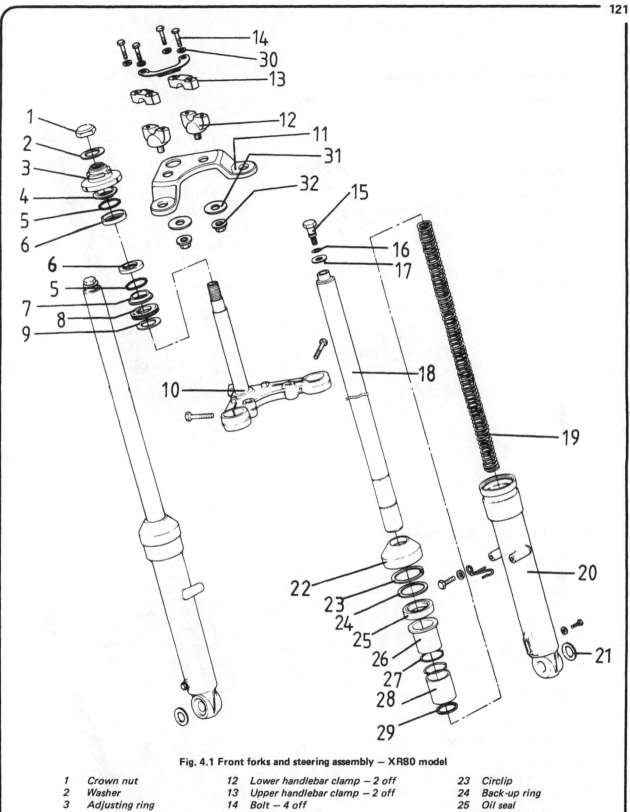

Fig. 4.1 Front forks and steering assembly — XR80 model

1	Crown nut	12	Lower handlebar clamp — 2 off	23	Circlip
2	Washer	13	Upper handlebar clamp — 2 off	24	Back-up ring
3	Adjusting ring	14	Bolt — 4 off	25	Oil seal
4	Upper cone	15	Top bolt	26	Shouldered guide
5	Bearing race	16	O-ring	27	Stopper ring
6	Bearing cup	17	Washer	28	Piston
7	Lower cone	18	Stanchion	29	Circlip
8	Dust seal	19	Spring	30	Washer — 4 off
9	Washer	20	Lower leg	31	Washer — 2 off
10	Steering stem	21	Washer	32	Nut — 2 off
11	Upper fork yoke	22	Dust seal		

Fig. 4.2 Front forks – XL100 S and XL125

1 Lower leg
2 Stanchion
3 Main spring
4 Top bolt
5 O-ring
6 Damper rod
7 Piston ring
8 Rebound spring
9 Dust seal
10 Damper rod seat
11 Wire clip
12 Oil seal
13 Sealing washer
14 Drain bolt
15 Sealing washer
16 Allen screw

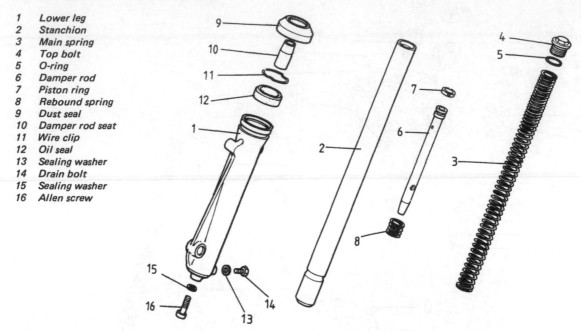

Fig. 4.3 Front fork – 185 and 200 models

1 Lower leg
2 Stanchion
3 Gaiter – XR models
4 Main spring
5 Top bolt
6 O-ring
7 Piston ring
8 Damper rod
9 Rebound spring
10 Gaiter retaining clip
11 Screw
12 Allen screw
13 Sealing washer
14 Drain bolt
15 Sealing washer
16 Oil seal seat – XR models
17 Oil seal
18 Wire clip
19 Damper rod seat
20 Dust seal
21 Clamp
22 Bolt
23 Nut
24 Upper spring – XR200
25 Spring spacer – XR200

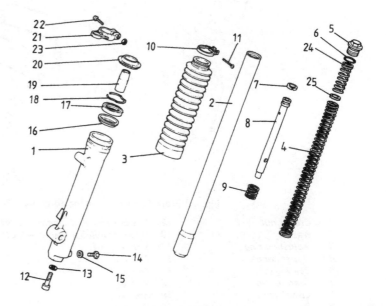

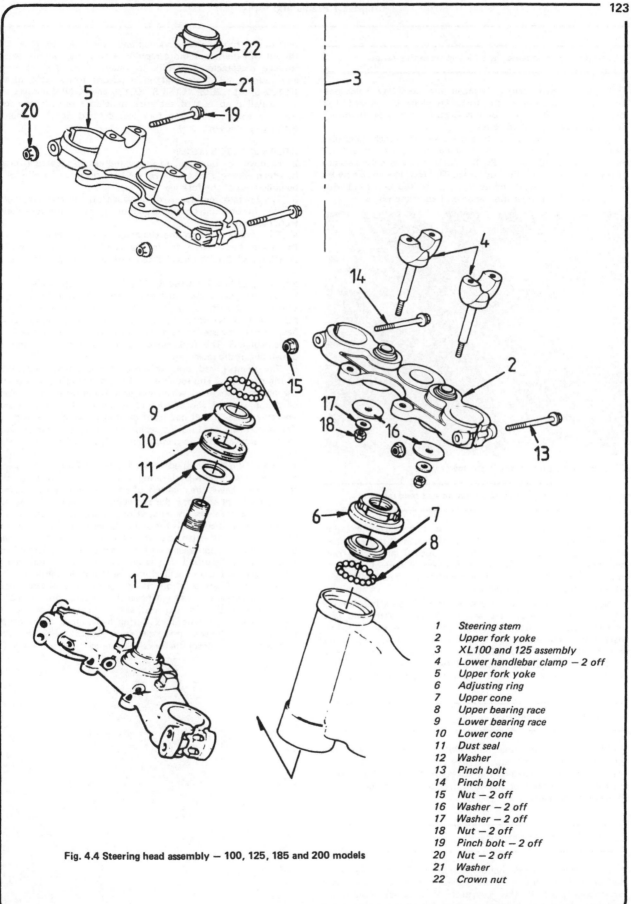

1   Steering stem
2   Upper fork yoke
3   XL100 and 125 assembly
4   Lower handlebar clamp − 2 off
5   Upper fork yoke
6   Adjusting ring
7   Upper cone
8   Upper bearing race
9   Lower bearing race
10 Lower cone
11 Dust seal
12 Washer
13 Pinch bolt
14 Pinch bolt
15 Nut − 2 off
16 Washer − 2 off
17 Washer − 2 off
18 Nut − 2 off
19 Pinch bolt − 2 off
20 Nut − 2 off
21 Washer
22 Crown nut

Fig. 4.4 Steering head assembly — 100, 125, 185 and 200 models

4   Front forks: removing the fork legs from the yokes

1   It is not necessary to remove the complete headstock assembly if attention to the fork legs alone is required. The instructions in Section 2 of this Chapter should be followed, then proceed as described below.

2   Loosen the fork top bolts and slacken the pinch bolts that clamp the fork stanchions in the upper and lower yokes. Note that on the XR80 and XL80 S models, the top bolts also act as retaining bolts for the top yoke. The fork legs may now be removed individually. If necessary, tap the top bolt head with a soft-faced mallet to jar the stanchion free of the yokes.

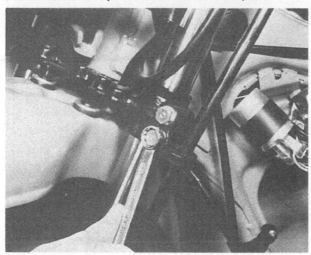

4.2 Slacken the pinch bolts in the lower yoke

5   Steering head bearings: examination and renovation

1   Before reassembly of the forks is commenced, examine the steering head races. The ball bearing tracks of the respective cup and cone bearings should be polished and free from indentations or cracks. If wear or damage is evident, the cups and cones must be renewed as a complete set. They are a tight press fit and should be drifted out of position.

2   Note that when each head race bearing has its full complement of ball bearings they are not packed tightly and that sufficient room is left to accommodate one extra ball. This spacing is essential because if the bearings are packed tightly against each other, they will skid rather than roll, thus greatly accelerating the rate of wear. Ball bearings are cheap; if the originals are marked or discoloured, they should be renewed.

6   Fork yokes: examination

1   To check the top yoke for accident damage, push the fork stanchions through the bottom yoke and fit the top yoke. If it lines up, it can be assumed the yokes are not bent. Both yokes must also be checked for cracks. If they are damaged or cracked, fit new replacements.

7   Front fork legs: dismantling

1   It is advisable to dismantle each fork leg separately, using an identical procedure. There is less chance of unwittingly exchanging parts if this approach is adopted. Commence operations by draining the oil from each fork leg. Unscrew the

bolt at the top of each fork leg and, upending the leg so that the oil will drain out more rapidly, empty the contents into a suitable receptacle. The fork leg may be pumped to expel any residual oil. Note that on the front forks fitted to the XL100 S, XL125 S, XL185 S, XR185 and XR200 models, the spring will drop out of the fork stanchions directly the fork leg is upended. With the fork legs drained of oil dismantle each leg as follows.

XR80 and XL80 S models

2   Remove the dust seal to reveal the circlip located beneath it, which when detached allows the back-up ring and oil seal to be pulled out of the fork leg.

3   The bottom fork leg is then tapped off the fork stanchion, leaving the bearings on the stanchion. Remove the fork spring from inside the stanchion.

4   The sliding bush on the stanchion is easily removed whilst the fixed bush needs to have a circlip removed first. The stanchion is left with two circlips which position the bearings.

XL100 S, XL125 S, XL185 S, XR185 and XR200 models

5   Remove the rubber dust seal. Clamp the fork lower leg in a vice fitted with soft jaws, or prevent damage to the leg by padding it before inserting it in the vice. Unscrew the socket head screw, recessed into the housing which carries the front wheel spindle. The fork stanchion (upper tube) can now be drawn out of the lower leg.

6   The damper rod unit will now lift out from the centre of the stanchion. The lower end of the main spring rests on the damper rod assembly, which has an integral piston and a form of piston ring. A short rebound spring fits below the piston assembly. All will pull out from the upper end of the stanchion. The piston ring need not be separated from the piston attached to the damper rod. The damper rod unit will be followed by its seat.

7   The oil seal is retained inside the lower fork leg by a wire clip. Remove the clip by levering it out with the flat of a small screwdriver. Lever the seal out of position in a similar manner, taking care not to mark the surrounding surface. The oil seal may need a considerable amount of persuasion to come out of the fork leg, particularly if the seal is damaged. If the screwdriver proves to be ineffective, try using a tyre lever, preferably with the levering end somewhat curled to act as a drift. The oil seal must only be disturbed if it is worn or damaged, indicated by oil leakage, as the seal will almost certainly be damaged when being prised from position. It is however, considered good practice to renew the seals whenever the forks are dismantled.

8   Note that the damper rod seat is fitted to prevent an instantaneous hydraulic lock during heavy fork compression, which could otherwise make the forks 'freeze' solid. The taper on the restrictor allows the forks to come to a stop gradually.

7.2a On 80 models, remove the front fork leg dust seal ...

7.2b ... followed by the circlip and back-up ring ...

7.2c ... and finally the oil seal

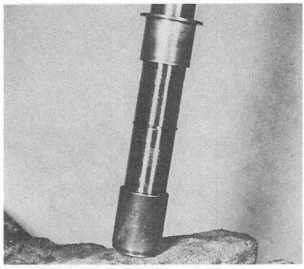

7.3 The bearings will remain on the stanchion

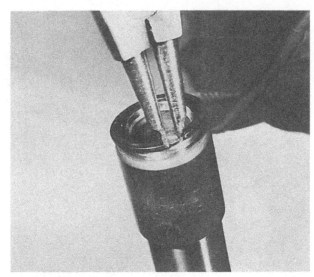

7.4a Remove the circlip ...

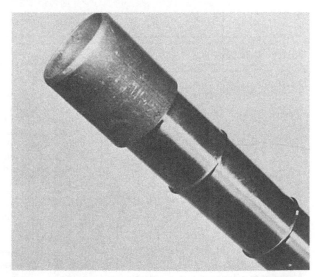

7.4b ... to allow the fixed bearing to be slid off the stanchion

7.4c Note that the damping mechanism is sealed and cannot be dismantled

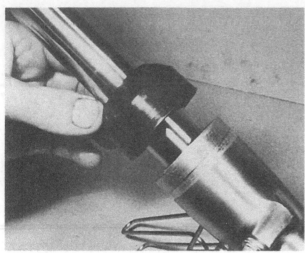

7.5a On all except the 80 models, remove the fork leg dust seal ...

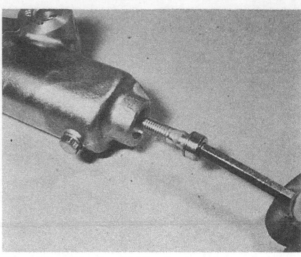

7.5b ... followed by the socket head screw

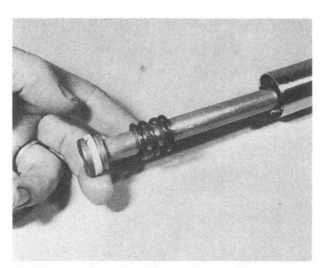

7.6a Withdraw the damper rod assembly from the stanchion ...

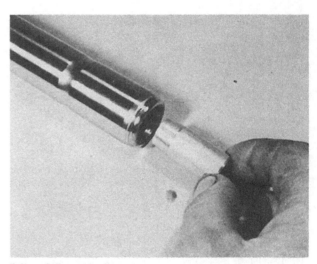

7.6b ... followed by its seat

7.7 Remove the oil seal retaining clip

8 Front forks: examination and renovation

XR80 and XL80 S models

1   The type of front forks fitted to these machines are equipped with bushes. Some indication of the extent of wear of these bushes can be gained before the machine is dismantled. If the front wheel is gripped between the knees and the handlebars rocked to and fro, the amount of wear will be magnified by the leverage at the handlebar ends. Cross-check by applying the front brake and pulling and pushing the machine backwards and forwards.

2   The fork bushes are left on the stanchion. The lower one will press off if the circlip is removed; the upper one can be slid off in a similar fashion. The new top bush will slide on to be retained by the circlip. Follow up with the lower bush which can be pushed on, and retained in the same manner.

3   Check the lower fork leg and the stanchion for any score marks or excessive wear. In extreme cases the damaged component will have to be renewed.

XL100 S, XL125 S, XL185 S, XR185 and XR200 models

4 The front forks fitted to these machines are not provided with bushes. The fork legs slide directly against the hard chrome surface of the stanchions and if wear occurs, the stanchion and/or the lower fork leg will have to be renewed. Wear normally occurs only after a very considerable mileage has been covered and can be detected by a juddering sensation when the front brake is applied. A slack steering head assembly will give the same effect, so this should always be checked first and adjusted if necessary, before condemning the forks.

5 Wear is often visually apparent in the form of scuffing or break-through of the chrome surface of the fork stanchions. If evidence of damage of this nature is apparent, the stanchion in question must be renewed. In extreme cases the fork lower leg will have to be renewed as well.

6 If damping action is lost, the piston ring around the damper piston should be renewed; it is catalogued as a separate item. Check that the small holes in the damper tube are not blocked and if no substantial improvement is shown when the forks are reassembled and refilled, renew the complete damper assembly.

All models

7 If the fork legs have shown a tendency to leak oil or if there is any other reason to suspect the condition of the oil seals, now is the time to renew them. Details of seal removal are given in Section 7 of this Chapter.

8 Renew the O-ring fitted to the top bolt of each fork leg. Inspect the dust seal for signs of deterioration and damage, and renew if necessary. It is recommended that this seal is renewed whenever the forks are dismantled. If the drain plug at the bottom of the lower fork leg has been removed for any reason, the drain plug sealing washer should be renewed.

9 It is rarely possible to straighten forks which have been badly damaged in an accident, especially if the correct jigs are not available. It is always best to err on the side of safety and fit new replacements, especially since there is no easy means of checking to what extent forks have been overstressed. The fork stanchions can be checked for straightness by rolling them on a flat surface. Any misalignment will immediately be obvious.

10 The fork springs may show signs of compression after lengthy service, in which case they can be replaced to advantage. The correct spring lengths are given in the Specifications Section of this Chapter. Note that the main fork springs have closer coiled springing at one end (except XR80 and XL80 S models). Upon refitting, the end with the closer pitches must be at the top of the stanchion.

9 Front forks: reassembly

1 Reassembling the front forks is essentially a reverse of the dismantling procedure given in Section 7 of this Chapter, noting the following points.

2 It is essential that all fork components are thoroughly washed in solvent and wiped clean with a lint-free cloth before assembly takes place. Any trace of dirt inside the fork leg assembly will quickly destroy the oil seal or score the stanchion to outer fork leg bearing surfaces.

3 When fitting the new oil seal, the seal should be coated with the recommended fork oil on its inner and outer surfaces. This serves to make the fitting of the seal into the lower fork leg easier and also reduces the risk of damage to the seal when the fork stanchion is passed through it. Great care should be taken when fitting the stanchion through the seal.

4 Honda recommend the use of a service tool with which to drive the seal into the fork leg recess. With the seal located partially in its recess and the fork stanchion passed through it, the tool, which takes the form of a short length of metal tube approximately 3 in long, with an inner diameter just greater than the outer diameter of the stanchion and an outer diameter just less than that of the outer diameter of the oil seal, may be passed over the stanchion and used to tap the seal home by using it as a form of slide hammer. If this tool is not readily available it can, of course, be fabricated from a piece of metal tubing of the appropriate dimensions. Care should be taken however, to ensure that the end of the tube that makes contact with the seal is properly chamfered, free of burrs and absolutely square to the fork stanchion. Alternatively, a suitable socket may be used to drive the seal into position. Ensure that the seal is driven home squarely.

5 On models where protective gaiters are fitted, the gaiters should be closely inspected for signs of deterioration and any small holes or splits in the rubber where dust or water may be allowed to enter. If necessary, renew the gaiters. It is well worth considering at this point, fitting gaiters even if they were not originally fitted. It is a fact that the life of the oil seal can be considerably lengthened by doing this, with the additional advantage that the lower part of the fork stanchion is also protected. Several manufacturers provide gaiters to fit the type of forks fitted to these machines, if the genuine Honda replacement cannot be obtained.

6 On the XL100 S, XL125 S, XL185 S, XR185 and XR200 models, ensure that when the socket head screw is refitted into the base of the lower fork leg, it is torque loaded to the recommended figure given in the Specifications Section of this Chapter. Apply a locking compound to the threads of each bolt before it is fitted.

10 Refitting the forks in the frame

1 If it has been necessary to remove the fork assembly completely from the frame, refitting is accomplished by following the dismantling procedure in reverse. Check that none of the balls are misplaced whilst the steering head stem is passed through the head lug. It has been known for a ball to be displaced, drop down and wear a groove or even jam the steering, so be extremely careful in this respect. The balls and the bearing surfaces of the cup and cone races must be liberally coated with grease of the recommended type before installation. This will serve to help retain the balls during the refitting procedure.

2 Tighten the steering head carefully, so that all play is eliminated without placing undue stress on the bearings. The adjustment is correct if all play is eliminated and the handlebars will swing to full lock of their own accord when given a push on one end.

3 It is possible to place several tons pressure on the steering head bearings if they are overtightened. The usual symptom of overtight bearings is a tendency for the machine to roll at low speeds, even though the handlebars may appear to turn quite freely.

4 The fork legs may be filled with the correct quantity and grade of damping fluid whilst they are fitted in the fork yokes and before the top bolts are fitted. Check that both the drain bolts and the socket bolts in the bottom of the lower legs have been refitted and tightened before refilling the legs with oil.

5 Ensure that after the front mudguard, speedometer, tachometer, indicator lamp, headlamp and handlebar assemblies have been fitted, all control cables, electrical wires, etc are correctly routed, refitted and reconnected. The headlamp beam height should be adjusted; see the relevant Section in Chapter 6. The handlebar must be refitted in the top yoke clamps with the punch marks in the correct positions; that is with the punch mark on the handlebar in line with the rear mating face of the lower clamp and with the punch mark on the upper half of the clamp facing forward. Torque load the retaining bolts to the figure given in the Specifications Section of this Chapter; front bolts first, followed by the rear bolts.

6 When refitting the front wheel, follow the procedure given in Chapter 5 of this Manual.

7   When everything is in place, bounce the forks a few times so that they settle into their natural position. Finally, fully tighten the pinch bolts and top bolts. Note that this operation must be done last, only after the front wheel has been refitted and the steering head bearings adjusted. The top yoke (except XR80 and XL80 S models) and bottom yoke pinch bolts should be torque loaded to the figure given in the Specifications.

8   Before finally tightening the pinch bolts, ensure that the top of each fork leg stanchion is flush with the upper face of the top yoke (XL100 S, XL125 S, XL185 S, XR185 and XR200 models only). Refit and tighten the top bolts, ensuring that the new O-ring has been fitted.

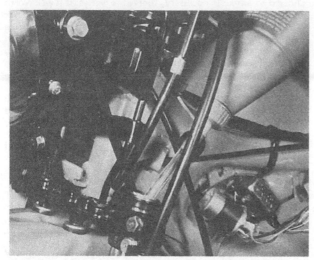

10.4 Refill the fork legs with the correct damping fluid

---

11 Frame assembly: examination and renovation

1   If the machine is stripped for a complete overhaul, this affords a good opportunity to inspect the frame for cracks or other damage which may have occurred in service. Check the top of the front downtube and the front of the top tube where it joins the steering head, the two points where fractures are most likely to occur. The straightness of the tubes concerned will show whether the machine has been involved in a previous accident.

2   Check carefully areas where corrosion has occurred on the frame. Corrosion can cause a reduction in the material thickness and should be removed by use of a wire brush and derusting agents.

3   If the frame is broken or bent, professional attention is required. Repairs of this nature should be entrusted to a competent repair specialist, who will have available all the necessary jigs and mandrels to preserve correct alignment. Repair work of this nature can prove expensive and it is always worthwhile checking whether a good replacement frame of identical type can be obtained at a reasonable cost.

4   Remember that a frame which is in any way damaged or out of alignment will cause, at the very least, handling problems. Complete failure of a main frame component could well lead to a serious accident.

---

12 Swinging arm: removal, examination, renovation and refitting

1   Any wear in the swinging arm pivot bushes will cause imprecise handling of the machine, with a tendency for the rear end of the machine to twitch or hop. Any wear in the bushes may be detected by placing blocks under the sump-guard or footrest assembly so as to raise the rear wheel clear of the ground, and pulling and pushing on the fork ends in a horizontal direction with one hand whilst holding the frame firmly in the other. Ensure the machine is blocked securely to prevent it toppling during this test. Any play in the bushes will be greatly magnified by the leverage effect of the fork legs. No play at all should be detected during this test. If play is apparent then the swinging arm assembly should be removed from the machine and the bushes renewed.

2   Remove the rear wheel assembly as described in Chapter 5 and disconnect the chain, but do not pull it right off. Leave it on the gearbox sprocket.

3   Remove the two rear suspension units from their lower locating points and then remove the chainguard (if considered necessary). The swinging arm fork may now be detached from the frame mounting point and serviced as follows.

XR80 and XL80 S models

4   Remove the pivot bolt locknut from the right-hand side of the machine then withdraw the bolt from the left. It should pull out of position quite easily if the pivot has been greased regularly. Support the swinging arm with the right hand as the pivot bolt is withdrawn.

5   In order to renew the bushes, it is necessary to fabricate a tool with which to press the bushes out of their housings. It is suggested that a tool similar to the one shown in the accompanying diagram is made, utilising a short length of thick-walled tube, the inside diameter of which is slightly larger than the outside diameter of the bush, a high tensile bolt and nut, and two thick plate washers, one of which has an outer diameter slightly smaller than that of the bush. Note that, as a means of removal, attempting to drive the bushes out will probably prove unsuccessful, because the rubber will effectively damp out the driving force, and damage to the lugs may occur.

6   If, due to corrosion between the mating faces of the bush and swinging arm lug, the bushes are reluctant to move, even using this method, it is recommended that the unit be returned to a Honda Service Agent whose expertise can be brought to bear on the problem.

7   New bushes may be driven in using a tubular drift against the outer sleeve, or by reversing the removal operation, using the fabricated puller. Whichever method is adopted, the outer sleeve should be lubricated sparingly, and care must be taken to ensure that the bush remains square with the housing bore. The swinging arm must be properly supported whilst doing this.

XL100 S, XL125 S, XL185 S, XR185 and XR200 models

8   Remove the locknut and plain washer from the left-hand end of the pivot bolt and withdraw the bolt from the right. If the pivot has been greased regularly, the bolt should pull out of position quite easily. Support the swinging arm with the left hand as the bolt is withdrawn.

9   With the swinging arm placed on a work surface, remove the nylon chain slider from the end of the pivot housing, taking with it the outer dust seal. Remove the inner dust seal. The two pivot bushes may now be removed by tapping them out from the opposite end using a long soft-metal drift and hammer, after having first withdrawn the collar from the pivot housing. No difficulty should be experienced when removing these bushes as they are a very light interference fit in the housing.

10 Clean the inside bearing faces of the pivot housing and press the new bushes into position. It may be necessary to tap the bushes home to ensure the shoulder of the bush is firmly against the end of the pivot housing. To do this, place a wooden block over the end of the bush and with the swinging arm properly supported, use a soft-faced mallet to tap the bush home.

11 Measure the outside diameter of the collar and compare it with the figure given in the Specifications Section of this Chapter. If the collar is worn beyond the service limit given, it must be renewed; if not, then it should be lubricated with the recommended grease and refitted.

12 Inspect the two dust seals before refitting them; if they are in any way damaged or have deteriorated, then they should be renewed. Both seals should be lightly greased. It is very unlikely that any wear will have taken place to the chain slider until a considerable number of miles has been covered by the machine. If the unit is seen to be badly worn however, then it must be renewed.

All models

13 Reassemble and refit the swinging arm fork by reversing the dismantling procedure. Ensure that before the pivot bolt is inserted, it is coated liberally with the recommended type of grease. Torque load the pivot bolt nut to the figure given in the Specifications. The rear suspension unit lower securing nuts or bolts should also be torque loaded to the recommended figure.

14 On all except the XR80 and XL80 S models, apply a grease gun to the nipple provided on the swinging arm pivot and pump grease into the assembly until it can be seen to be forced out of the ends of the pivot bushes. Wipe off any excess lubricant.

15 Refit the rear wheel, using the procedure given in Chapter 5. Refer also to the Sections in Chapter 5 for adjustment of the rear brake and chain tension. Finally, check the routing of any breather pipes that pass through the swinging arm assembly.

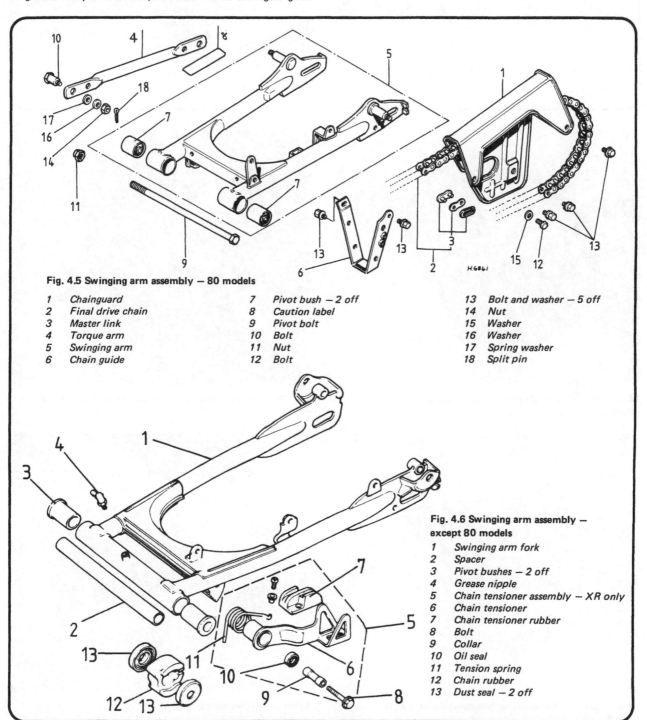

Fig. 4.5 Swinging arm assembly — 80 models

1	Chainguard	7	Pivot bush — 2 off	13	Bolt and washer — 5 off
2	Final drive chain	8	Caution label	14	Nut
3	Master link	9	Pivot bolt	15	Washer
4	Torque arm	10	Bolt	16	Washer
5	Swinging arm	11	Nut	17	Spring washer
6	Chain guide	12	Bolt	18	Split pin

Fig. 4.6 Swinging arm assembly — except 80 models

1 Swinging arm fork
2 Spacer
3 Pivot bushes — 2 off
4 Grease nipple
5 Chain tensioner assembly — XR only
6 Chain tensioner
7 Chain tensioner rubber
8 Bolt
9 Collar
10 Oil seal
11 Tension spring
12 Chain rubber
13 Dust seal — 2 off

Fig. 4.7 Swinging arm bush removal tool

1	Nut
2	Washer
3	Tube
4	Swinging arm
5	Washer
6	Bolt
7	Bush

H.6840

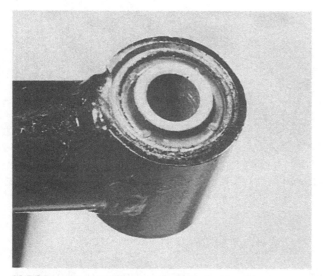

12.5 80 models have 'Silentbloc' type bushes fitted in the swinging arm

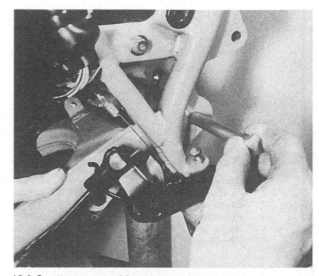

12.8 On all except the 80 models, withdraw the pivot bolt ...

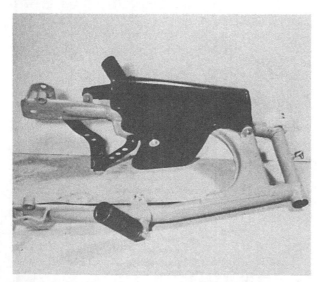

12.9a ... position the swinging arm assembly on a work surface ...

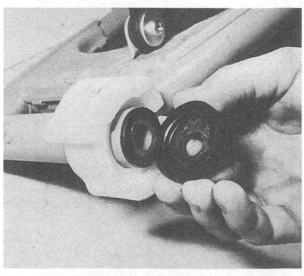

12.9b ... and remove the outer dust seal ...

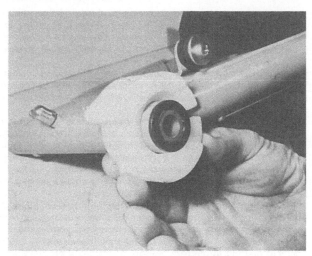

12.9c ... followed by the nylon chain slider ...

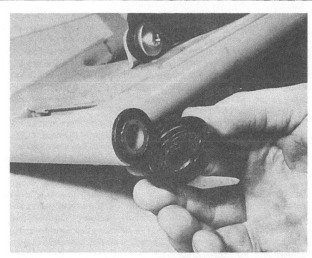

12.9d ... and inner dust seal (one either end of the pivot housing)

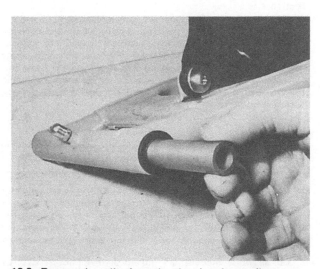

12.9e Remove the collar from the pivot housing to allow removal of the bushes

12.13a Torque load the pivot bolt nut ...

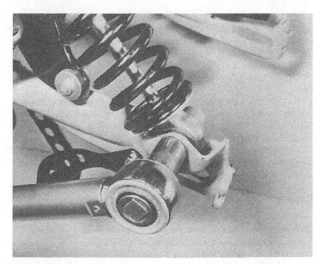

12.13b ... and rear suspension lower securing nuts or bolts to the recommended figure

13 Rear suspension units: removal, examination, renovation and refitting

1   All the machines covered by this Manual have the same basic type of rear suspension; that is, a swinging arm rear fork assembly supported by two suspension units. These suspension units do, however, differ between the model types. The types of suspension unit and the models to which they are fitted are as follows. Always service the units as a pair, never individually.

XR80 and XL80 S models
2   The suspension units fitted to these models are removed from the frame and swinging arm attachment studs simply by removing the upper and lower cap nut and plain washer and pulling the unit out away from the machine.
3   Each unit comprises a hydraulic damper, effective primarily on rebound, and a concentric spring. It is secured to the frame and swinging arm by means of rubber-bushed lugs at either end of the unit. The units are provided with an adjustment of the spring tension, giving five settings. The settings can be easily altered without removing the units from the machine

by using a C-spanner or a screwdriver in the holes directly below the springs. Turning clockwise will increase the spring tension and stiffen the rear suspension, turning anti-clockwise will lessen the spring tension and therefore soften the ride. As a general guide the softest setting is recommended for road use only, when no pillion passenger is carried. The hardest setting should be used when a heavy load is carried, and during high-speed riding either on or off road. The intermediate positions may be used as conditions dictate.

4 There is no means of draining the units or topping up, because the dampers are built as a sealed unit. If the damping fails or if the units start to leak, the complete damper assembly must be renewed. This applies equally if the damper rod has become bent.

5 The damping efficiency of the units can best be judged after removal of the springs. The compression springs can be removed by holding each unit upright on the workbench with it set on the softest setting. With an assistant pulling down on the top of the spring so that it clears the top mounting lug plate, insert an open-ended spanner over the locknut situated underneath the mounting lug. The mounting lug may now be unscrewed by passing a bar through the stud bush and turning it against the spanner. Note that it is essential an assistant keeps the spring pressure from acting on the mounting lug plate during this operation, because once the plate reaches the end of its thread spring pressure will cause it to fly off. If difficulty is experienced whilst trying to remove the springs from the units, it is advisable to return the units to an official Honda Service Agent who will have the necessary equipment to dismantle the units.

6 Inspect the damper unit for signs of leakage, corrosion or damage to the chromed surface of the piston shaft, deterioration of the seal, rubber buffer and rubber mounting bushes, and damage to the piston housing. Renew any components as considered necessary.

7 Place the unit upright with the lower bush lug gripped in a vice, temporarily refit the upper mounting lug (spring removed) and compress and extend the damper. Little resistance should be felt when compressing the damper, whereas heavy damping should be apparent as the unit is extended.

8 If the damper springs have become weakened or are of differing lengths they should be renewed as a matched pair. The spring free lengths and the maximum allowable shortening of the springs are given in the Specifications at the beginning of this Chapter.

9 Reassembly and refitting of the units is a reversal of the removal and dismantling procedures, noting the following points. Ensure that the units have been thoroughly cleaned before reassembly, especially around the area of the chromed piston shaft and seal. Do not grease the chromed shaft. Apply a locking compound to the threads of the locknut and mounting lug plate before refitting them and ensure that the plate and locknut are fully tightened. With the units fitted to the machine, torque load the retaining nuts to the figure given in the Specifications Section of this Chapter and check that both units are on the same setting.

XL100 models

10 The suspension units fitted to this model are similar to those fitted to the XR80 and XL80 S machines, the one main difference being the fitting of a second concentric spring. This spring is located below the main spring and is separated from it by a spacer plate and seat plate. Another seat plate is located between the base of the spring and the adjuster ring.

11 All the servicing procedures listed in the previous paragraphs of this Section apply to these suspension units.

XL125, XL185, XR185 and XR200 models

12 The suspension units fitted to these models are of the gas-filled type, the gas being under high pressure. Because of this, some care must be taken when handling these units and it is recommended that the following safety precautions are observed:-

a) *Do not subject the dampers to a naked flame of high heat source because this may expand the gas beyond the limits of the container, thereby causing an explosion.*

b) *Do not damage or deform the damper cylinder in any way.*

c) *Do not discard the damper carelessly where it may present a danger to others. Return it to an official Honda Service Agent who will have the facilities to dispose of the unit safely.*

13 Removal of these units is similar to that procedure given in paragraph 2 of this Section. Remove the spring from the unit by inverting the unit, placing it upright on a work surface and pulling the top of the spring downwards so that the spring seat may be withdrawn from between the end of the spring and the removable rubber-bushed lug. The spring can now be lifted off the damper unit along with the upper spring seat.

14 Note that the units fitted to XR185 and XR200 models have a set ring and plate fitted above the upper spring seat. This provides a method of adjusting the spring tension, whilst the unit is dismantled, by moving the set ring to a lower groove in the damper housing to increase the spring tension, or moving the set ring to a higher groove to decrease the spring tension. A general guide for spring setting is contained in paragraph 3 of this Section.

15 Examine the units as described in paragraphs 4, 6, 7 and 8 of this Section. Reassembly and refitting of the units is a reversal of the removal and dismantling procedures, noting the points made in paragraph 9 of this Section. Note also that the spring must be refitted with the closer pitched coils toward the removable rubber-bushed lug. If this lug has been removed it should be tightened, to the torque figure given in the Specifications, against the locknut.

14 Footrest assembly: removal, examination and refitting

1 On the XR80 and XL80 S models, the footrest assembly is in the form of a bar passing underneath the engine; a pivoting metal footrest being attached to either end of the bar by means of a clevis pin, plain washer and split-pin. The complete assembly may be removed from the machine by unscrewing the four retaining bolts.

2 On all other models, the individual metal footrests pivot on brackets which are part of the frame itself. They are also retained in position by means of a clevis pin, plain washer and split-pin.

3 If the footrests become damaged in an accident, it may be possible to straighten them. The area around the deformed portion should be heated to a dull red before any attempt is made to bend the footrest back into shape. If required, the hinged portion of the footrest may be separated from the frame or bar bracket after removing the split-pin and clevis pin.

4 If there is evidence of failure of the metal either before or after straightening, it is advised that the damaged component is renewed. If a footrest breaks in service, loss of machine control is almost inevitable.

5 Refitting of the footrest assemblies is a reversal of the removal procedure. When refitting the clevis pin, ensure that the return spring is correctly located and a new split-pin fitted.

15 Rear brake pedal: examination and renovation

1 On the XR80 and XL80 S models, the rear brake pedal pivots on a plain shaft attached to a lug on the lower right-hand side of the frame, just behind the engine. It is retained on the shaft by a plain washer and split-pin.

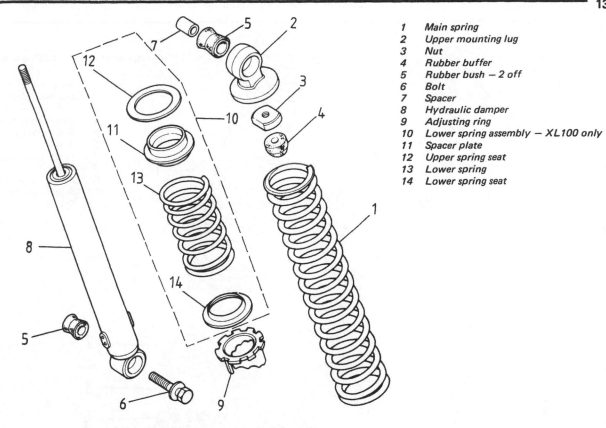

1 Main spring
2 Upper mounting lug
3 Nut
4 Rubber buffer
5 Rubber bush — 2 off
6 Bolt
7 Spacer
8 Hydraulic damper
9 Adjusting ring
10 Lower spring assembly — XL100 only
11 Spacer plate
12 Upper spring seat
13 Lower spring
14 Lower spring seat

Fig. 4.8 Rear suspension unit — 80 and 100 models

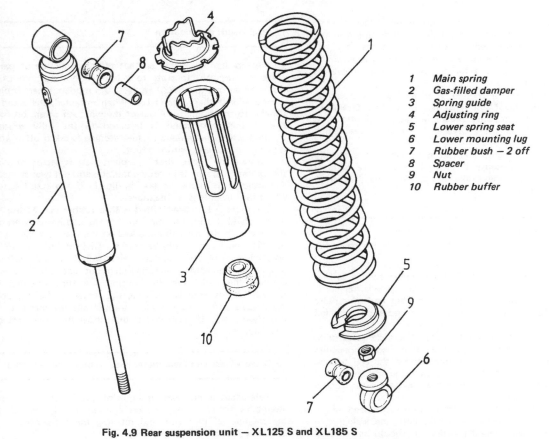

1 Main spring
2 Gas-filled damper
3 Spring guide
4 Adjusting ring
5 Lower spring seat
6 Lower mounting lug
7 Rubber bush — 2 off
8 Spacer
9 Nut
10 Rubber buffer

Fig. 4.9 Rear suspension unit — XL125 S and XL185 S

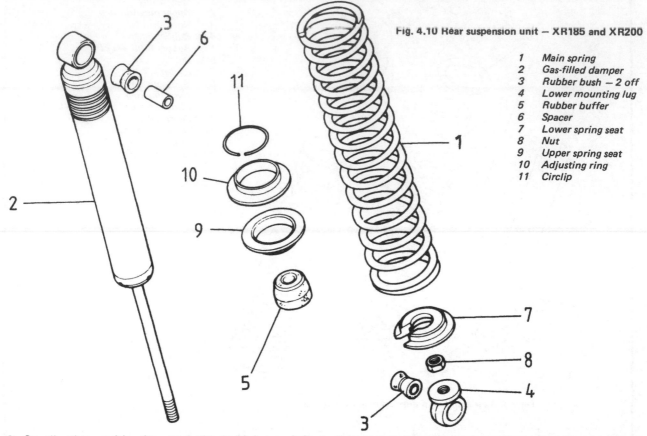

Fig. 4.10 Rear suspension unit — XR185 and XR200

1   Main spring
2   Gas-filled damper
3   Rubber bush — 2 off
4   Lower mounting lug
5   Rubber buffer
6   Spacer
7   Lower spring seat
8   Nut
9   Upper spring seat
10   Adjusting ring
11   Circlip

2   On all other models, the rear brake pedal has a shaft attached and this shaft passes through a mounting lug in the frame. The pedal is retained in position by means of a nut held in position by a spring clip. A coil spring around the pivot ensures the pedal is returned to its normal operating position after braking.

3   If the pedal is bent or twisted in an accident, it should be removed from the pivot and clamped in a vice. Straighten the pedal using the method recommended for footrests in Section 14 of this Chapter. Grease the shaft before fitting the lever.

4   The warning relating to footrest breakage applies equally to the brake pedal because it follows that failure is most likely to occur when the brake is applied firmly, which is when it is required most.

16 Kickstart lever: examination and renovation

1   The kickstart lever is splined and is secured to its shaft by means of a pinch bolt. The kickstart crank swivels so that it can be tucked out of the way when the engine is started. It is held in position on the swivel by a washer and spring clip (XR80 and XL80 S models) or by a pin and spring clip (all other models).

2   A spring-loaded ball bearing locates the kickstart arm in either the operating or folded position; if the action becomes sloppy it is probable that the spring behind the ball bearing needs renewing. It is advisable to remove the spring clip and washer or pin occasionally, so that the kickstart crank can be detached and the swivel greased.

3   It is unlikely that the kickstart crank will bend in an accident unless the machine is ridden with the kickstart in the operating and not folded position. It should be removed and straightened, using the same technique as that recommended for the footrest assembly.

17 Prop stand: examination

1   A prop stand is fitted so that the machine can be parked without difficulty. It bolts to a lug on the underside of the frame on all but the XR80 and XL80 S models, where it pivots on the end of the footrest bar. When retracted, the stand lies parallel to the frame, well out of the way. An extension spring ensures the prop stand is returned to the fully retracted position when the weight of the machine is taken off it and it is pushed backwards with the foot.

2   Periodically check that the pivot bolt is secure and that the extension spring is in good condition and not over-stretched. An accident is more or less inevitable if the stand extends whilst the machine is on the move.

3   On those XL models fitted with a rubber pad at the base of the stand, Honda recommend that the pad be renewed once it is worn to the wear line marked on the rearmost face of the pad. The new pad should be marked 'BELOW 259 LB ONLY', no other type of pad is suitable. Removal of the pad is achieved simply by unscrewing the retaining nut and bolt and withdrawing the pad from its housing. Refitting the pad is a reversal of this process, ensuring the arrow on the pad points downwards and faces outwards away from the machine. The pad should also be renewed if the rubber has deteriorated in any way.

18 Speedometer and tachometer head: removal and refitting

1   References to the various types of speedometer/tachometer assemblies fitted to the models covered in this Manual, and the method of removing and refitting these assemblies, may be found in Section 3 of this Chapter.

2   Apart from defects in either the drive or the cable, a speedometer or tachometer that malfunctions is difficult to

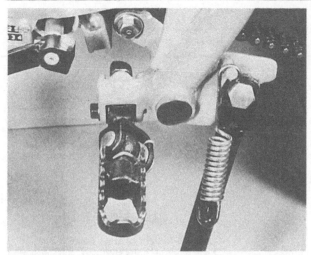

17.2 Check the condition of the prop stand extension spring and pivot bolt

repair. Fit a replacement or, alternatively, obtain one from a crash repair specialist who may have perfectly good instruments with scratched cases and glass etc.

3 Remember that a speedometer in correct working order is a statutory requirement in the UK. Apart from this legal necessity, reference to the odometer reading is the most satisfactory means of keeping pace with the maintenance schedules.

19 Speedometer and tachometer cables: examination and renovation

1 It is advisable to detach the speedometer and tachometer drive cables from time to time in order to check whether they are adequately lubricated and whether the outer cables are compressed or damaged at any point along their run. A jerky or sluggish movement at the instrument head can often be attributed to a cable fault.

2 To grease the cable, uncouple both ends and withdraw the inner cable. (On some model types this may not be possible in which case a badly seized cable will have to be renewed as a complete assembly). After removing any old grease, clean the inner cable with a petrol soaked rag and examine the cable for broken strands or other damage. Do not check the cable for broken strands by passing it through the fingers or palm of the hand, this may well cause a painful injury if a broken strand snags the skin. It is best to wrap a piece of rag around the cable and pull the cable through it, any broken strands will snag on the rag.

3 Regrease the cable with high melting point grease, taking care not to grease the last six inches closest to the instrument head. If this precaution is not observed, grease will work into the instrument and immobilise the sensitive movement.

4 If the cable breaks, it is usually possible to renew the inner cable alone, provided the outer cable is not damaged or compressed at any point along its run. Before inserting the new inner cable, it should be greased in accordance with the instructions given in the preceding paragraph. Try to avoid tight bends in the run of the cable because this will accelerate wear and make the instrument movement sluggish.

20 Seat: removal and refitting

1 The seat is secured in position by two bolts towards the rear which pass through or into frame lugs close to the seat base. The front of the seat is located in a rubber buffered retainer.

2 Removal of the seat is straightforward; unscrew the bolts and pull the seat to the rear so that it disengages at the front. The seat can then be lifted away.

3 Refitting the seat is a reversal of the removal procedure. It is important that the seat is correctly engaged on the forward mounting point and that the rear mounting bolts are fully tightened. Should the seat become detached whilst the machine is in motion, the resulting loss in balance of the rider could well prove disastrous.

21 Steering lock: removal and fitting - XL100 S, XL125 S and XL185 S models

1 The steering lock is situated at the front of the machine, on the bottom fork yoke. It is advisable to use it whenever the machine is left unattended, even for a short time.

2 To remove the lock, remove the countersunk screws holding the lock body to the bottom yoke bracket. Fitting the lock is the reverse procedure to removal. When the lock has been renewed, ensure a new key is obtained and carried with you when the machine is used.

22 Cleaning the machine

1 After removing all surface dirt with a rag or sponge washed frequently in clean water, the application of car polish or wax will give a good finish to the machine.

2 The plated parts should require only a wipe with a damp rag, but if they are badly corroded, as may occur during the winter when the roads are salted, it is permissible to use one of the proprietary chrome cleaners. These often have an oily base which will help to prevent corrosion from recurring.

3 If the engine parts are particularly oily, use a cleaning compound such as Gunk or Jizer. Apply the compound whilst the parts are dry and work it in with a brush so that it has an opportunity to penetrate and soak into the film of oil and grease. Finish off by washing down liberally, taking care that water does not enter the carburettors, air cleaners or the electrics.

4 If possible, the machine should be wiped down immediately after it has been used in the wet, so that it is not garaged under damp conditions that will promote rusting. Make sure the chain is wiped and re-oiled, to prevent water from entering the rollers and causing harshness with an accompanying rapid rate of wear. Remember there is less chance of water entering the control cables and causing stiffness if they are lubricated regularly as described in the Routine Maintenance Section.

23 Cleaning the plastic mouldings

1 The moulded plastic cycle parts, which include the front and rear mudguards, the sidepanels and number plate assemblies, need treating in a different manner than normal metal cycle parts.

2 These plastic parts will not respond to cleaning in the same way as painted metal parts; their construction may be adversely affected by traditional cleaning and polishing techniques, and lead as a result, to the surface finish deteriorating. It is best to wash these parts with a household detergent solution, which will remove oil and grease in a most effective manner.

3 Avoid the use of scouring powder or other abrasive cleaning agents because this will score the surface of the mouldings making them more receptive to dirt, and permanently damaging the surface finish.

24 Fault diagnosis: frame and forks

Symptom	Cause	Remedy
Machine veers either to the left or the right with hands off handlebars	Bent frame Twisted forks Wheels out of alignment	Check, and renew. Check, and renew. Check and realign.
Machine rolls at low speed	Overtight steering head bearings	Slacken until adjustment is correct.
Machine judders when front brake is applied	Slack steering head bearings Worn fork stanchion or lower legs - XL100 S, XL125 S, XL185 S, XR185 and XR200 models Worn fork bushes - XR80 and XL80 S models	Tighten, until adjustment is correct. Dismantle forks and renew worn components.
Machine pitches on uneven surfaces	Ineffective fork dampers Ineffective rear suspension units Suspension too soft	Check oil content of front forks. Check whether units still have damping action. Raise suspension unit adjustment one notch.
Fork action stiff	Fork legs out of alignment (twisted in yokes)	Slacken yoke clamps, and fork top bolts. Pump fork several times then retighten from bottom upwards.
Machine wanders. Steering imprecise. Rear wheel tends to hop	Worn swinging arm pivot	Dismantle and renew bushes and pivot shaft.

Chapter 5 Wheels, brakes and tyres

For information relating to 1981 — 1987 models, see Chapter 7

Contents

Specifications

	XR80 and XL80 S	XL100 S	XL125 S, XL185 S, XR185 and XR200
Wheels			
Type	Chromed steel rims and steel spokes		
Size:			
Front	16 inch	19 inch	21 inch
Rear	14 inch	17 inch	18 inch
Rim runout (service limit)	2.0 mm (0.1 in)	2.0 mm (0.1 in)	2.0 mm (0.1 in)
Brakes			
Front and rear:			
Type	Internally expanding, single leading shoe, drum		
Lining thickness	4 mm (0.2 in)	4 mm (0.2 in)	4 mm (0.2 in)
Service limit	2 mm (0.1 in)	2 mm (0.1 in)	2 mm (0.1 in)
Brake drum ID	110 mm (4.33 in)	110 mm (4.33 in)	110 mm (4.33 in)
Service limit	111 mm (4.37 in)	111 mm (4.37 in)	111 mm (4.37 in)
Tyres			
Size:			
Front	2.50 - 16 - 4PR	2.75 - 19 - 4PR	2.75 - 21 - 4PR
Rear	3.00 - 14 - 4PR	3.25 - 17 - 4PR	4.10 - 18 - 4PR
Pressures:	**XR80**	**XL80 S and 100 S**	**All others**
Front	18 psi (1.2 kg/cm^2)	21 psi (1.5 kg/cm^2)	14 psi (1.0 kg/cm^2) - XR models only 21 psi (1.5 kg/cm^2) - XL models only
Rear	20 psi (1.4 kg/cm^2)	21 psi (1.5 kg/cm^2)	14 psi (1.0 kg/cm^2) - XR models only 21 psi (1.5 kg/cm^2) - XL models only

Torque wrench settings kgf m (lbf ft)	XR80 and XL80 S	XL100 S, XL125 S, XL185 S, XR185 and XR200
Front wheel spindle nut	4.0 - 5.5 (29 - 39.8)	4.0 - 5.0 (29 - 36)
Rear wheel spindle nut	4.0 - 5.5 (29 - 39.8)	6.0 - 8.0 (43 - 58)
Rear wheel sprocket pin nuts	—	5.5 - 6.5 (40 - 47)
Rear brake torque arm nut	1.0 - 2.0 (7.3 - 14.5)	1.8 - 2.5 (13 - 18)
Front and rear brake arm nuts	0.8 - 1.2 (6 - 9)	0.8 - 1.2 (6 - 9)

1 General description

All the models covered in this Manual utilise the traditional type of wheel, that is, a chromed steel rim laced to the hub by steel spokes. The wheel sizes vary between the model types and reference should be made to the Specifications Section of this Chapter for both wheel and tyre sizes.

The type of brake fitted to all machines is a drum brake of standard single leading shoe type. The front brake on the XR80, XL80 S and XL100 S models is of the full-width hub design; a half-width hub is fitted to all other models in the range. The rear brake remains of the full-width hub design throughout the model range.

2 Front wheel: examination and renovation

1 Position blocks underneath the sumpguard plate so that the front wheel is raised clear of the ground. Spin the wheel and check the rim alignment; this should be no more than 2.0 mm (0.1 in) out of true. Small irregularities in alignment can be corrected by tightening the spokes in the affected area although a certain amount of experience is necessary to prevent over-correction. Any flats in the wheel rim will be evident at the same time. These are more difficult to remove and in most cases it will be necessary to have the wheel rebuilt on a new rim. Apart from the effect on stability, a flat will expose the tyre bead and walls to greater risk of damage if the machine is run with a deformed wheel.

2 Check for loose and broken spokes. Tapping the spokes is the best guide to tension. A loose spoke will produce a quite different sound and should be tightened by turning the nipple in an anti-clockwise direction. Always check for runout by spinning the wheel again. If the spokes have to be tightened by an excessive amount, it is advisable to remove the tyre and tube as detailed in Section 16 of this Chapter. This will enable the protruding ends of the spokes to be ground off, thus preventing them from chafing the inner tube and causing punctures.

3 Front wheel: removal

1 Place blocks underneath the engine sumpguard so that the front wheel is raised clear of the ground. Detach the speedometer drive cable at the wheel hub by removing the single crosshead retaining screw and pulling the cable clear.

2 Remove the front brake cable from the brake backplate attachment points by screwing both the handlebar lever and brake backplate cable adjusters fully in, detaching the cable inner from the handlebar lever and detaching the cable inner from the brake backplate lever. The cable and adjuster may now be detached from the brake backplate, take care to retain the large return spring as the cable is removed.

3 Remove the split-pin from the wheel spindle nut and unscrew the nut. The spindle may now be withdrawn with the aid of a tommy bar and the wheel lowered clear of the forks.

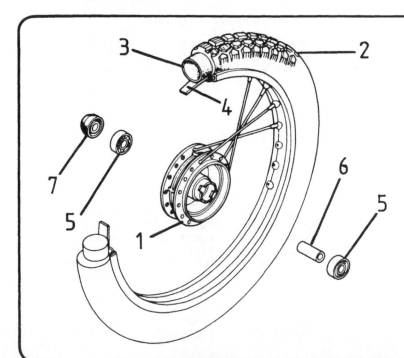

Fig. 5.1 Front wheel -- 80 and 100 models

1 Hub
2 Tyre
3 Inner tube
4 Rim band
5 Bearing — 2 off
6 Spacer
7 Dust seal

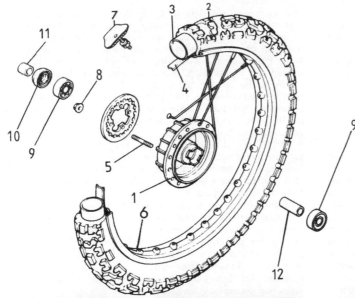

Fig. 5.2 Front wheel — 125, 185 and 200 models

1 Hub
2 Tyre
3 Inner tube
4 Rim band
5 Stud — 4 off
6 Rim
7 Security bolt
8 Nut
9 Bearing — 2 off
10 Dust seal
11 Spacer
12 Spacer

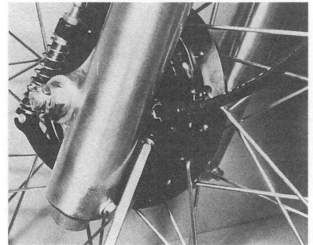

3.1 Detach the speedometer drive cable by removing the single crosshead retaining screw

3.2a Screw the handlebar cable adjuster ...

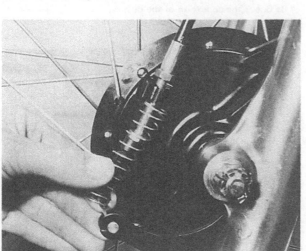

3.2b ... and the brake backplate cable adjuster fully in before detaching the cable

4 Front brake assembly: dismantling, examination, renovation and reassembly

1 The front brake assembly complete with the brake backplate can be withdrawn from the front wheel hub after removing the front wheel from the forks. With the wheel laid on a work surface, brake backplate uppermost, the brake backplate may be lifted away from the hub. It will come away quite easily, with the brake shoe assembly attached to its back.

2 Examine the condition of the brake linings. If they are thin or unevenly worn, the brake shoes should be renewed. The linings are bonded on and cannot be supplied separately. The linings are 4 mm (0.2 in) thick when new and should receive attention when worn to the wear limit thickness of 2 mm (0.1 in).

3 If fork oil or grease from the wheel bearings has badly contaminated the linings, the brake shoes should be renewed. There is no satisfactory way of degreasing the lining material.

4 Examine the drum surface for signs of scoring or oil contamination. Both of these conditions will impair braking efficiency. Remove all traces of dust, preferably using a brass wire brush, taking care not to inhale any of it, as it is of an asbestos nature, and consequently harmful. Remove oil or

grease deposits, using a petrol soaked rag.

5 If deep scoring is evident, due to the linings having worn through to the shoe at some time, the drum must be skimmed on a lathe, or renewed. Whilst there are firms who will undertake to skim a drum whilst fitted to the wheel, it should be borne in mind that excessive skimming will change the radius of the drum in relation to the brake shoes, therefore reducing the friction area until extensive bedding in has taken place. Also full adjustment of the shoes may not be possible. If in doubt about this point, the advice of one of the specialist engineering firms who undertake this work should be sought.

6 Note that it is a false economy to try to cut corners with brake components; the whole safety of both machine and rider being dependent on their good condition.

7 Removal of the brake shoes is accomplished by folding the shoes together so that they form a 'V'. With the spring tension relaxed, both shoes and springs may be removed from the brake backplate as an assembly.

8 Before fitting the brake shoes, check that the brake operating cam is working smoothly and is not binding in its pivot. The cam can be removed by withdrawing the retaining bolt on the operating arm and pulling the arm off the shaft. Before removing the arm, it is advisable to mark its position in relation to the shaft, so that it can be relocated correctly.

The shaft and arm should be already marked with a manufacturer's punch mark to indicate the correct relative positions of the two components. Lightly grease both the shaft and the faces of the operating cam and pivot prior to reassembly. Note that where a wear indicator plate is fitted over the splined end of the shaft, the plate should be aligned with the master spline on the shaft. Check the condition of the dust seal located beneath the indicator plate and renew it if considered necessary. Note that the arm clamp bolt and nut should be torque loaded to the figure given in the Specifications Section of this Chapter.

9 Before refitting existing shoes, roughen the lining surface sufficiently to break the glaze which will have formed in use. Glasspaper or emery cloth is ideal for this purpose but take care not to inhale any of the asbestos dust that may come from the lining surface.

10 Fitting the brake shoes and springs to the brake backplate is a reversal of the removal procedure. Some patience will be needed to align the assembly with the pivot and operating cam whilst still retaining the springs in position; once they are correctly aligned though, they can be pushed back into position by pressing downwards in order to snap them into position. Do not use excessive force, or there is risk of distorting the brake shoes permanently.

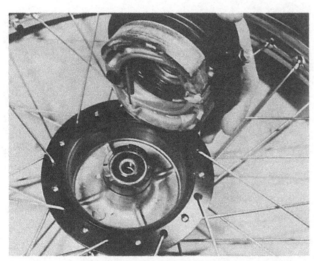

4.1 The brake backplate may be lifted away from the hub

4.8a Check the condition of the dust seal ...

4.8b ... before fitting the brake operating shaft ...

4.8c ... followed by the wear indicator plate

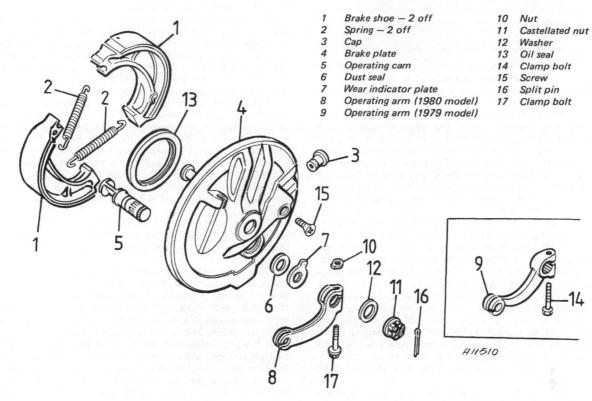

Fig. 5.3 Front brake assembly — 80 models

1	Brake shoe — 2 off	10	Nut
2	Spring — 2 off	11	Castellated nut
3	Cap	12	Washer
4	Brake plate	13	Oil seal
5	Operating cam	14	Clamp bolt
6	Dust seal	15	Screw
7	Wear indicator plate	16	Split pin
8	Operating arm (1980 model)	17	Clamp bolt
9	Operating arm (1979 model)		

H11510

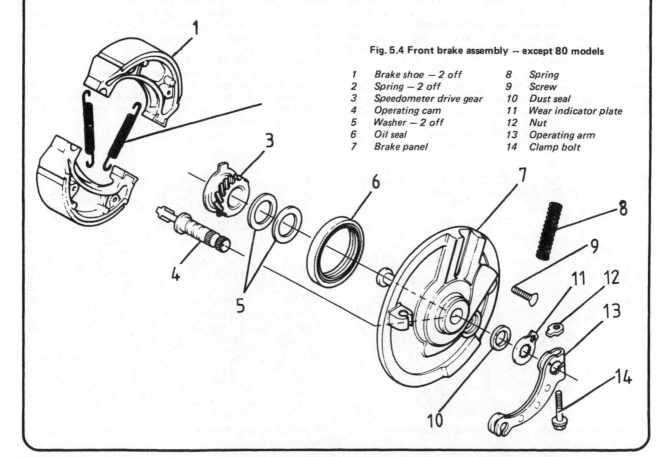

Fig. 5.4 Front brake assembly — except 80 models

1	Brake shoe — 2 off	8	Spring
2	Spring — 2 off	9	Screw
3	Speedometer drive gear	10	Dust seal
4	Operating cam	11	Wear indicator plate
5	Washer — 2 off	12	Nut
6	Oil seal	13	Operating arm
7	Brake panel	14	Clamp bolt

5 Speedometer drive gear: removal, examination and refitting

1 The speedometer drive assembly is contained within the front wheel brake backplate and should be examined and re-packed with grease whenever work is carried out on the wheel bearings or brake assembly.

2 To remove the main drive gear from its housing, place the brake backplate assembly, inner side uppermost, on a work surface. Remove the brake shoes and springs, as described in Section 4 of this Chapter, and inspect the large dust seal for signs of damage and deterioration. To renew this seal, carefully lever it out of position using the flat of a screwdriver; great care must be taken not to damage the surrounding alloy casting. The new seal may be fitted after removal, examination and fitting of the drive gear assembly.

3 The main drive gear may now be withdrawn from the brake backplate housing; some difficulty being experienced in with-drawal due to the gear being engaged with the worm drive gear and to the retentive qualities of the grease around the base of the gear. Note the position of the two thrust washers below the gear and remove them from the housing.

4 Remove all old grease from the drive gear, thrust washers, brake backplate housing and worm gear by wiping the com-ponents with a clean rag. Inspect the gears for broken teeth and signs of excessive wear due to lack of lubrication. If it is considered necessary to renew the worm drive gear, then the brake backplate assembly should be returned to an official Honda Service Agent who will be able to remove and insert a new gear assembly.

5 Refit the two thrust washers into the brake backplate housing, followed by the main drive gear. Pack the housing with the recommended grease as these components are fitted. Fit the new dust seal, if required, into its recess in the brake backplate by pressing it into position evenly and squarely. The brake shoes and springs may now be refitted and the brake backplate assembly re-inserted into the wheel hub. Ensure that the speedometer drive tabs are aligned with the corresponding slots in the wheel hub boss.

6 Front wheel bearings: removal, examination and refitting

1 There are two bearings in the front wheel. If the wheel has any side play when fitted to the machine, or any roughness, the wheel bearings need to be renewed.

2 Before bearing removal can be undertaken, the brake back plate assembly must be removed from the wheel hub and the wheel laid flat on a work surface, supported by wooden blocks so that enough clearance is left beneath the wheel to drive the bearing out. Ensure the blocks are placed as close to the bearing as possible, to lessen the risk of distortion occurring to the hub casting whilst the bearings are being removed or fitted.

3 With the brake backplate side of the wheel hub facing uppermost, place the end of a small flat-ended drift against the upper face of the lower bearing and tap the bearing down-wards out of the wheel hub. The spacer located between the two bearings may be moved sideways slightly in order to allow the drift to be positioned against the face of the bearing. Move the drift around the face of the bearing whilst drifting it out of position, so that the bearing leaves the hub squarely. The end spacer and dust seal will be driven out along with the bearing.

4 With the one bearing removed, the wheel may be lifted and the spacer withdrawn from the hub. Invert the wheel and remove the second bearing, using a similar procedure to that used for the first.

5 Wash the bearings thoroughly in clean petrol to remove all traces of the old grease. Check the bearing tracks and balls for wear or pitting or damage to the hardened surfaces. A small amount of side movement in the bearing is normal but no radial movement should be detectable. If wear or damage is found the bearing in question should be renewed.

6 If the original bearings are to be refitted, then they should be repacked with the recommended grease before being fitted into the hub. New bearings must also be packed with the recommended grease. Ensure that the bearing recesses in the hub are clean and both bearing and recess mating surfaces lightly greased. The two bearings and central spacer may now be fitted. With the hub well supported by the wooden blocks, drift the bearing into the brake backplate side of the wheel hub using a length of metal tube, or a socket, of suitable diameter and a soft-faced mallet. Invert the wheel, insert the spacer and fit the second bearing using the same procedure as that given for the first.

7 Before fitting the dust seal into its recess in the hub, inspect it for signs of damage or deterioration and renew it if considered necessary. The seal will press quite easily into its recess in the hub; ensure it is fitted with the lip facing outwards. Lightly grease the end spacer and insert it into position through the seal.

5.2 Carefully remove the seal with the flat of a screwdriver

5.5a Refit the two thrust washers into the brake backplate housing ...

5.5b ... followed by the main drive gear

6.4 With the wheel hub well supported, drift out the bearings

6.6a Insert the spacer into the wheel hub ...

6.6b ... followed by the second bearing

6.7a Press the seal into the hub recess ...

6.7b ... and lightly grease the end spacer before inserting it through the seal

7 Front wheel: refitting

1   Before refitting the wheel into the front forks, ensure that the brake backplate assembly has been inserted into the wheel hub with the speedometer drive tabs aligned with the corresponding slots in the hub boss. Check also that the end spacer is correctly located in the bearing dust seal and is not displaced during the fitting of the wheel.

2   Roll the wheel into position beneath the front forks and align the slot in the brake backplate with the spigot cast in the base of the left-hand fork. Lift the wheel so that the spigot locates correctly into the slot and insert the wheel spindle. Note that if the brake backplate is allowed to rotate, due to its not being retained to the front fork by the correct location of the spigot in the slot, the wheel will lock on the first application of the front brake, with disastrous consequences. Fit the spindle retaining nut and tighten it to the torque figure given in the Specifications Section of this Chapter. Insert a new split-pin through the spindle and nut and lock it in position by bending out both of its legs.

3   Push the speedometer drive cable end into position in the brake backplate housing, ensuring that the spigot on the inner cable locates correctly in the end of the worm drive gear, and retain it with the single crosshead retaining screw.

4   Refit the brake operating cable by inserting the adjuster through the retainer cast in the brake backplate; refit the large return spring and reconnect the cable inner to the brake lever. Reconnect the cable inner to the handlebar lever.

5   The front brake should now be correctly adjusted by turning the cable lower adjuster to give 10 - 20 mm (0.39 - 0.75 in) on XL80S models, 20 - 30 mm (0.75 - 1.25 in) on all other models, of free play at the end of the handlebar lever. Any minor adjustments necessary may then be made with the cable adjuster at the handlebar lever bracket. Lock both adjusters in position by tightening the locknuts and check the brake for correct operation by spinning the wheel and applying the brake lever. There should be no indication of the brake binding as the wheel is spun. If the brake shoes are heard to be brushing against the surface of the wheel drum, back off on the cable adjuster slightly until all indication of binding disappears. The brake may be readjusted after a period of bedding-in has been allowed for the brake shoes.

8 Rear wheel: examination

1   Position blocks underneath the sumpguard plate so that the rear wheel is raised clear of the ground.

2   Check the wheel rim alignment and look for damage to the rim or loose or broken spokes, by following the procedure adopted for the front wheel in Section 2 of this Chapter.

9 Rear wheel: removal

1   With the machine securely positioned on blocks placed underneath the engine sumpguard, so that the rear wheel is raised clear of the ground, remove the rear wheel as follows.

2   Slacken the rear brake adjuster nut and remove it to release the brake operating rod. Reassemble the brake rod components, after the wheel has been removed, to avoid loss.

3   Turn the rear wheel until the joining link in the drive chain is accessible, then, using pointed-nose pliers, prise off the spring link, and separate the chain. Loosely reassemble the joining link and store it safely to avoid loss. Lift the chain off the wheel sprocket and lay the ends on a piece of clean rag or paper placed underneath the machine to prevent any grit or dirt attaching itself to the chain.

4   Remove the split-pin which retains the torque arm nut, then remove the nut and disengage the torque arm from the brake plate.

5   Slacken off the two wheel spindle drawbolt adjusters, having made a note of their position against the graduations on the fork ends. Pull out the split-pin from the castellated wheel spindle nut, and remove the nut. The spindle may now be withdrawn and the wheel lowered clear of the frame.

10 Rear brake assembly: dismantling, examination, renovation and reassembly

1   With the rear wheel removed from the frame as described in the previous Section of this Chapter the brake backplate assembly can be pulled clear of the drum for examination. Follow the procedure given for the front brake assembly as detailed in Section 4 of this Chapter.

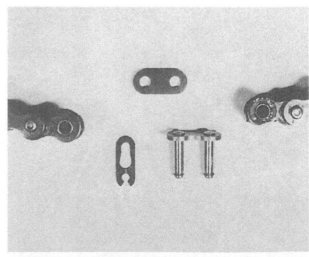

7.2 The fork leg spigot must locate correctly in the brake backplate slot

9.3 The joining link components should be stored safely to avoid loss

10.1a Note the positioning of the brake shoes and springs ...

10.1b ... and the position of the wear indicator plate

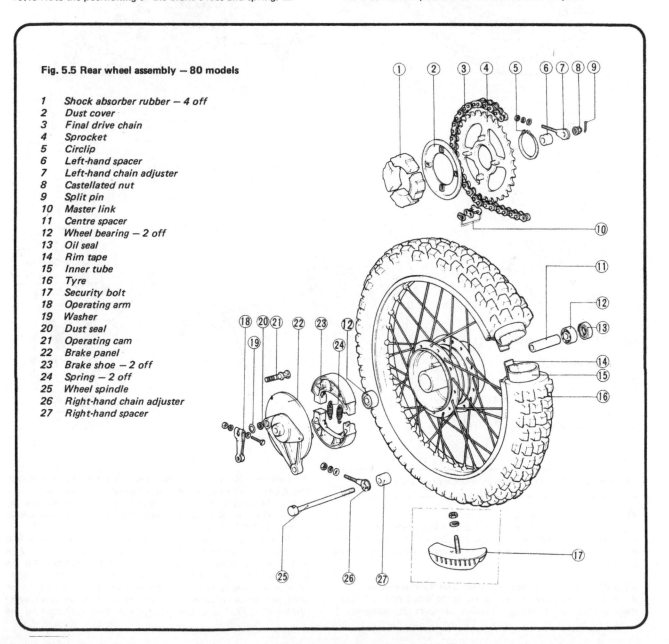

Fig. 5.5 Rear wheel assembly — 80 models

1 Shock absorber rubber — 4 off
2 Dust cover
3 Final drive chain
4 Sprocket
5 Circlip
6 Left-hand spacer
7 Left-hand chain adjuster
8 Castellated nut
9 Split pin
10 Master link
11 Centre spacer
12 Wheel bearing — 2 off
13 Oil seal
14 Rim tape
15 Inner tube
16 Tyre
17 Security bolt
18 Operating arm
19 Washer
20 Dust seal
21 Operating cam
22 Brake panel
23 Brake shoe — 2 off
24 Spring — 2 off
25 Wheel spindle
26 Right-hand chain adjuster
27 Right-hand spacer

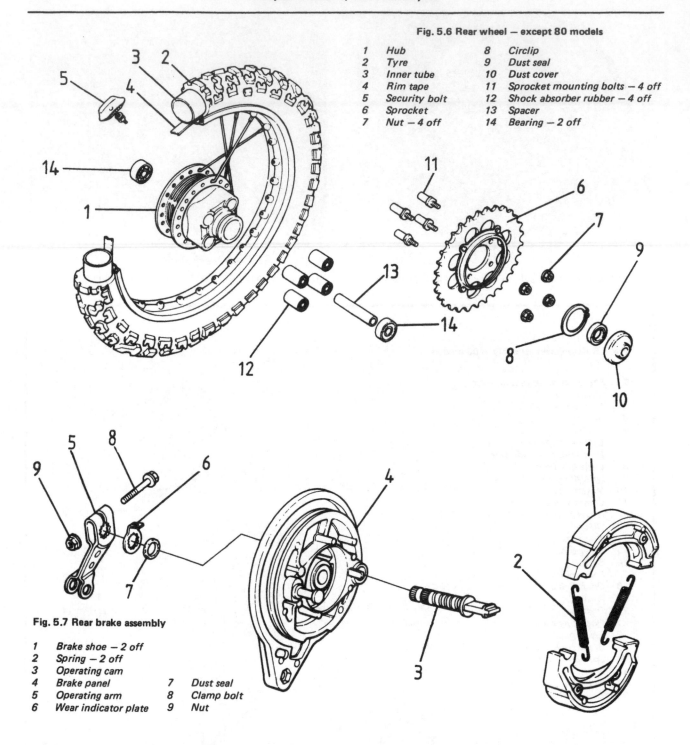

Fig. 5.6 Rear wheel — except 80 models

1	Hub	8	Circlip
2	Tyre	9	Dust seal
3	Inner tube	10	Dust cover
4	Rim tape	11	Sprocket mounting bolts — 4 off
5	Security bolt	12	Shock absorber rubber — 4 off
6	Sprocket	13	Spacer
7	Nut — 4 off	14	Bearing — 2 off

Fig. 5.7 Rear brake assembly

1	Brake shoe — 2 off		
2	Spring — 2 off		
3	Operating cam		
4	Brake panel	7	Dust seal
5	Operating arm	8	Clamp bolt
6	Wear indicator plate	9	Nut

11 Rear wheel bearings: removal, examination and refitting

XR80 and XL80 S models

1 On the type of rear wheel fitted to these machines, bearing examination, removal and refitting is carried out in a similar manner to that given for the front wheel in Section 6 of this Chapter.

XL100 S, XL125 S, XL185 S, XR185 and XR200 models

2 If the wheel bearings are found to be worn or rough, then they should be renewed. Bearing removal is, however, a difficult operation due to the fact that the spacer between the bearings cannot be moved sideways in order to allow a drift to be placed on the inner face of the bearing, thus making it impossible to drift the bearing out of position as described in Section 6 of this Chapter.

3 With the brake backplate withdrawn from the wheel hub and the sprocket removed from the wheel (see Section 12 of this Chapter), place the wheel on the work surface with the brake backplate side of the wheel uppermost. The only method of bearing removal found possible was by using a slide hammer with two outward facing feet, as shown in the accompanying photograph. With the feet of the tool firmly located in the

recess between the bearing and spacer, and the wheel held firmly down on the bench surface, pull the bearing out of position by sliding the hammer up the tool shaft so that it butts against the end of the tool with enough impact to jar the bearing out of its housing in the wheel hub. If no tool is available, it is recommended that the wheel be returned to an official Honda Service Agent who will have at his disposal the necessary tools required for bearing removal.

4 With the one bearing removed, the wheel may be inverted to allow the spacer to drop out of the wheel hub. Re-invert the wheel and support it on wooden blocks placed beneath the hub with enough clearance left between the hub and work surface to allow the second bearing to be driven out of position by using a drift placed against the bearing face and a soft-faced mallet.

5 Examination and refitting of the bearings and spacer is as listed in paragraphs 5 and 6 of Section 6 of this Chapter.

All models

6 It should be noted that the oil seal fitted above the bearing on the sprocket side of the wheel, will be drifted out of position as the bearing is removed. This seal should be examined for signs of deterioration or damage and renewed if necessary. Before fitting the seal, lightly grease its inner and outer surfaces. The seal may be pushed into position in the hub recess and tapped home by using a block of wood and a soft-faced mallet. Ensure the seal remains square whilst fitting it into the recess.

12 Rear wheel sprocket and shock absorbers: removal, examination and refitting

XR80 and XL80 S models

1 The rear wheel sprocket fitted to these machines is held onto the hub by a circlip. Once this circlip has been removed the sprocket, dust cover and shock absorber rubbers can be removed.

2 The shock absorber rubbers will remain in the wheel hub and should be checked for any damage or deterioration. All oil or grease should be wiped away as this may cause premature deterioration.

XL100 S, XL125 S, XL185 S, XR185 and XR200 models

3 The rear wheel sprocket fitted to these machines is mounted on four pins which engage in bonded rubber bushes pressed into the hub casting. It is retained by a large circlip, which, when removed, allows the sprocket to be pulled free complete with the four driving pins. On older machines, it is possible that the pins may become jammed in place in the bonded bush centres, in which case, the four nuts may be removed to allow the sprocket to be detached.

4 Should the above condition arise, or if the bushes are worn, the wheel should be taken to a Honda Service Agent who will have the equipment necessary to extract the old bushes, and fit the new ones. It is extremely unlikely that this operation can be performed at home due to the tight fit of the bushes. In all probability, any attempt to dislodge them will result in the inner metal sleeve tearing out of the rubber, making subsequent removal difficult.

All models

5 With the sprocket removed from the wheel, clean it thoroughly and examine it closely, paying particular attention to the condition of the teeth. The sprocket should be renewed if the teeth are hooked, chipped, broken or badly worn. It is considered bad practice to renew one sprocket on its own; both drive sprockets should be renewed as a pair, preferably with a new final drive chain. If this recommendation is not observed, rapid wear resulting from the running of old and new parts together will necessitate even earlier replacement on the next occasion.

6 Refit the sprocket and shock absorber assemblies to the rear wheel by using a reverse procedure to that given for removal. Ensure that the sprocket retaining circlip is correctly located in its groove.

7 On the XL100 S, XL125 S, XL185 S, XR185 and XR200 models, the four sprocket pins must be correctly aligned with the sprocket; that is, the flats on the pin must be seated against the inner boss on the sprocket and follow the line of the sprocket circumference. The pin retaining nuts should be torque loaded to the setting given in the Specifications Section of this Chapter. Do not omit to refit the metal dust cover over the circlip.

11.3 The use of a slide hammer is required to remove the first wheel bearing (all except 80)

11.6 Lightly grease the seal before fitting it into the hub recess

12.1a On 80 models, remove the circlip to release the sprocket ...

12.1b ... followed by the dust cover

12.2a Inspect the shock absorber rubbers for damage or deterioration ...

12.2b ... and renew the seal if necessary

12.3a On all except the 80 models, remove the dust cover ...

12.3b ... followed by the circlip ...

12.3c ... to allow the sprocket to be pulled free complete with the driving pins

12.3d The pins may be removed from the sprocket if necessary

13 Rear wheel: refitting

1 Position the rear wheel between the swinging arm fork ends and retain it in position by inserting the wheel spindle, with a chain adjuster bolt fitted under the spindle head (except XR200 models), from the right-hand side. Fit the second chain adjuster bolt over the left-hand spindle end and fit the spindle retaining nut, finger-tight.
2 Feed the rear brake operating rod through the brake lever trunnion and fit the adjusting nut onto the end of the rod to retain it in position. Ensure that the spring is correctly positioned between the butt flange on the operating rod and the forward facing face of the lever trunnion.
3 Refit the end of the torque arm to the brake backplate stud. Retain the arm in position by fitting and tightening the retaining nut to the recommended torque and locking it in position by fitting the split-pin. The torque arm must be properly attached to the brake backplate. Failure to ensure this will mean that on the first application of the rear brake, the wheel will lock, with disastrous consequences.
4 With the wheel moved fully forward, refit the chain over the rear wheel sprocket and connect the ends by inserting the joining link. Note that the link securing clip must be fitted with the closed end facing the direction of rotation of the chain.
5 The final drive chain tension should now be correctly adjusted as described in Section 14 of this Chapter. Tighten the wheel spindle to the torque figure given in the Specifications Section of this Chapter and fit the new split-pin to retain it in position. Finally, adjust the rear brake pedal free play as described in Section 15 of this Chapter before lowering the machine off its supporting blocks.

14 Final drive chain: examination, adjustment and lubrication

1 The final drive chain is exposed for most of its travel and has only a lightweight chainguard to protect the forward part of the upper run. No provision is made for lubricating the chain.
2 Periodically the tension will need to be adjusted, to compensate for wear. This is accomplished as follows.

XR80 and XL80 S models
3 With the rear wheel raised clear of the ground and the gear lever placed in the neutral position, depress the chain at a mid-point between the sprockets on its lower run and measure the amount of slack. If the chain tension is correct, this slack should measure 20 mm (0.75 in).

XL100 S, XL125 S and XL185 S models
4 Position the machine on its side stand and select neutral. Select a point on the chain midway between the sprockets on its lower run and move the chain up and down at this point. The amount of slack present in the chain should measure 25 - 35 mm (1.0 - 1.4 in) for XL100 S models or 30 - 40 mm (1.25 - 1.60 in) for XL125 S and XL185 S models.

XR185 models
5 Place the machine on its side stand and select neutral. Before inspecting the chain tension, check the degree of wear on the chain tensioner slider block. If wear extends to the wear line on the slider then the block should be renewed.
6 Note the chain tension gauge marks on the chainguard plate just behind and to the rear of the chain tensioner. If the upper surface of the chain aligns with the upper chain mark on the gauge, then it should be adjusted down to align with the bottom chain mark.
7 If the gauge marks have become obliterated during the life of the machine, then the following check should be made. Measure the distance between the bottom of the swinging arm fork and the top of the chain at a point on the chain run just forward of the tensioner slider block. If this distance is less than the service limit of 21 mm (0.9 in), then it should be corrected to 31 mm (1.25 in) by adjusting the chain tension.

XR200 models
8 The method of examining chain tension on this machine is similar to that given for the XR185 model, except that the chain tension gauge on the chainguard plate takes the form of two rivets. The chain should be adjusted so that the upper surface of the chain aligns with the upper edge of the lower rivet. The chain must be adjusted before its upper surface reaches the upper edge of the upper rivet. There are no alternative measurements given because the rivets are a permanent fixture.

All models
9 The method of drive chain adjustment is the same for all the model types covered in this Manual and is as follows.
10 Remove the split-pin from the wheel spindle nut and loosen the nut just enough to allow the wheel to be drawn backwards by means of the two drawbolt adjusters. Before carrying out chain adjustment, rotate the wheel so that the chain is moved to its tightest point, as a chain rarely wears evenly during service.

11 Always adjust the drawbolts an equal amount in order to preserve wheel alignment. The fork ends are clearly marked with a series of horizontal lines above or below the adjusters, to provide a simple, visual check. If desired, wheel alignment can be checked by running a plank of wood parallel to the machine, so that it touches the side of the rear tyre. If wheel alignment is correct, the plank will be equidistant from each side of the front wheel tyre, when tested on both sides of the rear wheel. It will not touch the front wheel tyre because this tyre is of smaller cross section. See the accompanying diagram.

12 On completion of chain adjustment, tighten the wheel spindle nut to the torque figure given in the Specifications Section of this Chapter and fit a new split-pin. Check that the rear wheel rotates freely and check the rear brake pedal free play.

13 Do not run the chain overtight to compensate for uneven wear. A tight chain will place excessive stresses on the gearbox and rear wheel bearings, leading to their early failure. It will also absorb a surprising amount of power.

14 After a period of running, the chain will require lubrication. Lack of oil will accelerate the rate of wear of both chain and sprockets and will lead to harsh transmission. The application of engine oil will act as a temporary expedient, but it is preferable to remove the chain and immerse it in a molten lubricant such as Linklyfe or Chainguard after it has been cleaned in a

paraffin bath. These latter lubricants achieve better penetration of the chain links and rollers and are less likely to be thrown off when the chain is in motion. A recommended alternative to using engine oil as a temporary lubricant is the aerosol type chain lubricant of which there are many makes available. This type of lubricant is very sticky and, thus, is less likely to be flung off the moving chain.

15 To check whether the chain is due for replacement, lay it lengthwise in a straight line and compress it endwise so that all the play is taken up. Anchor one end and measure the length. Now, pull the chain with one end anchored firmly, so that the chain is fully extended by the amount of play in the opposite direction. If there is a difference of more than ¼ inch per foot in the two measurements, the chain should be replaced in conjunction with the sprockets. Note that this check should be made after the chain has been washed out, but before any lubricant is applied, otherwise the lubricant may take up some of the play.

16 When replacing the chain, make sure that the spring link is seated correctly, with the closed end facing the direction of travel.

17 An equivalent British-made chain of the correct size is available from Renold Limited. When ordering a new chain always quote the size (length and width of each pitch), the number of links and the machine to which it is fitted.

13.1 Insert the wheel spindle through the chain adjuster bolt (except XR200)

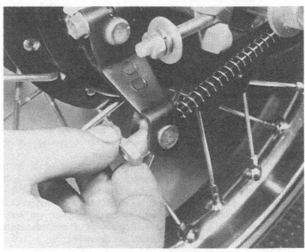

13.2 Loosely fit the brake rod adjusting nut

13.3 The torque arm retaining nut must be properly locked in position

13.4 The link securing clip must be fitted with the closed end facing the direction of travel

Tyre changing sequence - tubed tyres

 Deflate tyre. After pushing tyre beads away from rim flanges push tyre bead into well of rim at point opposite valve. Insert tyre lever adjacent to valve and work bead over edge of rim.

Use two levels to work bead over edge of rim. Note use of rim protectors

 Remove inner tube from tyre

When first bead is clear, remove tyre as shown

 When fitting, partially inflate inner tube and insert in tyre

Work first bead over rim and feed valve through hole in rim. Partially screw on retaining nut to hold valve in place.

 Check that inner tube is positioned correctly and work second bead over rim using tyre levers. Start at a point opposite valve.

Work final area of bead over rim whilst pushing valve inwards to ensure that inner tube is not trapped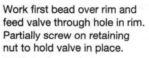

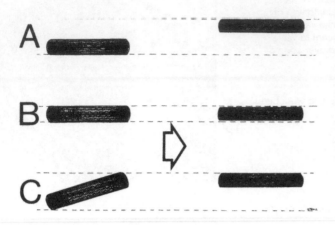

Fig. 5.8 Wheel alignment

A & C — Incorrect   B — correct

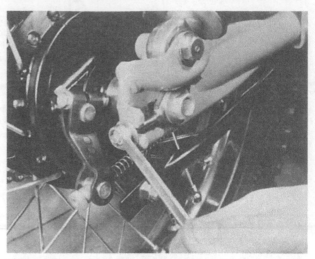

14.11a Adjust the drawbolts an equal amount to preserve wheel alignment

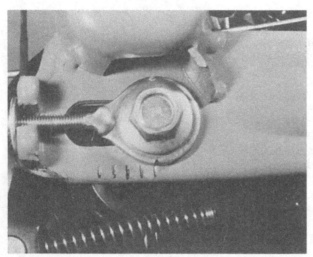

14.11b The fork ends are clearly marked to provide a simple visual check on wheel alignment

15.2 The amount of rear brake pedal travel is controlled by the adjuster nut

15 Adjusting the rear brake

1   If the adjustment of the rear brake is correct, the rear brake pedal will have 20 - 30 mm (0.75 - 1.25 in) of free play at the footplate end of the lever before the brake begins to operate.
2   The length of travel is controlled by the adjuster at the end of the brake operating rod, close to the brake operating arm. If the nut is turned clockwise, the amount of travel is reduced and vice-versa. Always check that the brake is not binding after adjustments have been made.
3   Note that it may be necessary to re-adjust the height of the stop lamp switch if the pedal height has been changed to any marked extent. The body of the switch is threaded, so that it can be raised or lowered, after the locknuts have been slackened. If the stop lamp lights too soon, the switch should be lowered and vice-versa.

16 Tyres: removal and refitting

1   At some time or other the need will arise to remove and replace the tyres, either as the result of a puncture or because

a replacement is required to offset wear. To the inexperienced, tyre changing represents a formidable task yet if a few simple rules are observed and the technique learned, the whole operation is surprisingly simple.
2   To remove the tyre from either wheel, first detach the wheel from the machine by following the procedure in Section 3 or Section 9 of this Chapter, depending on whether the front or rear wheel is involved. Deflate the tyre by removing the valve insert and when it is fully deflated, push the bead of the tyre away from the wheel rim on both sides so that the bead enters the centre well of the rim. Remove the locking cap and push the tyre valve into the tyre itself.
3   Insert a tyre lever close to the valve and lever the edge of the tyre over the outside of the wheel rim. Very little force should be necessary; if resistance is encountered it is probably due to the fact that the tyre beads have not entered the well of the wheel rim all the way round the tyre.
4   Once the tyre has been edged over the wheel rim, it is easy to work around the wheel rim so that the tyre is completely free on one side. At this stage, the inner tube can be removed.
5   Working from the other side of the wheel, ease the other edge of the tyre over the outside of the wheel rim that is furthest away. Continue to work around the rim until the tyre is free completely from the rim.

6 If a puncture has necessitated the removal of the tyre, reinflate the inner tube and immerse it in a bowl of water to trace the source of the leak. Mark its position and deflate the tube. Dry the tube and clean the area around the puncture with a petrol soaked rag. When the surface has dried, apply the rubber solution and allow this to dry before removing the backing from the patch and applying the patch to the surface.

7 It is best to use a patch of the self-vulcanising type which will form a very permanent repair. Note that it may be necessary to remove a protective covering from the top surface of the patch, after it has sealed in position. Inner tubes made from synthetic rubber may require a special type of patch and adhesive if a satisfactory bond is to be achieved.

8 Before fitting the tyre, check the inside to make sure the agent which caused the puncture is not trapped. Check also the outside of the tyre, particularly the tread area, to make sure nothing is trapped that may cause a further puncture.

9 If the inner tube has been patched on a number of past occasions, or if there is a tear or large hole, it is preferable to discard it and fit a replacement. Sudden deflation may cause an accident, particularly if it occurs with the front wheel.

10 To fit the tyre, inflate the inner tube sufficiently for it to assume a circular shape but only just. Then push it into the tyre so that it is enclosed completely. Lay the tyre on the wheel at an angle and insert the valve captive in its correct location.

11 Starting at the point furthest from the valve, push the tyre bead over the edge of the wheel rim until it is located in the central well. Continue to work around the tyre in this fashion until the whole of one side of the tyre is on the rim. It may be necessary to use a tyre lever during the final stages.

12 Make sure there is no pull on the tyre valve and again commencing with the area furthest from the valve, ease the other bead of the tyre over the edge of the rim. Finish with the area close to the valve, pushing the valve up into the tyre until the locking cap touches the rim. This will ensure the inner tube is not trapped when the last section of the bead is edged over the rim with a tyre lever.

13 Check that the inner tube is not trapped at any point. Reinflate the inner tube, and check that the tyre is seating correctly around the wheel rim. There should be a thin rib moulded around the wall of the tyre on both sides which should be equidistant from the wheel rim at all points. If the tyre is unevenly located on the rim, try bouncing the wheel when the tyre is at the recommended pressure. It is probable that one of the beads has not pulled clear of the centre well.

14 Always run the tyres at the recommended pressures and never under or over-inflate. See Specifications for recommended pressures.

15 Tyre fitting is aided by dusting the side walls, particularly in the vicinity of the beads, with a liberal coating of French chalk. Washing-up liquid can also be used to good effect, but this has the disadvantage of causing the inner surfaces of the wheel rim to rust.

16 Never fit the inner tube and tyre without the rim tape in position. If this precaution is overlooked there is a good chance of the ends of the spoke nipples chafing the inner tube and causing a crop of punctures.

17 Never fit a tyre which has a damaged tread or side walls. Apart from the legal aspects there is a very great risk of a blow-out, which can have serious consequences on any two wheel vehicle.

18 Tyre valves rarely give trouble but it is always advisable to check whether the valve itself is leaking before removing the tyre. Do not forget to fit the dust cap which forms an effective second seal.

17 Security bolt: function and fitting

1 If the drive from a high-powered engine is applied suddenly to the rear wheel of a motorcycle, wheel spin will occur with an initial tendency for the wheel rim to creep in relation to the tyre and inner tube. Under these circumstances there is risk of the valve being torn from the inner tube, causing the tyre to deflate rapidly, unless movement between the rim and tyre can be restrained in some way. A security bolt fulfils this role in a simple and effective manner, by clamping the bead of the tyre to the well of the wheel rim so that any such movement is no longer possible.

2 Only the XR80, XL185 S, XR185 and XR200 models have a single security bolt fitted as standard to the rear wheel. Before attempting to remove or fit a tyre, the security bolt must be slackened off completely so that the clamping action is released. The inside edge of the wheel rim is normally ribbed to help hold the tyre firmly in position.

18 Tyre valve dust caps

1 Tyre valve dust caps are often left off when a tyre has been fitted, despite the fact that they serve an important two-fold function. Firstly, they prevent dirt or other foreign matter from entering the valve and causing the valve to stick open when the tyre pump is next applied. Secondly, they form an effective second seal so that in the event of the tyre valve sticking, air will not be lost.

'Fault diagnosis: wheels, brakes and tyres' on page 154

19   Fault diagnosis: wheels, brakes and tyres

Symptom	Cause	Remedy
Ineffective brake	Worn brake linings	Replace.
	Foreign bodies on brake lining surface	Clean.
	Incorrect engagement of brake arm serration	Reset correctly.
	Worn brake cam	Replace.
Handlebars oscillate at low speeds	Buckle or flat in wheel rim, most likely front wheel	Check rim alignment by spinning wheel. Correct by retensioning spokes or building on new rim.
	Tyre not straight on rim	Check tyre alignment.
Machine lacks power, poor acceleration	Brakes binding	Warm brake drum provides best evidence. Re-adjust brakes.
Brakes grab when applied gently	Ends of brake shoes not chamfered	Chamfer with file.
	Elliptical brake drum	Lightly skim on lathe.
Brake pull-off spongy	Brake cam binding in housing	Free and grease.
	Weak brake shoe springs	Renew if springs have not become displaced.
Harsh transmission	Worn or badly adjusted final drive chain	Adjust or renew.
	Hooked or badly worn sprockets	Renew as a pair.
	Loose rear sprocket	Check bolts.
	Worn damper rubbers	Renew rubber inserts.

Chapter 6 Electrical system

For information relating to 1981 — 1987 models, see Chapter 7

Contents

Specifications

	XL80 S, XL100 S, XL125 S and XL185 S
Battery	
Capacity	6V, 4Ah
Electrolyte specific gravity	1.260 - 1.280 at 20°C (68°F)
Earth	Negative
Rectifier	Silicon diode
Fuse	10A rating

	XL80 S	XL100 S, XL125 S and XL185 S	XR185 and XR200
Flywheel generator			
Output	58W @ 5000 rpm	90W @ 5000 rpm	45W @ 5000 rpm

	XL80 S	XL100 S	XL125 S and XL185 S	XR185 and XR200
Bulbs				
Headlamp	6V, 15/15W	6V, 35/36.5W	6V, 35/35W	6V, 25/25W
Tail/stop lamp	6V, 5.3/25W	6V, 5.3/25W	6V, 5/21W	6V, 3W
Direction indicators	6V, 17W	6V, 17W	6V, 21W	—
Indicator warning	6V, 1.7W	6V, 1.7W	6V, 1.7W	—
Speedometer light	6V, 1.7W	6V, 3W	6V, 3W	6V, 3W
Neutral indicator	6V, 3W	6V, 3W	6V, 3W	—
High beam indicator	6V, 1.7W	6V, 1.7W	6V, 1.7W	—
Pilot lamp	—	—	6V, 4W	—

1 General description

The XR80 models are fitted with a flywheel generator containing a single power coil which provides ignition source power. No provision for lighting is made on this machine.

The XR185 and XR200 models are fitted with a flywheel generator containing two separate power coils; one to provide ignition source power and the other to provide a lighting current for the headlamp, stop lamp and speedometer light. These two coils are not interconnected in any way, and for the purposes of testing and fault isolating may be considered as separate component systems.

Electrical power on the XL80 S, XL100 S, XL125 S and XL185 S models is provided by a multi-coil flywheel generator. Alternating current (ac) provided by one of these coils is converted to direct current (dc) to enable the battery to be charged and the instrument lights and tail lamp to be fed. A silicon diode rectifier is used to convert the current. Ac is also provided, direct to the headlamp, by another of the coils (two in the four-coil set up of the XL125 S and XL185 S models). The remaining coil provides ignition source power.

2 Checking the electrical system: general

Many of the test procedures applicable to motorcycle electrical systems require the use of test equipment of the multimeter type. Although the tests themselves are quite straightforward, there is a real danger, particularly on alternator systems, of damaging certain components if wrong connections are made. It is recommended, therefore, that no attempt be made to investigate faults in the charging system, unless the owner is reasonably experienced in the field. A qualified Honda Service Agent will have in his possession the necessary diagnostic equipment to effect an economical repair.

3 Flywheel generator: checking the output - XL80 S, XL100 S, XL125 S and XL185 S models only

1 An ammeter and a voltmeter, both of the centre zero and moving coil type, are required for this test. The voltmeter should have a 0 - 10 volt range and the ammeter a range of 0 - 5 amps. Before carrying out the following check, ensure the battery is fully charged (specific gravity of the electrolyte 1.260 - 1.280 at 20°C).
2 Disconnect the battery leads at their bullet connectors. Connect the lead from the positive (+) terminal of the battery to the positive lead of the voltmeter and to the negative lead of the ammeter. Connect the lead from the negative (—) terminal of the battery to the negative lead of the voltmeter and to the green lead connector. To complete the test circuit, connect the positive lead of the ammeter to the red lead connector.
3 Start the engine and measure the battery voltage and charging current; they should correspond with the figures given in the following table:

Model	Charging start	4000 rpm	8000 rpm
XL80 S	1600 rpm (main switch on)	0.7A min/ 8.0V	2.2A max/ 9.0V
XL100 S	1800 rpm max	1.2A min/ 8.0V	2.5A max/ 9.0V
XL125 S and XL185 S	900 rpm max (lighting switch off)	2.4A min/ 8.0V	4.5A max/ 9.0V
	1800 rpm max (lighting switch on - main beam)	1.2A min/ 8.0V	3.4A max/ 9.0V

If the results of the test are not similar to the figures given in the above table, it is probable that the flywheel generator coils have failed or a loose connection has occurred between the flywheel generator and battery. Check all connections visually and clean them of dirt and corrosion if necessary. The condition of the coils in the stator may be checked, by determining whether continuity exists, by carrying out the following test procedure. Removal of the stator from the machine is not necessary.
4 Disconnect the leads from the flywheel generator at the block connector. Using a multimeter, set to the resistance function, check the resistance between the yellow and the pink leads; this should be 0.58 ohm. Check also the resistance between the white lead (XL100 S models) or white/yellow lead (XL125 S and XL185 S models) and earth; this should be 0.47 ohm. If the readings shown on the multimeter do not correspond with those given, do not consign the stator to the scrap bin but place it in the hands of an official Honda Service Agent or an auto-electrician who may have the necessary equipment and facilities for carrying out further tests and rectification. On XL80 S models no figures are available for stator testing. A test can be made, however, with the stator in place on the machine, for continuity between the black/white wire and earth, the blue/white or yellow wire and earth, and also the white wire and earth. In all cases good continuity should exist.
5 If the tests on the flywheel generator charging current and battery voltage prove to be unsatisfactory, yet the test on the stator proves satisfactory, the rectifier should be checked as described in Section 5 of this Chapter.

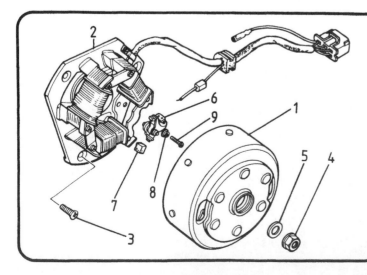

Fig. 6.1 Flywheel generator — XL100 S (80 models similar)

1 Flywheel generator rotor
2 Stator
3 Screw
4 Nut
5 Washer
6 Contact breaker assembly
7 Felt lubricating wick
8 Washer
9 Screw

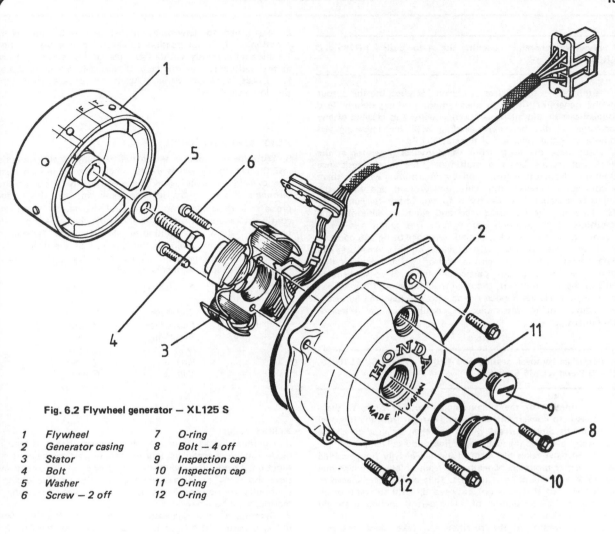

Fig. 6.2 Flywheel generator — XL125 S

1	Flywheel	7	O-ring
2	Generator casing	8	Bolt — 4 off
3	Stator	9	Inspection cap
4	Bolt	10	Inspection cap
5	Washer	11	O-ring
6	Screw — 2 off	12	O-ring

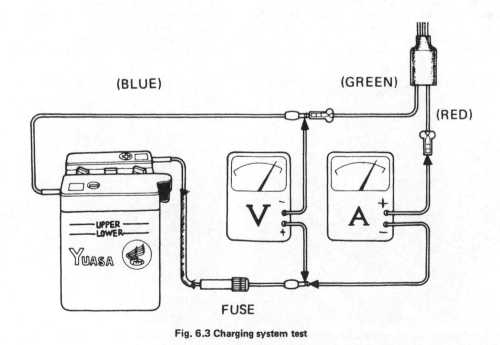

(BLUE) (GREEN) (RED)

UPPER LOWER

YUASA

FUSE

Fig. 6.3 Charging system test

4   Flywheel generator: checking the stator coils - XR185 and XR200 models

1   Although no test figures are given for checking the output of the generator fitted to these models, the generator lead connections should all be checked visually and cleaned of any corrosion or dirt before proceeding with the following test to determine the condition of the stator coils.
2   Disconnect the leads from the flywheel generator at the block connector. Using a multimeter, set to the resistance function, check the lighting coil by determining whether there is continuity between the white/yellow lead and earth; the coil is serviceable if continuity is found. Check for continuity also between the black/red lead and earth to determine the condition of the ignition source coil. The coil is serviceable if continuity is found; the specified resistance being 245 ohms.
3   If this test proves satisfactory, yet problems are still experienced within the lighting and/or ignition systems that cannot be traced to any particular individual component or fault in the wiring circuit, then the machine should be returned to an official Honda Service Agent who will have the necessary test equipment to determine the conditon of the generator component parts.

5   Rectifier: removal, testing and fitting - XL80 S, XL100 S, XL125 S and XL185 S models

1   The function of the silicon diode rectifier fitted to these machines is to convert the alternating current from the flywheel generator to direct current which can then be used to charge the battery and feed the instrument lights and tail lamp.
2   The rectifier takes the form of a small heavily finned sealed unit which is mounted beneath the frame plate on which the fuel tank rear mounting is located. It is deliberately placed in this location so that it is not exposed directly to water or oil and yet has free circulation of air to permit cooling. It should be kept clean and dry.
3   Before removal of the rectifier can take place, the seat and fuel tank must first be removed to expose the retaining bolt head. Details of seat and tank removal and refitting are given in Chapters 2 and 4 of this Manual. Remove the rectifier by disconnecting it at the block connector and unscrewing the retaining nut and bolt.
4   If a pocket multimeter is possessed, then a simple test can

be performed to determine whether or not the unit is malfunctioning. It is not possible to check whether the rectifier is functioning correctly without the use of test equipment, such as the multimeter. In this case, if the performance of the unit is suspect, a Honda Service Agent or auto-electrical expert should be consulted.

XL100 S, XL125 S and XL185 S

5   With the multimeter set to the resistance testing function, test the continuity across the terminals of the block connector as shown in the table below. Place the positive probe on those terminals shown in the left-hand column, and the negative probe on those in the right-hand column. In these tests continuity should exist on all except the pink to yellow terminals where no continuity should exist. Repeat this test but this time reversing the position of the probes. Non-continuity should exist in all of these tests. If any test is incorrect, the rectifier must be renewed.

Test probe	Positive (+)	Negative (−)
Terminal	*Red/White*	*Pink*
	Red/White	*Yellow*
	Red/White	*Green*
	Pink	*Green*
	Pink	*Yellow*
	Yellow	*Green*

XL80 S model

6   On these models only a single-diode rectifier is fitted. The diode can be checked using a multimeter, set to the resistance function. Continuity should be found only in the direction of flow shown by the arrow embossed on the rectifier casing. If continuity or non-continuity is found in both directions the rectifier must be renewed.
7   Symptoms of a damaged rectifier are constantly blowing bulbs, persistent flat batteries or persistent overcharging. Note that if the battery is connected up incorrectly, a damaged rectifier is very likely to result.
8   Refitting the rectifier is a direct reversal of the removal sequence. Check that the block connector is clean and free of corrosion before the two halves are pushed together and take great care not to overtighten the retaining bolt and nut.

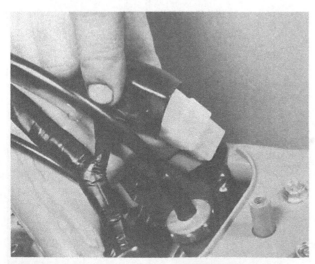

5.3a The rectifier fitted to the XL models may be removed by pulling apart the block connector halves ...

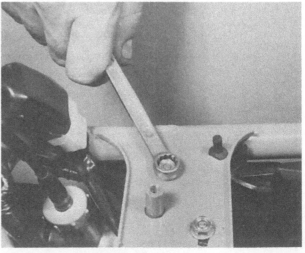

5.3b ... and unscrewing the retaining bolt

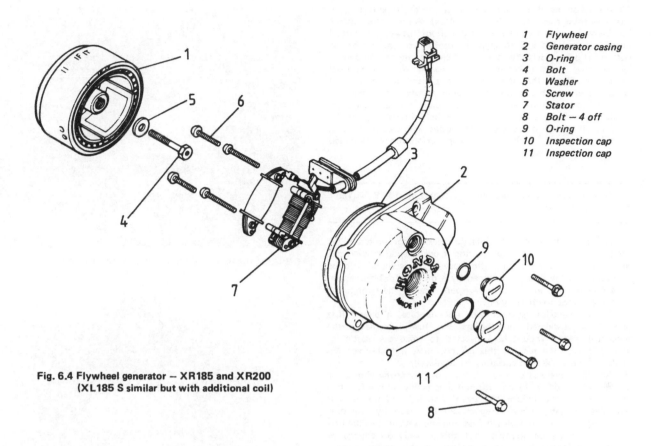

1 Flywheel
2 Generator casing
3 O-ring
4 Bolt
5 Washer
6 Screw
7 Stator
8 Bolt — 4 off
9 O-ring
10 Inspection cap
11 Inspection cap

Fig. 6.4 Flywheel generator — XR185 and XR200
(XL185 S similar but with additional coil)

6 Battery: removal, charging procedure, maintenance and refitting - XL80 S, XL100 S, XL125 S and XL185 S models

1 Whilst the machine is used on the road it is unlikely that the battery will require attention other than routine maintenance because the generator will keep it fully charged. However, if the machine is used for a succession of short journeys only, mainly during the hours of darkness when the lights are in full use, it is possible that the output from the generator may fail to keep pace with the heavy electrical demand, especially if the machine is parked with the lights switched on. Under these circumstances it will be necessary to remove the battery from time to time to have it charged independently.

2 The battery is located behind the right-hand sidepanel within a plastic cover. This cover has an inspection cut-out so that the electrolyte level may be noted without necessitating the removal of the sidepanel or cover. A padded metal bracket, hinged at one end, retains the battery in its case.

3 To remove the battery from the machine, unclip the sidepanel from its three frame attachment points, release the battery cover by unclipping its lower attachment spigot and hinging it upwards to free it from its upper locating points, disconnect the battery leads at their bullet connectors, undo the bracket securing nut to allow the bracket to be hinged clear of the battery and release the vent pipe to allow the battery to be lifted clear of its tray.

4 Maintenance is normally limited to keeping the electrolyte level between the prescribed upper and lower limits and making sure the vent tube is not blocked. The lead plates and their separators are visible through the transparent case, a further guide to the general condition of the battery.

5 Unless acid is spilt, as may occur if the machine falls over, the electrolyte should always be topped up with distilled water to restore the correct level. If acid is spilt onto any part of the machine, it should be neutralised with an alkali such as washing soda or baking powder and washed away with plenty of water, otherwise serious corrosion may occur. Top up with sulphuric acid of the correct specific gravity (1.260 to 1.280) only when spillage has occurred. Check that the vent pipe end is well clear of the frame or any of the other cycle parts.

6 If the terminals are corroded, scrape away the deposits with a sharp knife and remove the few remaining traces by wiping with a rag soaked in a strong solution of bicarbonate of soda. Do not allow any of the solution to enter the battery, since it reacts violently with the acid and may cause permanent damage to the plates. Dry off the terminals, then give them a light coating with a petroleum jelly such as Vaseline (not grease) to obviate the risk of corrosion recurring.

7 It is seldom practicable to repair a cracked battery case because the acid present in the joint will prevent the formation of an effective seal. It is always best to replace a cracked battery, especially in view of the corrosion which will be caused if the acid continues to leak.

8 If the machine is not used for a period, it is advisable to remove the battery and give it a 'refresher' charge every six weeks or so from a battery charger. If the battery is permitted to discharge completely, the plate will sulphate and render the battery useless.

9 Before charging the battery, ensure that the cap is removed from each cell so that the battery may vent whilst charging. On no account allow sparks or naked lights near the battery when it is on charge. Turn the power off and on at the charger, never at the battery terminals.

10 The recommended charging current for the type of battery fitted to these machines is 0.4 amp (max). A more rapid charge can be given in an emergency but this should be avoided if at all possible because it will shorten the useful working life of the battery. It is not advisable to exceed a charge rate of 1 ampere.

11 Refitting of the battery is a reversal of the removal procedure. Check that the vent pipe is correctly routed and is not likely to chafe on any frame components, that the electrical connections are clean, the cell caps are correctly tightened and the battery is securely held in its tray by the bracket. If in doubt when reconnecting the electrical leads, consult the wiring diagram for the appropriate model contained at the end of this Chapter.

7 Fuse: location and replacement - XL80 S, XL100 S, XL125 S and XL185 S models

1 A fuse within a moulded plastic case is incorporated in the electrical system to give protection from a sudden overload, such as may occur during a short circuit. It is found in close proximity to the battery, retained by a moulded rubber carrier. The fuse is rated at 10 amps. A spare fuse of the same rating is clipped inside the battery cover.

2 If the electrical system will not operate, a blown fuse should be suspected, but before the fuse is replaced the electrical system should be inspected to trace the reason for the failure of the fuse. If this precaution is not observed the replacement will almost certainly blow as well.

3 At least one spare fuse of the correct rating should be carried at all times. In an extreme emergency and only when the cause of the failure has been rectified and no spare is available, a get-you-home repair can be made by wrapping silver paper around the blown fuse and re-inserting it in the fuse holder. It must be stressed that this is only an emergency measure and the fuse should be replaced at the earliest possible opportunity, as it affords no protection at all to the electrical system when bridged in this fashion.

8 Headlamp: bulb renewal and beam adjustment

XL80 S, XL100 S, XL125 S and XL185 S models

1 In order to gain access to the headlamp bulbs fitted within the conventional type of headlamp unit fitted to these models, it is first necessary to remove the headlamp rim, complete with the reflector and headlamp glass. The rim is retained in position by two screws passing through the headlamp shell into lugs projecting from the rim. These screws are in the 8 o'clock and 4 o'clock positions, viewed from the front of the machine.

2 UK models are fitted with a headlamp bulb and a pilot bulb. The headlamp bulb holder is retained in position in the reflector unit by a single spring which must be detached from either the reflector unit or bulb holder before the holder and bulb can be withdrawn. To release the bulb, push it inwards slightly and twist it anti-clockwise; the bulb may now be pulled out of the holder. The pilot bulb holder is a straight push-fit in the reflector unit, the bulb having the same fitting in the holder as that of the headlamp bulb.

3 US models have a sealed beam headlamp unit with no provision for a pilot bulb. If one filament blows, the complete unit must be renewed. The unit is secured to the headlamp rim by two panhead screws, at the 12 o'clock and 6 o'clock positions. With these screws removed and the rim detached, disconnect the leads from the back of the unit at their respective terminals, noting the colour coding of the leads for reference when refitting.

4 Fitting of the bulbs/sealed beam unit is a reversal of the removal procedure. Ensure that all bulb and electrical connections are free from contamination.

6.3 Undo the battery retaining bracket securing nut

6.11 Instructions for routing the vent pipe are contained within the battery cover. Note also the spare fuse

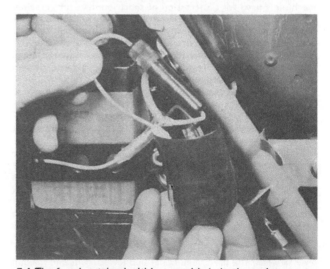

7.1 The fuse is retained within a moulded plastic carrier

8.1 On the XL models, remove the headlamp rim retaining screws ...

8.2a ... to gain access to the headlamp bulb holder ...

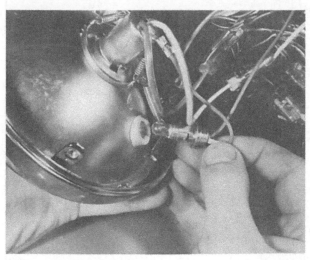

8.2b ... and pilot bulb holder

5 Beam height is effected by pivoting the headlamp about the two mounting bolts. These bolts should be loosened slightly before adjustment is carried out. Horizontal beam adjustment is provided by the adjuster screw which passes through the headlamp rim at the 9 o'clock position.

6 UK lighting regulations stipulate that the lighting system must be arranged so that the light will not dazzle a person standing in the same horizontal plane as the vehicle at a distance greater than 25 feet from the lamp, whose eye level is not less than 3 feet 6 inches above that plane. It is easy to approximate this setting by placing the machine 25 feet away from a wall, on a level road, and setting the dipped beam height so that it is concentrated at the same height as the distance from the centre of the headlamp to the ground. The rider must be seated normally during this operation and also the pillion passenger, if one is carried regularly.

7 For the US, the headlamp beam should be adjusted as specified by the State laws and regulations.

XR185 and XR200 models

8 The combined headlamp/number plate unit fitted to these machines must first be detached from the machine before bulb removal is possible. This is achieved by removing the three retaining bolts, one at the bottom centre of the plate and two at the top.

9 The headlamp unit is retained to the number plate by two securing screws, the attachment bracket being hinged to the reflector case. If damaged, the unit may be removed from the plate by unscrewing these two screws.

10 Bulb removal is achieved by unplugging the rubber-covered socket from the back of the reflector unit and then removing the bulb from the socket. Fitting the replacement bulb and refitting the headlamp/number plate unit to the machine is a reversal of the removal procedure.

11 Beam height is effected by screwing in or out the adjuster screw located directly beneath the headlamp lens. This serves to pivot the headlamp unit about its hinged mounting bracket. No provision for horizontal beam adjustment is made. The beam should be set, as far as is possible, within the laws stated in paragraphs 6 and 7 of this Section.

---

9 Tail and stop lamp: bulb renewal

XL80 S, XL100 S, XL125 S and XL185 S models

1 The combined tail and stop lamp unit fitted to these models is fitted with a double filament bulb having offset pins to prevent its unintentional reversal in the bulb holder. The lamp unit serves a two-fold purpose — to illuminate the rear of the machine and the rear number plate, and to give visual warning when the rear brake is applied.

2 To gain access to the bulb, remove the three screws holding the red plastic lens in position. The bulb is released by pressing inwards and with an anti-clockwise turning action, the bulb will now come out.

XR185 and XR200 models

3 The purpose of the tail lamp unit fitted to these models is to illuminate the rear of the machine; no facility for a stop lamp is provided.

4 To gain access to the bulb, remove the two screws that hold the red plastic lens in position. The bulb may be released from its holder by pushing it into the holder, turning it anti-clockwise and releasing it to allow it to spring clear.

All models

5 Refitting of the bulb and lens is a reversal of the removal procedure. Check the inside of the bulb holder for any signs of corrosion or moisture, ensuring that the contacts are free to move when depressed. When refitting the lens securing screws, take care not to overtighten them as it is possible that the lens may crack.

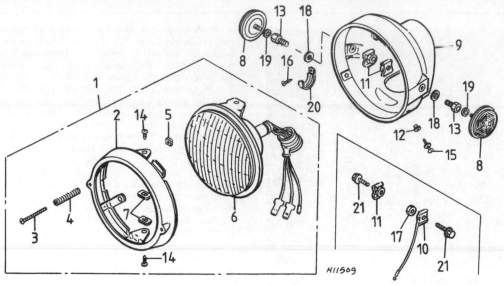

Fig. 6.5 Headlamp – XL models

1	Headlamp	8	Shell reflector – 2 off	15	Screw and washer – 2 off
2	Rim	9	Shell	16	Screw
3	Beam adjusting screw	10	Nut with attached lead	17	Nut
4	Spring	11	Nut	18	Washer – 2 off
5	Nut	12	Collar – 2 off	19	Spring washer – 2 off
6	Reflector unit	13	Screw – 2 off	20	Cable clip
7	Nut – 2 off	14	Screw – 2 off	21	Bolt – 2 off

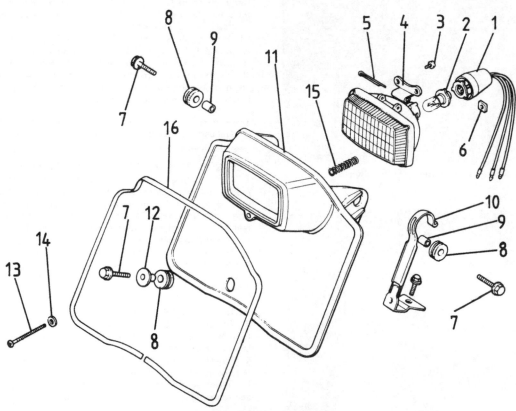

Fig. 6.6 Headlamp – XR models

1	Bulb holder	7	Bolt – 3 off	12	Shouldered spacer
2	Bulb	8	Grommet – 3 off	13	Adjustment screw
3	Screw	9	Spacer – 2 off	14	Washer
4	Headlamp unit	10	Spring clip	15	Spring
5	Split pin	11	Headlamp shroud	16	Headlamp shroud trim
6	Nut				

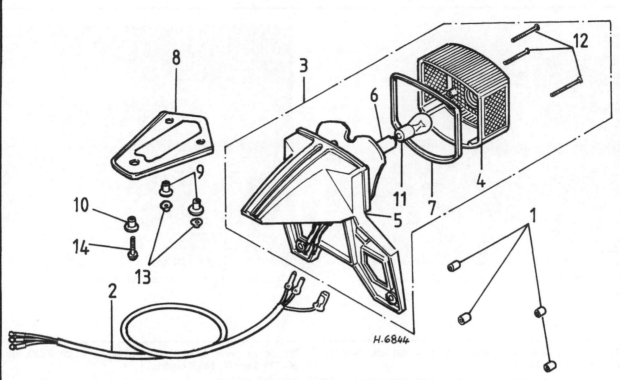

Fig 6.7 Tail and stop lamp — XL models

1	Collar — 4 off	6	Reflector unit	11	Bulb
2	Wiring lead	7	Rubber seal	12	Screw — 3 off
3	Tail lamp assembly	8	Mounting bracket	13	Nut — 2 off
4	Lens	9	Collar — 2 off	14	Bolt
5	Tail lamp body	10	Collar		

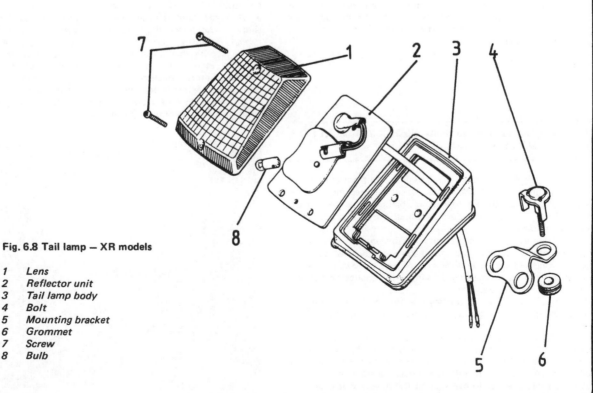

Fig. 6.8 Tail lamp — XR models

1   Lens
2   Reflector unit
3   Tail lamp body
4   Bolt
5   Mounting bracket
6   Grommet
7   Screw
8   Bulb

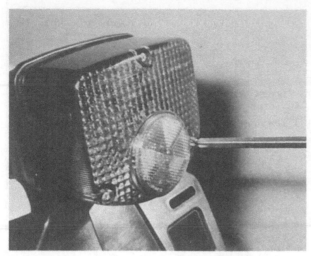

9.2a On the XL models, remove the three lens securing screws ...

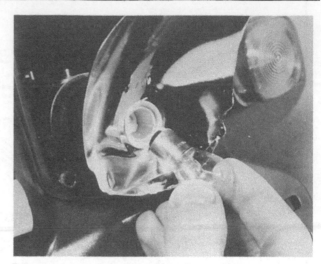

9.2b ... to gain access to the tail/stop lamp bulb

10 Instrument illumination and warning lights: bulb renewal - all models except XR80

1 The illuminating bulbs fitted to the speedometer and tachometer heads are rubber mounted and press into the base of each instrument. To remove the bulbs, pull out the rubber bulb holder and push in and twist the bulb to remove it from the holder.
2 The same method of bulb attachment is used for the warning light bulbs fitted in the separate console. Where this console is mounted between a speedometer and tachometer and forms part of a complete assembly, the top of this assembly must first be removed by unscrewing the four crosshead retaining screws so that access may be gained to the base of the console and the bulbs removed.

11 Flashing indicator lamps: bulb renewal - XL80 S, XL100 S, XL125 S and XL185 S models

1 Flashing indicator lamps are fitted to the front and rear of these machines. The front lamps may be either mounted on stalks, either side of the headlamp shell, or mounted on the handlebar. The rear lamps are mounted on stalks attached to the rear mudguard.
2 With the earlier round type of lens, the lens may be removed to give access to the bulb by unscrewing the two retaining screws. On the later rectangular type of lens, removal is achieved either by the above method or simply by unclipping the lens from the indicator case.
3 The bulbs fitted are of the bayonet type and may be released by pushing in, turning anti-clockwise and pulling out of the holder.
4 Refitting of the bulb and lens is a reversal of the removal procedure. Check the inside of the bulb holder for any signs of corrosion or moisture, ensuring that the contact is free to move when depressed. When refitting lenses retained by securing screws, take care not to overtighten the screws as it is possible to crack the lens by doing so.

12 Flasher unit: location and renewal - XL80 S, XL100 S, XL125 S and XL185 S models

1 The flasher relay unit is located underneath the fuel tank, just behind the headstock on the left-hand side. It is cylindrical in profile and retained to the frame by a rubber mounting band.
2 If the flasher unit is functioning correctly, a series of audible clicks will be heard when the indicator lamps are in action. If the unit malfunctions and all the bulbs are in working order, the usual symptom is one initial flash before the unit goes dead; it will be necessary to replace the complete unit if the fault cannot be attributed to any other cause.
3 If replacement is necessary, removal of the existing unit is simple. Disconnect the leads and release the unit from the rubber mounting band. Take care when disconnecting the leads, to note their location on the unit terminals so that they may be reconnected to the replacement unit in the same manner. Before condemning the unit, make sure a defective bulb or connection is not the cause of the trouble.
4 When fitting a new unit, or handling the existing one, take great care; it is easily damaged if dropped.

10.1 The instrument illuminating bulbs are rubber mounted

10.2 Gain access to the warning light bulbs by removing the top of the console

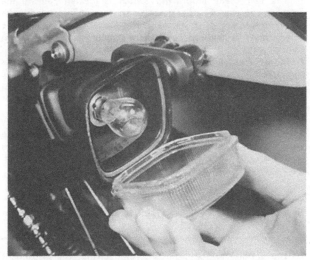

11.2 On the XL models, access to the flashing indicator bulbs may be gained simply by unclipping the lens

12.1 The flasher relay unit fitted to the XL models is located underneath the fuel tank

13 Switches: general

1 Generally speaking, the switches fitted to the machines covered in this Manual should give little trouble, but if necessary they may be tested as described in the following Sections.
2 To test the switches, a multimeter, set to the resistance function, will be required to carry out the various continuity checks described.
3 On models fitted with a battery, always disconnect the battery before removing any of the switches, to prevent the possibility of a short circuit. Most troubles are caused by dirty contacts, but in the event of the breakage of some internal part, it will be necessary to renew the complete switch.

14 Lighting and headlamp dip switch: testing - XR185 and XR200 models

1 The lighting and headlamp dip switch is incorporated in the handlebar controls and is located next to the left-hand handlebar rubber.
2 The switch may be tested by checking for continuity between the white/yellow, brown, blue and white wire terminals located behind the headlamp/number plate unit. Remove this unit, disconnect the aforementioned terminals and carry out continuity checks between them in the following sequence. Ensure that the lighting switch is moved to the ON position for all of these checks.
3 Move the dip switch to the 'Hi' position and check for continuity between the white/yellow and brown wire terminals, between the white/yellow and blue wire terminals, and between the brown and blue wire terminals. Now move the switch to the 'Lo' position and check for continuity between the white/yellow and brown wire terminals, between the white/yellow and white wire terminals, and between the brown and white wire terminals.
4 With the switch positioned mid-way between the 'Hi' and 'Lo' positions, continuity should be found between any combination of the four wire terminals.
5 If continuity is found in all of the tests, the switch is serviceable. If either one of these tests shows non-continuity, then the switch must be renewed, as any failure that occurs when changing from one beam to the other can plunge the lighting system into darkness, with disastrous consequences.

15 Lighting switch: testing - XL125 S and XL185 S models

1 The lighting switch fitted to these models is incorporated in the engine stop switch casing located next to the handlebar throttle twistgrip and may be tested for continuity by removing the headlamp reflector unit, disconnecting the black, brown/white, blue/white and white/yellow wire terminals, and checking for continuity between the terminals in the following sequence.
2 With the switch in the 'P' position, check for continuity between the black and brown/white wire terminals. Turn the switch to the 'H' position and check for continuity between the black and brown/white wire terminals and between the blue/white and white/yellow wire terminals.
3 If continuity is found in both of the tests, the switch is serviceable. If either one of these tests shows non-continuity then the switch must be renewed.

15.1 The right-hand switch unit fitted to the XL125 S model incorporates the lighting and engine stop switches ...

16 Headlamp dip switch: testing - XL80 S, XL100 S, XL125 S and XL185 S

1 The headlamp dip switch is incorporated in the switch assembly located next to the left-hand handlebar grip rubber and may be tested for continuity at the six-point block connector and blue/white terminal situated within the headlamp shell.
2 Remove the headlamp reflector unit and separate the two halves of the block connector. Disconnect the blue/white wire terminal. With the switch turned to the 'Hi' position, check for continuity between the blue and blue/white wire terminals. Turn the switch to the 'Lo' position and check for continuity between the blue/white and white wire terminals. Now move the switch to a position mid-way between the 'Hi' and 'Lo' positions and check the continuity between the blue and blue/white wire terminals, between the blue and white wire terminals, and between the blue/white and white wire terminals.
3 If continuity is found in all three of the tests, the switch is serviceable. If any one of these tests shows non-continuity, then the switch must be renewed, as any failure that occurs when changing from one beam to the other can plunge the lighting system into darkness, with disastrous consequences.

16.1 ... whereas the left-hand unit incorporates the headlamp dip and direction indicator switches as well as the horn push

17 Front brake stop lamp switch: testing - XL80 S, XL100 S, XL125 S and XL185 S models

1 The front brake stop lamp switch is incorporated in the handlebar front brake lever assembly. No means of switch adjustment is provided.
2 To test the switch for continuity, gain access to the black and green/yellow wire terminals. On the XL80 S and XL100 S models, these terminals are located behind the headlamp reflector. On the XL125 S and XL185 S models the seat and fuel tank must be removed to expose the terminals. Disconnect the terminals and check for continuity between them with the handlebar lever pulled back. If continuity exists, the switch is serviceable. If no continuity is found, the switch must be renewed as it is not possible to effect a satisfactory repair.

18 Rear brake stop lamp switch: adjustment and testing - XL80 S, XL100 S, XL125 S and XL185 S models

1 The rear brake stop lamp switch is located beneath the right-hand sidepanel and is operated by movement of the brake pedal transmitted to the base of the switch via a spring. The switch body is secured to a frame bracket by means of an adjuster nut.
2 Before testing the switch for continuity, check that it is correctly adjusted by holding the switch body and with the correct size spanner, screwing the adjuster nut up or down to raise or lower the switch thus making it operate sooner or later. Never adjust too tightly; the light should come on just as the first braking pressure is felt.
3 With the sidepanel removed, disconnect the switch wire terminals and check for continuity between the terminals with the switch fully extended. If no continuity is found, the switch must be renewed.

19 Ignition switch: testing - XL80 S, XL100 S, XL125 S and XL185 S models

1 The ignition switch is incorporated within the instrument console and may be tested for continuity at the block connector within the headlamp shell.
2 Remove the headlamp reflector unit and identify the block connector by the black, red, black/white and green wires running into it from the switch. Disconnect the terminals by pulling apart the two halves of the block. With the switch in the 'On' position, check for continuity between the black and red wire terminals. With the switch in the 'Off' position, check for continuity between the black/white and green wire terminals. If continuity exists between both of these tests, the switch is serviceable. If one or both of these tests show non-continuity, the switch should be renewed.

20 Engine stop switch: testing - all models

1 The engine stop switch is incorporated in the handlebar controls and takes the form of a single off - on switch located next to the throttle twistgrip.
2 The switch may be tested by gaining access to the black/white and green wire terminals. On the XR80 models, these terminals are located just to the rear of the front number plate. On all other models, they are located behind the headlamp reflector. Disconnect the terminals and check for continuity between them with the switch in the 'Off' position. If continuity exists, the switch is serviceable.

3 If no continuity is found, the switch must be removed from the handlebar and renewed, as it is rarely possible to effect a satisfactory repair. To remove the switch, separate the two halves which form a split clamp around the handlebar and release the clips holding the leads to the handlebar.

21 Direction indicator switch: testing - XL80 S, XL100 S, XL125 S and XL185 S models

1 The direction indicator switch is incorporated in the switch assembly located next to the left-hand handlebar grip rubber and may be tested for continuity at the six-point block connector situated within the headlamp shell.
2 Remove the headlamp reflector unit and separate the two halves of the block connector. With the switch turned to the 'R' position, check for continuity between the light blue and grey wire terminals. Turn the switch to the 'L' position and check for continuity between the grey and orange wire terminals. If continuity is found in both of these tests, then the switch is serviceable. If one or both of these tests show non-continuity, the switch should be renewed.

22 Horn push: testing - XL80 S, XL100 S, XL125 S and XL185 S models

1 The horn push button is incorporated in the switch assembly located next to the left-hand handlebar grip rubber and may be tested for continuity at the six-point block connector situated within the headlamp shell.
2 Remove the headlamp reflector unit and separate the two halves of the block connector. With the button pushed fully in, check for continuity between the black and light green wire terminals. If continuity exists the switch is serviceable. If no continuity is found, the switch must be renewed.

23 Horn: adjustment - XL80 S, XL100 S, XL125 S and XL185 S models

1 The horn is situated just below and to the right of the head-lamp, being secured to the headlamp and instrument retaining bracket by a single flange-headed bolt.
2 A volume adjustment screw and locknut are incorporated into the centre of the rear of the horn and may need to be reset from time to time to compensate for wear inside the horn.

3 To adjust the horn, slacken the locknut, depress the horn button (with the ignition on) and screw the adjuster in or out to obtain maximum horn volume. Re-tighten the locknut when the adjustment is correct.

23.3 Horn adjustment on the XL models is carried out on the centre screw

24 Wiring: layout and examination

1 The cables of the wiring harness are colour-coded and will correspond with the accompanying wiring diagram.
2 Visual inspection will show whether any breaks or frayed outer coverings are giving rise to short circuits which will cause the main fuse to blow. Another source of trouble is the snap connectors and spade terminals, which may make a poor connection if they are not pushed home fully, or if corrosion has occurred.
3 Intermittent short circuits can sometimes be traced to a chafed wire passing through, or close to, a metal component, such as a frame member. Avoid tight bends in the cables or situations where the cables can be trapped or stretched, especially in the vicinity of the handlebars or steering head.

25 Fault diagnosis: electrical system

Symptom	Cause	Remedy
Complete electrical failure	Blown fuse	Check wiring and electrical components for short circuit before fitting new 10 amp fuse. Check battery connections, also whether connections show signs of corrosion.
Dim tail lamp and instrument lights, horn inoperative	Discharged battery	Re-charge battery with battery charger. Check whether generator is giving correct output. Check rectifier.
Constantly blowing bulbs	Vibration, poor earth connection	Check security of bulb holders. Check earth return connections.
	Damaged rectifier	Renew rectifier.
Persistent overcharging of battery	Damaged rectifier	Test and renew.

Right-hand view of the XL125R-C

Right-hand view of the XL125R-F

Chapter 7 The 1981 to 1987 models

Contents

Specifications

Except where entered overleaf, specifications for the models covered in this Chapter are the same as those given for the earlier models at the beginning of each Chapter. For information on XR80R models refer to that given for the XR80, for XR100, XR100R models refer to that given for the XL100S, for XL125R models refer to that given for the XL125S and for the XR200R refer to that given for the XR200. For the XL200R, refer either to XL185S or XR200 information as shown. Note that UK models are identified by their suffix letter, US models by their production year.

Note: *The 1984 and 1985 XR200R models are powered by a new 199cc RFVC (4-valve) engine/gearbox unit and are not covered in this Manual*

Model dimensions and weights

	XL80S 1983 on	XR80 1983 on	XR80R	XL100S-B
Overall length	1723 mm (67.8 in)	1695 mm (66.7 in)	1730 mm (68.1 in)	1920 mm (75.6 in)
Overall width	755 mm (29.7 in)	750 mm (29.5 in)	755 mm (29.7 in)	805 mm (31.7 in)
Overall height	970 mm (38.2 in)	1010 mm (39.8 in)	955 mm (37.6 in)	1060 mm (41.7 in)
Wheelbase	1135 mm (44.7 in)	1140 mm (45.0 in)	1195 mm (47.1 in)	1310 mm (51.6 in)
Ground clearance	195 mm (7.7 in)	195 mm (7.7 in)	210 mm (8.3 in)	255 mm (10.0 in)
Seat height	720 mm (28.3 in)	725 mm (28.5 in)	725 mm (28.5 in)	780 mm (30.7 in)
Dry weight	72 kg (159 lb)	66.5 kg (147 lb)	64 kg (141 lb)	80 kg (176 lb)

	XL100S '81 '82	XL100S '83, '84, '85	XR100 '81, '82	XR100 '83, '84
Overall length	1920 mm (75.6 in)	1893 mm (74.5 in)	1820 mm (71.7 in)	1830 mm (72.1 in)
Overall width	805 mm (31.7 in)	805 mm (31.7 in)	805 mm (31.7 in)	805 mm (31.7 in)
Overall height	1060 mm (41.7 in)	1060 mm (41.7 in)	1050 mm (41.3 in)	1050 mm (41.3 in)
Wheelbase	1225 mm (48.2 in)	1225 mm (48.2 in)	1225 mm (48.2 in)	1225 mm (48.2 in)
Ground clearance	255 mm (10.0 in)	315 mm (12.4 in)	255 mm (10.0 in)	255 mm (10.0 in)
Seat height	800 mm (31.5 in)	800 mm (31.5 in)	800 mm (31.5 in)	800mm (31.5 in)
Dry weight	80 kg (176 lb)	81 kg (179 lb)	73.5 kg (162 lb)	74 kg (163 lb)

	XR100 ,	XL125S-B,C	XL125R-C	XL125R-F
Overall length	1855 mm (73.0 in)	2080 mm (81.9 in)	2045 mm (80.5 in)	2045 mm (80.5 in)
Overall width	800 mm (31.5 in)	840 mm (33.1 in)	840 mm (33.1 in)	840 mm (33.1 in)
Overall height	1030 mm (40.6 in)	1110 mm (43.7 in)	1185 mm (46.7 in)	1185 mm (46.7 in)
Wheelbase	1255 mm (49.4 in)	1310 mm (51.6 in)	1355 mm (53.4 in)	1350 mm (53.2 in)
Ground clearance	265 mm (10.4 in)	270 mm (10.6 in)	265 mm (10.4 in)	265 mm (10.4 in)
Seat height	770 mm (30.3 in)	820 mm (32.3 in)	840 mm (33.1 in)	840 mm (33.1 in)
Dry weight	68 kg (150 lb)	106 kg (234 lb)	106 kg (234 lb)	108 kg (238 lb)

	XL185S-B	XL200R	XR200R '81, '82, '83 XR200R-C	XR200R '86, '87
Overall length	2080 mm (81.9 in)	2112 mm (83.2 in)	2080 mm (81.9 in)	2035 mm (80.1 in)
Overall width	855 mm (33.7 in)	844 mm (33.2 in)	815 mm (32.1 in)	865 mm (34.1 in)
Overall height	1110 mm (43.7 in)	1143 mm (45.0 in)	1210 mm (47.6 in)	1180 mm (46.5 in)
Wheelbase	1310 mm (51.6 in)	1373 mm (54.1 in)	1355 mm (53.4 in)	1360 mm (53.5 in)
Ground clearance	265 mm (10.6 in)	273 mm (10.8 in)	340 mm (13.4 in)	305 mm (12.0 in)
Seat height	820 mm (32.3 in)	851 mm (33.5 in)	890 mm (35.0 in)	885 mm (34.8 in)
Dry weight	107.5 kg (237 lb)	106 kg (234 lb)	101 kg (223 lb)	98 kg (216 lb)

Specifications relating to Chapter 1

XL200R .... Refer to XR200 information with the following exceptions

Engine
Maximum horsepower:
XR100 .... 9.34 bhp @ 9500 rpm
XL125R .... 13.6 ps @ 9000 rpm
XL200R .... 16.4 ps @ 8000 rpm
Maximum torque:
XR100 .... 0.71 kg m @ 7500 rpm
XL125R .... 1.20 kg m @ 7500 rpm
XL200R .... 1.59 kg m @ 7000 rpm
Compression ratio – XL200R 9.2:1

Valves
XR80R, XR100R .... as XL100S

Valve springs
XR80R, XR100R .... as XL100S

Valve timing – XL125R
Inlet valve opens .... 5° BTDC
Exhaust valve opens .... 35° BBDC
Inlet valve closes .... 30° ABDC
Exhaust valve closes .... 5° ATDC

Camshaft

	XR80R	XR100R
Cam height:		
Inlet:	28.017 – 28.197 mm	27.840 – 28.020 mm
	(1.1030 – 1.1101 in)	(1.0961 – 1.1032 in)
Service limit	27.950 mm (1.1004 in)	27.800 mm (1.0945 in)
Exhaust	27.835 – 28.015 mm	27.776 – 27.950 mm
	(1.0959 – 1.1029 in)	(1.0935 – 1.1004 in)
Service limit	27.750 mm (1.0925 in)	27.700 mm (1.0906 in)
Journal OD – XR80R	as XL100S	
Camshaft holder ID – XR80R	as XL100S	

Rockers
XR80R, XR100R .... as XL100S

Crankshaft – XR80R, XR100R
Runout tolerances .... as XR80
Big-end clearances .... as XL100S

Clutch

	All later 100 models	All later 185 models
Spring free length:		
Standard	26.1 mm (1.03 in)	37.9 mm (1.49 in)
Service limit	24.6 mm (0.97 in)	34.7 mm (1.37 in)
Friction plate thickness – XR80R, XR100R	as XR80	

Gearbox
Gear ratios – XL200R .... as XL185S
Final reduction ratio:
XL100S – all models .... 2.800 : 1 (42/15T)
XR100, XR100R – all models .... 3.521 : 1 (50/14T)
XL125S-B, C .... 3.357 : 1 (47/14T)
XL125R .... 3.533 : 1 (53/13T)
XL200R .... 3.461 : 1 (45/13T)
XR200R – 1986, 1987 .... 3.615 : 1 (47/13T)
Selector fork dimensions – XR80R, XR100R .... as XL100S
Selector drum groove width – XR80R, XR100R .... 7.05 – 7.15 mm (0.2776 – 0.2815 in)
Service limit .... 7.16 mm (0.2819 in)
Kickstart assembly – XR80R, XR100R .... as XL100S
Shaft and gear pinion dimensions – XR80R, XR100R:
Mainshaft OD – at bearing surfaces of rotating gears .. 16.966 – 16.984 mm (0.6680 – 0.6687 in)
Service limit .... 16.950 mm (0.6673 in)
Mainshaft 4th and 5th gear pinion ID .... 17.016 – 17.034 mm (0.6699 – 0.6706 in)
Service limit .... 17.050 mm (0.6713 in)
Layshaft OD – at left-hand end .... 16.966 – 16.984 mm (0.6680 – 0.6687 in)
Service limit .... 16.950 mm (0.6673 in)
Layshaft OD – at centre splined section .... 21.959 – 21.980 mm (0.8645 – 0.8654 in)
Service limit .... 21.930 mm (0.8634 in)
Layshaft OD at right-hand end .... 17.966 – 17.980 mm (0.7073 – 0.7079 in)
Service limit .... 17.950 mm (0.7067 in)
Layshaft 1st gear pinion ID – XR80R .... 17.022 – 17.043 mm (0.6702 – 0.6710 in)

Service limit	17.060 mm (0.6717 in)
Layshaft 1st gear pinion ID – XR100R	20.622 – 20.643 mm (0.8119 – 0.8127 in)
Service limit	20.660 mm (0.8134 in)
Layshaft 2nd gear pinion ID	19.520 – 19.541 mm (0.7685 – 0.7693 in)
Service limit	19.560 mm (0.7701 in)
Layshaft 3rd gear pinion ID – XR80R	17.016 – 17.034 mm (0.6699 – 0.6706 in)
Service limit	17.050 mm (0.6713 in)
Layshaft 3rd gear pinion ID – XR100R	18.016 – 18.034 mm (0.7093 – 0.7100 in)
Service limit	18.050 mm (0.7106 in)

Shaft and gear dimensions – XR200R – 1986, 1987:

Mainshaft OD – at bearing surfaces of rotating gears ..	19.959 – 19.980 mm (0.7858 – 0.7866 in)
Service limit	19.900 mm (0.7835 in)
Mainshaft kickstart driven gear ID	20.020 – 20.041 mm (0.7882 – 0.7890 in)
Service limit	20.070 mm (0.7902 in)
Mainshaft 5th and 6th gear pinion ID	23.020 – 23.041 mm (0.9063 – 0.9071 in)
Service limit	23.080 mm (0.9087 in)
Mainshaft 5th and 6th gear pinion bush ID	20.020 – 20.041 mm (0.7882 – 7890 in)
Service limit	20.070 mm (0.7902 in)
Mainshaft 5th gear pinion bush OD	22.959 – 22.980 mm (0.9039 – 0.9047 in)
Service limit	22.920 mm (0.9024 in)
Mainshaft 6th gear pinion bush OD	22.984 – 23.005 mm (0.9049 – 0.9057 in)
Service limit	22.940 mm (0.9032 in)
Layshaft OD – at 2nd gear pinion	19.974 – 19.987 mm (0.7864 – 0.7869 in)
Service limit	19.940 mm (0.7850 in)
Layshaft OD – at 3rd and 4th gear pinions	21.959 – 21.980 mm (0.8645 – 0.8654 in)
Service limit	21.900 mm (0.8622 in)
Layshaft OD – at 1st gear and kickstart idler gear	16.466 – 16.484 mm (0.6483 – 0.6490 in)
Service limit	16.410 mm (0.6461 in)
Layshaft kickstart idler gear ID	20.020 – 20.041 mm (0.7882 – 0.7890 in)
Service limit	20.070 mm (0.7902 in)
Layshaft kickstart idler gear bush ID	16.516 – 16.534 mm (0.6502 – 0.6509 in)
Service limit	16.560 mm (0.6520 in)
Layshaft kickstart idler gear bush OD	19.989 – 22.000 mm (0.7870 – 0.8661 in)
Service limit	19.960 mm (0.7858 in)
Layshaft 1st gear pinion ID	16.516 – 16.534 mm (0.6502 – 0.6509 in)
Service limit	16.560 mm (0.6520 in)
Layshaft 2nd gear pinion ID	23.020 – 23.041 mm (0.9063 – 0.9071 in)
Service limit	23.080 mm (0.9087 in)
Layshaft 2nd gear pinion bush ID	20.020 – 20.041 mm (0.7882 – 0.7890 in)
Service limit	20.070 mm (0.7902 in)
Layshaft 2nd gear pinion bush OD	22.984 – 23.005 mm (0.9049 – 0.9057 in)
Service limit	22.940 mm (0.9032 in)
Layshaft 3rd and 4th gear pinion ID	22.020 – 22.041 mm (0.8669 – 0.8678 in)
Service limit	22.080 mm (0.8693 in)

Kickstart
XR80R, XR100R	as XL100S

Torque wrench settings

Component	kgf m	lbf ft
Cylinder head 8 mm nuts – XL125R, all XL/XR200R models	2.8 – 3.0	20 – 22
Generator rotor nut or bolt:		
XL80S 1985, all XL100S, XR100 models	5.3 – 6.0	38 – 43
XR80R, XR100R	6.0 – 7.0	43 – 50.5
XL125R, all XL/XR200R models	4.5 – 5.5	33 – 40
Cam chain tensioner locknut:		
XL125R	0.8 – 1.2	6 – 9
XL/XR200R models	1.5 – 2.2	11 – 16
Engine mounting bolts:		
XR80R, XR100R – front mounting bolts	2.4 – 3.0	17 – 22
XR80R, XR100R – rear mounting bolts	3.0 – 4.0	22 – 29
8 mm – XR200R – 1986, 1987	3.0 – 4.0	22 – 29
8 mm – XL125R, all other XL/XR200R models	3.1 – 3.7	22.5 – 27
10 mm – XL125R, XL200R, XR200R 1986, 1987	5.5 – 6.5	40 – 47
10 mm – XR200R – 1981, 1982, 1983, XR200R-C ...	4.5 – 6.0	33 – 43
Kickstart lever clamp bolt:		
XR80R, XR100R	1.0 – 1.4	7 – 10
XR200R – 1986, 1987	2.0 – 3.5	14.5 – 25

Specifications relating to Chapter 2

Fuel tank

	Overall capacity	Reserve capacity
XL80S and XR80 – 1981, 1982 models	4.5 lit (1.19 US gal)	0.8 lit (0.21 US gal)

XL80S and XR80 – 1983 on models	6.0 lit (1.58 US gal)	1.4 lit (0.37 US gal)
XL100S-B, XL100S and XR100 – 1981 models	4.5 lit (0.99/1.19 Imp/US gal)	1.2 lit (0.26/0.32 Imp/US gal)
XL100S and XR100 – 1982 models	4.5 lit (1.19 US gal)	0.8 lit (0.21 US gal)
XL100S and XR100 – 1983 on models	6.0 lit (1.58 US gal)	1.4 lit (0.37 US gal)
XR80R, XR100R	6.5 lit (1.72 US gal)	0.9 lit (0.24 US gal)
XL125R, XL200R	8.0 lit (1.76/2.11 Imp/US gal)	1.4 lit (0.31/0.37 Imp/US gal)
XR200R – 1981 model	7.5 lit (1.98 US gal)	1.5 lit (0.40 US gal)
XR200R-C, XR200R – 1982, 1983 models	8.0 lit (1.76/2.11 Imp/US gal)	2.0 lit (0.44/0.53 Imp/US gal)
XR200R – 1986, 1987 models	9.0 lit (2.38 US gal)	1.5 lit (0.40 US gal)

Carburettor

ID number:

XL80S – 1981, 1982 models	PF65A-B
XL80S – 1983 on models	PF66A-A
XR80	PF10C-A
XR80R – 1985 model	PC10D-A
XR80R – 1986 model	PC20A-A
XR80R – 1987 model	PC20B-A
XL100S-B	PD38A-A
XL100S – 1981 model	PD37A-A
XL100S – 1982 model	PD37A-B
XL100S – 1983 on models	PD82A-A
XR100 – 1981 model	PD36A-A
XR100 – 1982 on models	PD36A-B
XR100R – 1985, 1986 models	PD36B-A
XR100R – 1987 model	PD80C-A
XL125S-B,C	PD21A-B
XL125S – 1981 model	PD13A-C
XL125S – 1982, 1984 models	PD13A-D
XL125R	PD52A-A
XL185S-B	PD15B-B
XL185S – 1981, 1982 models	PD13B-C
XL185S – 1983 model	PD13D-A
XL200R	PD61A-A
XR200 – 1981, 1982 models	PD32A-B
XR200 – 1983 on models	PD32A-E
XR200R – 1981, 1982, 1983 models, XR200R-C	PD28A-A
XR200R – 1986 model	PD97A-A
XR200R – 1987 model	PD97A-B

Venturi diameter – XL200R	24 mm

Main jet:

XR80R – 1985, 1986 models	92
XR80R – 1987 model	95
XL100S	98 (95 optional)
XR100	95
XR100R – 1985 model	100
XR100R – 1986, 1987 models	95
XL125R	102
XL200R	100
XR200R – 1981, 1982, 1983 models, XR200R-C	138
XR200R – 1986 model	112
XR200R – 1987 model	110

Pilot (slow) jet:

XL80S – 1983 on models	42
XR80, XR80R	35
XL100S and XR100 – 1982 on models, XR100R, XL125S – 1982 on models, XL125R	38
XL185S – 1983 on models	35
XL200R	40
XR200 – 1983 on models	35
XR200R – 1986, 1987 models	38

Pilot screw setting (turns out from fully screwed in):

XL80S – 1982, 1983, 1984 models	3
XL80S – 1985 model	$3\frac{3}{4}$
XR80R – 1985, 1986 models	$1\frac{1}{2}$
XR80R – 1987 model	$1\frac{3}{4}$
XL100S – 1981 model	$2\frac{5}{8}$
XL100S – 1982 on models, XL100S-B, XR100	$2\frac{1}{2}$
XR100R – 1985, 1986 models	$2\frac{5}{8}$
XR100R – 1987 model	$1\frac{3}{4}$
XL125S-B,C	$2\frac{1}{2}$
XL125R	$1\frac{5}{8}$
XL185S-B	$2\frac{1}{4}$

XL185S – 1981 on models	2
XL200R	$1\frac{1}{4}$
XR200R – 1981, 1982, 1983 models, XR200R-C	$2\frac{1}{2}$
XR200R – 1986, 1987 models	$1\frac{1}{8}$

Note – US XL models only: pilot screw setting is for initial assembly only and is preset by the factory – should not be altered except by qualified personnel

Needle clip position – grooves from top:

XR80R – 1987 models	2nd
XR100R	3rd
XR200R – 1981, 1982, 1983 models, XR200R-C	4th
XR200R – 1986, 1987 models	2nd
Float level – XL200R	14.0 mm (0.55 in)

Idle speed:

XL100S-B, XR100R	1400 ± 100 rpm
XL200R	1300 ± 100 rpm

Oil pump

XL200R	as XL185S

Specifications relating to Chapter 3

Flywheel generator

Output – @ 5000 rpm:

XR80R, XR100R	15W
XL125R	185W
XL200R	196W
XR200R – 1981, 1982, 1983 models, XR200R-C	50W
XR200R – 1986, 1987 models	108W

Ignition timing – dynamic

Initial:

XL80S – 1985 model, XR80R, XR100 – 1981 model	15° BTDC @ 1800 ± 100 rpm
XR100 – 1982 on models, XR100R	13° BTDC @ 2050 ± 150 rpm
XL200R	as XR200
XR200R – 1986, 1987 models	10° BTDC @ 1300 ± 100 rpm

Full advance:

XL80S – 1985 model	30° BTDC @ 4000 rpm
XR80R	30° BTDC @ 3400 rpm
XR100 – 1981 model	31.5 – 34.5° BTDC @ 4000 ± 200 rpm
XR100 – 1982 on models	26.5 – 29.5° BTDC @ 3650 ± 150 rpm
XR100R	28° BTDC @ 3650 rpm
XL200R	as XR200
XR200R – 1986, 1987 models	30° BTDC @ 3500 ± 150 rpm

Pulser generator coil resistance

All XL/XR200R models	30 – 200 ohm

Spark plug

Type:

All XL/XR80 and 100 models – 1982 on	NGK CR7HS or ND U22FSR-U
All XL/XR125, 185 and 200 models – 1982 on	NGK DR8ES-L or NDX24ESR-U

Ignition HT coil

Primary winding resistance:

XL80S – 1983 on models	1.0 – 1.2 ohm
XR80R, XR100R	1.3 – 1.7 ohm
XL125S and XR200 – 1984 models, XR200R – 1986, 1987 models	0.16 – 0.20 ohm

Secondary winding resistance:

XL80S – 1983 on models	5.0 – 5.5 K ohm
XR80R, XR100R	7.0 – 10.6 K ohm
XL125S and XR200 – 1984 models, XR200R – 1986, 1987 models	3.7 – 4.5 K ohm

Specifications relating to Chapter 4

Front forks

Travel:

All XL/XR100 models	126 mm (5.0 in)
XR80R, XR100R	140 mm (5.5 in)
XL125R	204 mm (8.1 in)
XL200R	220 mm (8.7 in)

XR200R – 1981, 1982, 1983 models, XR200R-C .... 249 mm (9.8 in)
XR200R – 1986, 1987 models .... 254 mm (10.0 in)

Spring free length:

	Standard	Service limit
XL80S and XR80 models – 1983 on	468.7 mm (18.45 in)	454.7 mm (17.90 in)
XR80R	525.2 mm (20.68 in)	514.7 mm (20.26 in)
XL100S and XR100 – 1981 models, XL100S-B	472.0 mm (18.58 in)	462.0 mm (18.19 in)
XL100S and XR100 – 1982 models	507.0 mm (19.96 in)	497.0 mm (19.57 in)
XL100S and XR100 – 1983 on models	526.5 mm (20.73 in)	510.8 mm (20.11 in)
XR100R	566.0 mm (22.28 in)	554.7 mm (21.84 in)
XL125R-C	549.1 mm (21.62 in)	538.1 mm (21.19 in)
XL125R-F	609.2 mm (23.98 in)	603.1 mm (23.74 in)
All later XL125S models, XL185S-B, XL185S – 1981 model – top spring	68.8 mm (2.71 in)	67.4 mm (2.65 in)
All later XL125S models, XL185S-B, XL185S – 1981 model – main spring	497.4 mm (19.58 in)	487.5 mm (19.19 in)
XL185S and XR200 – 1982 on models	565.5 mm (22.26 in)	554.2 mm (21.82 in)
XL200R – top spring	63.3 mm (2.49 in)	62.0 mm (2.44 in)
XL200R – main spring	539.8 mm (21.25 in)	529.0 mm (20.83 in)
XR200R – 1981 model	575.1 mm (22.64 in)	563.6 mm (22.19 in)
XR200R – 1982, 1983 models, XR200R-C	N/Av	N/Av
XR200R – 1986, 1987 models	596.0 mm (23.46 in)	590.0 mm (23.23 in)

Lower leg top bush ID – XL185S – 1982 on models .... 31.03 – 31.08 mm (1.222 – 1.224 in)
Service limit .... 31.43 mm (1.237 in)
Fork stanchion maximum runout .... 0.2 mm (0.008 in)
Fork oil capacity – per leg:
   XL80S and XR80 – 1983 on models .... 71 ± 2.5 cc (2.4 ± 0.1 US oz)
   XR80R .... 83 cc (2.8 US oz)
   XL100S-B, XL100S and XR100 – 1981 models .... 72 cc (2.5 Imp fl oz, 2.4 US oz)
   XL100S and XR100 – 1982 models .... 82 cc (2.8 US oz)
   XL100S and XR100 – 1983 on models .... 76 cc (2.6 US oz)
   XR100R .... 88 cc (3.0 US oz)
   XL125R-C .... 226 cc (8.0 Imp fl oz)
   XL125R-F .... 330 cc (11.6 Imp fl oz)
   All XL125S models, XL185S-B, XL185S – 1981 model ...... 160 cc (5.6 Imp fl oz, 5.4 US oz)
   XL185S – 1982 on models .... 165 cc (5.6 US oz)
   XR200 – 1982 on models .... 165 – 170 cc (5.6 – 5.8 US oz)
   XL200R .... 262 ± 2.5 cc (8.9 ± 0.1 US oz)
   XR200R – 1981 models .... 271 cc (9.2 US oz)
   XR200R-C, XR200R – 1982, 1983 models .... 320 cc (11.3 Imp fl oz, 10.8 US oz)
   XR200R – 1986, 1987 models .... 350 cc (11.8 US oz)
Oil level:
   XR80R .... 184 mm (7.24 in)
   XR100R .... 205 mm (8.07 in)
   XL125R-C .... 210 mm (8.27 in)
   XL125R-F .... 170 mm (6.69 in)
   XL200R .... 145 mm (5.71 in)
   XR200 – 1982 on models .... 173.5 mm (6.83 in)
   XR200R – 1981 model .... 162 mm (6.38 in)
   XR200R-C, XR200R – 1982, 1983 models .... 145 mm (5.71 in)
   XR200R – 1986, 1987 models .... 150 mm (5.91 in)
Oil level adjustable range:
   XR200 – 1982 on models .... 149 – 173.5 mm (5.87 – 6.83 in)
   XR200R – 1981 model .... 127 – 164 mm (5.00 – 6.46 in)
   XR200R-C, XR200R – 1982, 1983 models .... 135 – 175 mm (5.32 – 6.89 in)
   XR200R – 1986, 1987 models .... 140 – 180 mm (5.51 – 7.09 in)
Fork standard recommended air pressure:
   XL125R .... 0 – 2.8 psi (0 – 0.2 kg/cm^2)
   XL185S – 1982 on models .... 5.7 ± 1.4 psi (0.4 ± 0.1 kg/cm^2)
   XL200R .... 0 – 5.7 psi (0 – 0.4 kg/cm^2)
   XR200 – 1982 on models .... 4.3 – 7.2 psi (0.3 – 0.5 kg/cm^2)
   XR200R – 1981, 1982, 1983 models, XR200R-C .... 0 – 4.3 psi (0 – 0.3 kg/cm^2)
   XR200R – 1986, 1987 models .... 0 psi (0 kg/cm^2)
Maximum permissible air pressure – all models .... 14 psi (1.0 kg/cm^2)

Rear suspension

Type – XR80R, XR100R, XL125R, XL200R and XR200R .. Pivoted fork, operating on single coil-sprung, hydraulically damped suspension unit via rising rate linkage (Honda Pro-Link)

Travel:
   XR80R .... 110 mm (4.3 in)
   All XL/XR100 models .... 116 mm (4.6 in)
   XR100R .... 120 mm (4.7 in)

	Standard	Service limit
XL125R	175 mm (6.9 in)	
XL200R	190 mm (7.5 in)	
XR200R – 1981, 1982, 1983 models, XR200R-C	247 mm (9.7 in)	
XR200R – 1986, 1987 models	245 mm (9.6 in)	
Swinging arm pivot inner sleeve OD:	**Standard**	**Service limit**
XR80R, XR100R	14.966 – 14.984 mm	14.940 mm
	(0.5892 – 0.5899 in)	(0.5882 in)
XL125R-C	20.045 – 20.200 mm	19.993 mm
	(0.7892 – 0.7953 in)	(0.7871 in)
XR200R – 1981, 1982, 1983 models, XR200R-C	19.989 – 20.000 mm	19.915 mm
	(0.7870 – 0.7874 in)	(0.7841 in)
Swinging arm pivot bush ID – XR80R, XR100R	14.990 – 15.030 mm	15.200 mm
	(0.5902 – 0.5917 in)	(0.5984 in)
Suspension linkage pivot bearing dimensions – XL125R:		
Inner sleeve/bush maximum clearance – all bearings	N/App	0.194 mm (0.008 in)
Linkage rear arm/swinging arm pivot bush maximum ID	N/App	17.135 mm (0.675 in)
All other bushes maximum ID	N/App	15.135 mm (0.596 in)
Suspension linkage pivot bearing dimensions – XR80R, XR100R:		
Inner sleeve OD – all bearings	17.941 – 17.968 mm	17.910 mm
	(0.7063 – 0.7074 in)	(0.7051 in)
Pivot bush ID – all bearings	18.000 – 18.052 mm	18.250 mm
	(0.7087 – 0.7107 in)	(0.7185 in)
Rear suspension unit spring free length:		
XR80R	136.0 mm (5.35 in)	133.3 mm (5.25 in)
All XL/XR100 models	206.5 mm (8.13 in)	202.4 mm (7.97 in)
XR100R	136.5 mm (5.37 in)	133.8 mm (5.27 in)
XL125R-C	232.0 mm (9.13 in)	229.5 mm (9.04 in)
XL125R-F	N/Av	N/Av
XL200R	235.5 mm (9.27 in)	233.0 mm (9.17 in)
XR200R – 1981, 1982, 1983 models, XR200R-C	207.0 mm (8.15 in)	202.9 mm (7.99 in)
XR200R – 1986, 1987 models	190.0 mm (7.48 in)	186.0 mm (7.32 in)
Damper compression force – XL125R, XL200R, XR200R – 1981, 1982, 1983 models, XR200R-C	62 – 84 lbs (28 – 38 kg)	
Rear brake pedal pivot – XR80R, XR100R:		
Pedal pivot bore ID	17.300 – 17.327 mm	17.340 mm
	(0.6811 – 0.6822 in)	(0.6827 in)
Pivot sleeve OD	17.294 – 17.298 mm	17.270 mm
	(0.6809 – 0.6810 in)	(0.6799 in)

Torque wrench settings

Component	kgf m	lbf ft
Steering stem nut:		
All XL/XR100 models	6.0 – 7.0	43 – 50.5
XL125R, XL200R, XR200R – 1981, 1982, 1983 models, XR200R-C	8.0 – 12.0	58 – 87
XR200R – 1986, 1987 models	9.5 – 14.0	69 – 101
Steering adjuster nut:		
XL125R, XL200R, XR200R – 1986, 1987 models	0.1 – 0.2	0.5 – 1.5
XR200R – 1981, 1982, 1983 models, XR200R-C	0.55 – 0.65	3.5 – 4.5
Handlebar clamp bolts:		
XR80R, XR100R	1.0 – 1.4	7 – 10
All XL/XR100 models	0.8 – 1.2	6 – 9
XR200R – 1981, 1982, 1983 models, XR200R-C	1.8 – 3.0	13 – 22
XR200R – 1986, 1987 models	2.4 – 3.0	17 – 22
Handlebar holder nuts – all XL/XR100 models	3.0 – 4.0	22 – 29
Master cylinder clamp bolts – XL125R-F	0.8 – 1.2	6 – 9
Air valves – where fitted	0.4 – 0.7	3 – 5
Stanchion cap bolt or top plug:		
XL80S and XR80 – 1983 on models	4.0 – 5.0	29 – 36
XR80R, XR100R	1.5 – 3.0	11 – 22
XL100S-B, XL100S and XR100 – 1981 models	7.5 – 8.5	54 – 61.5
XL100S and XR100 – 1982 models	1.5 – 3.0	11 – 22
XR200R – 1986, 1987 models	2.5 – 3.5	18 – 25
Top yoke pinch bolts:		
XR80R, XR100R	0.9 – 1.3	6.5 – 9.5
XL125R-C, XL200R	2.0 – 2.5	14.5 – 18
XL125R-F, XR200R – 1986, 1987 models	2.5 – 3.0	18 – 22
Bottom yoke pinch bolts:		
XL80S and XR80 – 1983, 1984 models	1.5 – 2.5	11 – 18
XL80S – 1985 model, XL100S-B, XL100S and XR100 – 1981 model	3.0 – 4.0	22 – 29

Component	kgf m	lbf ft
XL100S and XR100 – 1982 models	2.5 – 3.5	18 – 25
XL100S and XR100 – 1983 on models	2.0 – 3.0	14.5 – 22
XR80R, XR100R	2.4 – 3.0	17 – 22
XR200R – 1981, 1982, 1983 models, XR200R-C	1.8 – 3.0	13 – 22
XL125R-F, XR200R – 1986, 1987 models	3.0 – 3.5	22 – 25
Headlamp bracket – bottom mounting bolt – XL80S and XR80 models 1983 on	1.0 – 1.4	7 – 10
Spring retaining threaded plug:		
XL80S and XR80 models – 1983 on	1.5 – 3.0	11 – 22
XL100S-B, XL100S and XR100 – 1981 models	1.5 – 2.0	11 – 14.5
Damper rod Allen bolt:		
XR80 – 1983 model	1.5 – 2.0	11 – 14.5
XL80S – 1983, 1984 models, XL100S and XR100 – 1983 on models, XR80R, XR100R, XL125R, XL200R, XR200R-C, XR200R – 1981, 1982, 1983 models	1.5 – 2.5	11 – 18
XL80S – 1985 model	0.8 – 1.2	6 – 9
XR200R – 1986, 1987 models	2.5 – 3.5	18 – 25
Fork drain plug – XL125R-F	0.6 – 0.9	4 – 6.5
Fuel tank mounting bolts:		
XR80R, XR100R	1.0 – 1.4	7 – 10
XR200R-C, XR200R – 1981, 1982, 1983 models	2.4 – 2.9	17 – 21
XR200R – 1986, 1987 models	0.8 – 1.2	6 – 9
Fuel tap mounting screws	0.5 – 0.9	3.5 – 6.5
Fuel tap filter bowl – where fitted	0.3 – 0.5	2 – 3.5
Footrest mounting bolts:		
XL80S – 1985 model, all XL100S and XR100 models, XL125R, XL200R, XR200R-C, XR200R – 1981, 1982, 1983 models	1.8 – 2.5	13 – 18
XR80R, XR100R	3.5 – 4.5	25 – 32.5
Swinging arm pivot bolt:		
XR80R, XR100R	5.5 – 7.0	40 – 50.5
All XL/XR100 models	3.0 – 4.0	22 – 29
XR200R – 1986, 1987 models	8.0 – 10.0	58 – 72
XL125R, XL200R, all other XR200R models	7.0 – 10.0	50.5 – 72
Rear suspension unit top mounting bolt:		
XL125R-C, XL200R, XR200R-C, XR200R – 1981, 1982, 1983 models	6.0 – 7.0	43 – 50.5
XL125R-F	6.0 – 7.5	43 – 54
XR200R – 1986, 1987 models	4.0 – 5.0	29 – 36
Rear suspension unit bottom mounting bolt:		
XR200R – 1986, 1987 models	4.0 – 5.0	29 – 36
XL125R, XL200R, all other XR200R models	3.8 – 4.8	27.5 – 34.5
Suspension linkage rear arm/swinging arm pivot bolt:		
XR80R, XR100R	4.0 – 5.0	29 – 36
XL125R, XL200R, XR200R	9.0 – 12.0	65 – 87
Suspension linkage rear arm/front arm pivot bolt:		
XR80R, XR100R, XR200R – 1986, 1987 models	4.0 – 5.0	29 – 36
XL125R, XL200R, all other XR200R models	6.0 – 7.5	43 – 54
Suspension linkage front arm/frame pivot bolt:		
XR80R, XR100R, XR200R – 1986, 1987 models	4.0 – 5.0	29 – 36
XL125R, XL200R, all other XR200R models	6.0 – 7.5	43 – 54

Specifications relating to Chapter 5

Wheels

Size – rear:
XL100S, XR100, XR100R	16 inch	
XR200R – 1986, 1987 models	17 inch	

Brakes

Brake drum ID – XR80R, XR100R	95.0 mm (3.74 in)	
Service limit	96.0 mm (3.78 in)	

Hydraulic disc brake – XL125R-F

Recommended fluid .... Good quality hydraulic brake fluid, DOT 4 specification

	Standard	Service limit
Disc thickness	3.5 mm (0.1378 in)	3.0 mm (0.1181 in)
Disc maximum warpage	N/App	0.3 mm (0.012 in)
Master cylinder bore ID	12.700 – 12.743 mm (0.5000 – 0.5017 in)	12.755 mm (0.5022 in)

	Standard	Service limit
Master cylinder piston OD	12.657 – 12.684 mm	12.640 mm
	(0.4983 – 0.4994 in)	(0.4976 in)
Caliper bore ID	25.400 – 25.405 mm	25.450 mm
	(1.0000 – 1.0002 in)	(1.0020 in)
Caliper piston OD	25.318 – 25.368 mm	25.300 mm
	(0.9968 – 0.9987 in)	(0.9961 in)

Tyres

	Front	Rear
Size:		
XR80, XR80R	2.50 x 16 – 4PR	3.60 x 14 – 4PR
XL100S, XR100, XR100R	2.50 x 19 – 4PR	3.00 x 16 – 4PR
XR200 – 1981 model	2.75 x 21 – 4PR	3.50 x 18 – 4PR
XR200 – 1982 on models	2.75 x 21 – 6PR	3.50 x 18 – 6PR
XR200R-C, XR200R – 1981, 1982, 1983 models	3.00 x 21 – 6PR	4.10 x 18 – 6PR
XR200R – 1986, 1987 models	80/100 – 21 51M	110/100 – 17 58M
Pressure:		
XR80R, XR100R	14 psi (1.00 kg/cm^2)	18 psi (1.25 kg/cm^2)
XL100S-B	21 psi (1.5 kg/cm^2)	32 psi (2.25 kg/cm^2)
XR100	18 psi (1.2 kg/cm^2)	20 psi (1.4 kg/cm^2)
XL125S-B,C, XL125R-F, XL185S-B – with passenger .	21 psi (1.5 kg/cm^2)	25 psi (1.75 kg/cm^2)

Torque wrench settings

Component	kgf m	lbf ft
Front wheel spindle or spindle nut:		
XL80S – 1983, 1984 models, XR80R, XL100S and		
XR100 – 1983 on models, XR100R	5.5 – 7.0	40 – 50.5
XL80S – 1985 model	4.0 – 5.0	29 – 36
XL125R, all XL/XR200R models	5.0 – 8.0	36 – 58
Front wheel spindle clamp nuts – XL125R-F, all XR200R		
models	1.0 – 1.4	7 – 10
Front hub spoke plate mounting nuts – where fitted	1.2 – 1.6	9 – 11.5
Front disc brake components – XL125R-F:		
Caliper/fork lower leg mounting bolts	2.0 – 3.0	14.5 – 22
Disc mounting screws	1.4 – 1.6	10 – 11.5
Brake hose union banjo bolt	3.0 – 4.0	22 – 29
Brake hose upper union	3.0 – 4.0	22 – 29
Brake hose upper union sleeve nut	1.2 – 1.5	9 – 11
Caliper bleed nipple	0.4 – 0.7	3 – 5
Brake pad retaining pins	1.5 – 2.0	11 – 14.5
Caliper axle bolts	2.0 – 2.5	14.5 – 18
Rear brake torque arm mountings – XL80S – 1985		
model	1.8 – 2.5	13 – 18
Rear wheel spindle nut:		
XL80S – 1985 model, XL100S-B, XL100S and		
XR100 – 1981, 1982 models	4.0 – 5.0	29 – 36
XL80S – 1983, 1984 models, XR80R, XL100S and		
XR100 – 1983 on models, XR100R	5.5 – 7.0	40 – 50.5
XL125R, all XL/XR200R models	8.0 – 11.0	58 – 80
Rear wheel sprocket mountings:		
XR80R, XR100R	3.0 – 3.5	22 – 25
XL125R-C, XL200R	2.8 – 3.4	20 – 24.5
XL125R-F	4.0 – 5.0	29 – 36
XR200R – 1986, 1987 models	2.7 – 3.3	19.5 – 24

Specifications relating to Chapter 6

Battery
Capacity – XL125R, XL200R	12V, 3Ah

Flywheel generator

Output @ 5000 rpm:	
XR80R, XR100R	15W
XL125R	185W
XL200R	196W
XR200R-C, XR200R – 1981, 1982, 1983 models	50W
XR200R – 1986, 1987 models	108W
Charging – XL200R:	
Minimum	17.5V/2.0A @ 3000 rpm
Maximum	18.5V/4.5A @ 8000 rpm

Note: above test conducted with regulator black wire disconnected, charging begins @ 1300 rpm

Lighting – XL200R (with AC regulator disconnected):
    Minimum .... 15V @ 2500 rpm
    Maximum .... 25V @ 10 000 rpm
Lighting – XL200R (between headlamp blue and white wires, dipswitch on Hi):
    Regulated voltage .... 13.5 – 14.5 V @ 5000 rpm
Lighting – XR200R – 1986, 1987 models:
    Regulated voltage .... 13.0 ± 0.5 V

Generator source coil resistance

Charging coil – White wire to earth:
    XL80S – 1983 on models .... 0.4 – 0.7 ohm
Charging coil – Yellow to Pink wires:
    XL100S – 1983 on models, XL125S – 1984 model,
    XL185S – 1983 model .... 0.58 ohm
    XL200R .... 0.2 – 1.0 ohm
Lighting coil – Yellow wire to earth:
    XL80S – 1983 on models .... 0.1 – 0.4 ohm
Lighting coil – White or White/Yellow wire to earth:
    XL100S – 1983 on models, XL125S – 1984 model,
    XL185S – 1983 model .... 0.47 ohm
    XL200R .... 0.2 – 1.0 ohm
Lighting coil – Blue wire to earth:
    XR200R – 1986, 1987 models .... 0.2 – 1.2 ohm
Ignition source coil – Black or Black/White wire to earth:
    XL80S – 1983 on models .... 1.7 – 2.6 ohm
    XR80R, XR100R .... 0.57 – 0.71 ohm
Ignition source coil – Black/Red wire to earth:
    XL125S – 1984 model, XL185S – 1983 model .... 241.5 ohm
    XL200R, XR200R – 1986, 1987 models .... 50 – 200 ohm
    All other XR200R models .... 200 – 500 ohm

Bulbs (all 12 volt)

	XL125R	XL200R	XR200R-C	XR200R '86, '87
Headlamp	35/35W	36.5/35W	35/35W	35W
Pilot lamp	4W	N/App	N/App	N/App
Stop/tail lamp	21/5W	27/8W	21/5W	3.4W
Turn signal lamps	21W	23W	N/App	N/App
Main beam warning lamp	1.7W	1.7W	N/App	N/App
Instrument illuminating and other warning lamps	3.4W	3.4W	N/App	N/App

1 Introduction

The first six chapters of this Manual describe the Honda XL/XR80-200 models sold between 1978 and 1980. This chapter covers the models from 1981 onwards, describing only those features which require a different working procedure. When working on one of these later models, check first with this Chapter to note any changes of specification or working procedure then refer to the relevant part of the main text. To assist the owner in identifying the machine exactly, a summary is given below of the major changes made to the various models, including the frame numbers and dates of import.

UK models

XL100S The XL100S-Z model continued until replaced in March 1981 by the XL100S-B, which differed in having a round speedometer only, the tachometer having been deleted. New front forks were fitted with the wheel spindle at the bottom of the lower leg instead of in front of it, and the fork movement was reduced. Restyled cycle parts were finished in dark red with matt black panels, and a one-piece exhaust system was fitted.

XL100S-B/HD05-5100001 on/March '81 to March '83

XL125S The XL125S-A model continued until replaced in May 1981 by the XL125S-B which was fitted with front fork gaiters. The cycle parts were finished in dark red with black panels. In February 1982 the XL125S-C was introduced; this can be identified only by the black-painted engine/gearbox unit. The only other changes were in minor modifications to the paintwork and graphics. Discontinued in March 1983.

XL125S-B/L125S-5200000 on/May '81 to Feb '82
XL125S-C/L125S-5300000 on/Feb '82 to March '83

XL125R Based on the XL125S model, this machine is fitted with Honda's own single suspension unit rising rate rear suspension system, known as 'Pro-Link'. The cycle parts were restyled in line with current trends and a 12 volt electrical system was employed. In February 1985 the XL125R-F was introduced to replace the original XL125R-C; the later model is easily distinguished by its disc front brake and square enduro-style headlamp housing/number plate. It also features revised front forks, a modified front wheel design, frame-mounted pillion footrests and new paintwork and graphics.

XL125R-C/JD04-5000885 on/April '82 to April '85
XL125R-F/JD04-5100006 on/Introduced Feb '85

XL185S The XL185S-A model continued until replaced in March 1981 by the XL185S-B. Fitted with front fork gaiters, an enduro-style headlamp housing/number plate, a spring-loaded chain tensioner and fixed chain guide, XR-style chain adjusters and slightly restyled cycle parts finished in dark red with matt black panels. Discontinued in November 1982.

XL185S-B/L185S-5200247 on/March '81 to Nov '82

XR200R Based on the XR200 model, this machine incorporated Pro-Link rear suspension and the other modifications outlined above for the XL125R-C model. Discontinued during 1983.

XR200R-C/MEO40*CK108702 on/April '82 to '83

US models

1981 The XL80S model was fitted with a slightly larger fuel tank and revised graphics, as was the XR80 model, which also featured a restyled seat and exhaust. The XL100S was fitted with new front forks with reduced travel and with the wheel spindle at the bottom of each fork lower leg instead of in front of it. A new model, the XR100, was introduced which was based on the XL100S but featured a more powerful engine and XR-style cycle parts with only the bare minimum of equipment. The XL125S and XL185S models were fitted with XR-style chain adjusters, rather than the previous drawbolt components, and were finished in a new paint scheme. The XR200 model was fitted with a larger front mudguard and crankcase bashplate and its lights and instruments were removed completely. A new model, the XR200R, was introduced which was based on the XR200 but was fitted with Pro-Link rear suspension, minimal 12 volt lighting and restyled cycle parts.

XL80S/HD040*BK100003 to BK106767/Oct '80 to Aug '81
XR80/HE010*BK200003 to BK217688/Oct '80 to June '81
XL100S/HD050*BK000011 to BK003939/Oct '80 to Aug '81
XR100/HE030*BK000016 to BK006122/Oct '80 to Aug '81
XL125S/JD020*BK200002 to BK206822/Oct '80 to Aug '81
XL185S/MD020*BK200002 to BK210000/Oct '80 to Aug '81
XR200/ME020*BK200001 to BK202164/Oct '80 to June '81
XR200R/ME040*BK000037 to BK004171/Oct '80 to Aug '81

1982 The XL80S and XL185S models were fitted with flexible odometer trip reset knobs, the XL185S and XR200 models were fitted with new front forks which incorporated valves in their top caps for the adjustment of air pressure and the XR200R models was fitted with revised front forks, wheel rims, tyre security bolts, a new rear suspension unit and fixed chain guides instead of the previous rollers. No other major changes were made.

XL80S/HD040*CK200004 to CK206263/Sept '81 to Sept '82
XR80/HE010*CK300004 to CK317722/July '81 to Sept '82
XL100S/HD050*CK100003 to CK104009/Sept '81 to Sept '82
XR100/HE030*CK100004 to CK109518/Sept '81 to Sept '82
XL125S/JD020*CK300003 to CK304770/Sept '81 to Sept '82
XL185S/MD020*CK300003 to CK307487/Sept '81 to Sept '82
XR200/ME020*CK300004 to CK303395/July '81 to Sept '82
XR200R/ME040*CK100001 to CK110976/Sept '81 to Sept '82

1983 The XL/XR80 and 100 models received larger, restyled fuel tanks, restyled seat and side panels, conical front wheel hubs and leading axle front forks. No XL125S model was imported for 1983. The XL185S and XR200 models received restyled fuel tanks, seats, side panels and rear mudguards, and the XR200R received an O-ring drive chain. A new model, the XL200R, was introduced which was based on the XR200R but with full lighting equipment etc for legal road use. Both the XL185S and XL200R models were fitted with the Evaporative Emission Control system for California only.

XL80S/HD040*DK300006 to DK303996/Oct '82 to Oct '83
XR80/HE010*DK400004 to DK412075/Oct '82 to Oct '83
XL100S/HD050*DK200001 to DK203665/Oct '82 to '83
XR100/HE030*DK200003 to DK205436/Oct '82 to Oct '83
XL185S/MD020*DK400003 on/Oct '82 to Oct '83
XL200R (49-state)/MD060*DK000001 to DK005845/Oct '82 to Oct '83
XL200R (California)/MD061*DK00001 to DK001036/Oct '82 to Oct '83
XR200/ME020*DK400001 to DK402563/Oct '82 to Oct '83
XR200R/ME040*DK200001 on/Oct '82 to Oct '83

1984 The XL125S model was reintroduced, being fitted with the Evaporative Emission Control system for California. The XL185S and XR200R models had been discontinued, the latter to be replaced by a new model fitted with the 4-valve RFVC engine unit. No major changes were made to any models.

XL80S/HD040*EK400001 to EK403140/Nov '83 to Sept '84
XR80/HE010*EK500001 on/Introduced Nov '83
XL100S/HD050*EK300001 to EK302070/Nov '83 to Sept '84
XR100/HE030*EK300001 on/Introduced Nov '83
XL125S/JD020*EK400001 on/Introduced Nov '83
XL200R/MD060*EK100001 on/Introduced Nov '83
XR200/ME020*EK500001 on/Introduced Nov '83

1985 The XL80S and XL100S models continued virtually unchanged, machines sold in California being fitted with the Evaporative Emission Control system. The XR80 and XR100 models were replaced by the XR80R and XR100R versions, these later models being fitted with a modified form of the Pro-Link rear suspension system. All larger capacity models were discontinued.

XL80S (49-state)/HD040*FK500001 on/Introduced Oct '84
XL80S (California)/HD041*FK500001 on/Introduced Oct '84
XR80R/HE010*FK600006 to FK612267/Oct '84 to Sept '85
XL100S (49-state)/HD050*FK400001 on/Introduced Oct '84
XL100S (California)/HD051*FK400001 on/Introduced Oct '84
XR100R/HE030*FK400005 to FK409365/Oct '84 to Sept '85

1986 The XL80S and XL100S models were discontinued and no significant changes were made to the XR80R and XR100R models. Having discontinued the more sophisticated RFVC-engined XR200R, Honda reintroduced the 195cc 2-valve engine unit in a machine which featured the full cradle frame and revised suspension components of its predecessor, thus providing better performance than the earlier 1983 model but at a lower cost than that of the RFVC machine.

XR80R/HE010*GK700001 to GK714676/Oct '85 to Sept '86
XR100R/HE030*GK500001 to GK504199/Jan '86 to Aug '86
XR200R/ME050*GK200004 on/Feb '86 to Aug '86

1987 No major changes made to any model

XR80R/HE010*HK800004 on/Introduced Oct '86
XR100R/HE030*HK600002 on/Introduced Sept '86
XR200R/ME050*HK300001 on/Introduced Sept '86

Note: *The digit indicated by the asterisk (*) in the new VIN numbers of US models varied from machine to machine.*

2 Fuel filter: general

1   On XL125S-C, XL125R, all US XL models from 1982 on and all US XR models (except the XR80R, XR100R and XR200R models) from 1983 on, an additional fuel filter is incorporated in the base of the tap body. To remove it, turn the tap to the 'Off' position and, using only a close-fitting spanner, unscrew the filter bowl. Carefully prise out the sealing O-ring and withdraw the filter gauze. Cleaning and renewal are as described in Routine Maintenance, and reassembly is the reverse of the dismantling procedure. Do not overtighten the filter bowl; note the specified torque setting of 0.3 – 0.5 kgf m (2 – 3.5 lbf ft). If the filter bowl is leaking, renew the sealing O-ring; overtightening the bowl in an attempt to cure a leak will merely worsen the problem.

3 Air filter: general

XL100S and XR100 models
1   The air filter assembly is now mounted behing the right-hand side panel and is removed and refitted as described in Routine Maintenance for the XR80 and XL80S models. It may

be necessary to remove the voltage regulator/rectifier unit and the turn signal relay to gain access to the filter casing.

XR80R and XR100R models

2 The air filter assembly is mounted behind the left-hand side panel. Withdraw the panel, detach the retaining strap and withdraw the filter cover, then unclip the second retaining strap and withdraw the element assembly.

3 The foam element can then be peeled off the metal frame and cleaned as described in Routine Maintenance.

XR200R – 1986 on model

4 Although the filter assembly is now box-shaped and is secured to the forward face of the filter casing, the removal, refitting and cleaning procedure is as described in Routine Maintenance for the earlier 125, 185 and 200 models.

4 Front forks: checking the oil level

1 Some later models, especially those with air-assisted front forks, have a specified oil level which must be set precisely to ensure optimum fork performance.

2 The level is measured from the top of the stanchion to the top of the oil when the fork top caps and springs are removed and the forks are fully compressed; a dipstick of suitable length can be made from welding rod or similar. It is essential that the level is at exactly the correct height and that the quantity in both legs is precisely the same; add or remove oil to achieve this. Note that the amount of fork oil in each leg is relatively unimportant; the oil level is the essential factor, as shown by the approximate amounts of fork oil specified for some models.

3 Note that for the XR200 and XR200R models it is permissible to add more or less than the specified amount of oil to vary the level within set limits. This is to enable the rider to alter the fork performance to suit his needs. The higher the level, the stiffer the fork springing will be, especially when the standard recommended air pressure is used.

5 Front forks: setting the air pressure

1 Most of the later 125, 185 and 200 models (see Specifications) are fitted with air valves in the stanchion top caps. This is to permit the adjustment of the air pressure inside the fork leg to alter the fork's effective spring rate; the higher the air pressure, the stiffer the fork springing will be.

2 Two tools are essential for setting fork pressure; a gauge capable of reading the low pressures involved, and a low-pressure pump. The gauge must be finely calibrated to ensure that both legs can be set to the same pressure, and must cause only a minimal drop in pressure whenever a reading is taken; as the total air volume is so small, an ordinary gauge, such as a tyre pressure gauge, will cause a large drop in pressure because of the amount of air required to operate it. Gauges for use on suspension components are now supplied by several companies and should be available through any good motorcycle dealer. The pump must be of the hand- or foot-operated type, a bicycle pump being ideal; aftermarket pumps are available for use on suspension systems and are very useful, but expensive. **Never** use a compressor-powered air line; it is all too easy to exceed the maximum recommended pressure which may cause damage to the fork oil seals and may even result in personal injury. Add air very carefully, and in small amounts at a time.

3 The air pressure must be set when the forks are cold, and with the machine securely supported so that its front wheel is clear of the ground, thus ensuring that the air pressure is not artificially increased.

4 Set the pressure to the required amount within the specified range, then be careful to ensure that each leg is at **exactly** the same pressure. This is essential as any imbalance in

pressures will impair fork performance and may render the machine unsafe to ride. Note that good quality aftermarket kits are now available to link separate air caps; the use of one of these, when correctly installed, will ensure that the pressures in the legs are equal at all times and will aid the task of setting the pressure in the future.

5 It is permissible to experiment with air pressures until the ideal setting is found, but this will require a great deal of care and patience. Do not exceed the maximum recommended pressure of 14 psi (1.0 kg/cm^2).

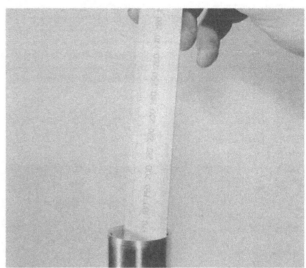

4.2a Measure oil level very carefully when rebuilding front forks or changing fork oil ...

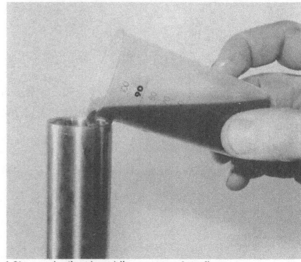

4.2b ... and adjust by adding or removing oil

6 Pro-Link rear suspension linkage: checking and lubricating

1 The Pro-Link suspension system consists of several highly-stressed bearing surfaces which will require regular checking and lubricating whenever the swinging arm pivot bearings are checked, as described in Routine Maintenance. Note that although the specified intervals are applicable to these models, it is recommended that the rear suspension is checked far more frequently, particularly if the machine is used extensively off-road.

2   To check the linkage components, two people are required; one to sit on the machine and bounce the rear suspension while the other watches closely the action of the various components. If any squeaks or other noises are heard, if the linkage appears stiff, or if any signs of dry bearings or other wear or damage are detected, the rear suspension should be removed as a complete assembly and dismantled for thorough cleaning, checking for wear, and lubrication.

3   The linkage should be greased at the interval specified in Routine Maintenance or whenever the machine has been used off-road. Where grease nipples are fitted, remove the weight from the suspension by placing the machine securely on a stand so that the rear wheel is clear of the ground, then inject grease into each bearing until all old grease is expelled and new grease can be seen issuing from each end of the bearing. If nipples are not fitted, the linkage must be dismantled so that the bearings can be cleaned and packed with new grease. Honda recommend the use of a grease which contains at least 45% molybdenum disulphide.

6.1 Pro-link suspension bearings need frequent attention

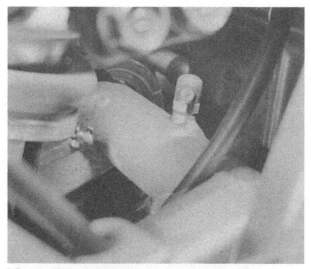

6.3a In addition to grease nipple on swinging arm pivot ...

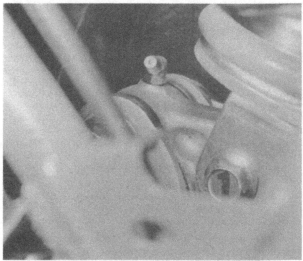

6.3b ... later models are fitted with grease nipples ...

6.3c ... on most suspension linkage pivot bearings

6.3d All bearings must be greased regularly and carefully

7 Pro-Link rear suspension: setting up the suspension unit – except XR80R and XR100R

1 The rear suspension unit spring preload is adjusted by slackening a slotted locknut and rotating the adjuster nut to alter the spring's fitted length. The standard settings are as follows:

XL125R-C .... 222.0 mm (8.74 in)
XL200R – all models .... 225.2 mm (8.87 in)
XR200R-C, XR200R – 1981,
1982 and 1983 .... 198.0 mm (7.80 in)
XR200R – 1986, 1987
models .... 185.5 mm (7.30 in)

2 On XR200R-1986 and 1987 models the spring can only be tensioned by shortening it to a maximum of 5 mm (0.20 in) less than the standard length, ie the spring final setting should always be within the range of 180.5 – 185.5 mm (7.11 – 7.30 in). On all other models, however, the spring can be adjusted through a range of 5 mm (0.20 in) shorter or longer than the set length; do not exceed these limits or the suspension unit may be damaged. Note that one full turn of the adjuster nut alters the spring length by 1.5 mm (0.06 in) on all models; to stiffen the suspension the spring must be shortened.

3 Depending on the tools being used, it will be necessary to remove one or both side panels. On XL125R, XL200R and XR200R-1981, 1982 and 1983 models it may also be necessary to remove the air filter hose; on XR200R-1986 and 1987 models it will be necessary also to remove the seat, the exhaust silencer and the air filter assembly. On all models, if the carburettor intake is exposed at any time, be careful to cover it with a piece of rag so that particles of dirt or debris do not drop into it. Be careful also to remove the rag before refitting the air filter.

4 When the suspension unit adjuster is accessible, use a slim C-spanner to rotate the slotted lock and adjuster nuts, and a ruler to measure the spring's length. The task is easier if the weight is taken off the suspension by raising the machine on to a stand so that the rear wheel is clear of the ground. Be careful to tighten the locknut securely when the required setting is reached; a torque setting of 8.0 – 10.0 kgf m (58 – 72 lbf ft) is specified for the locknut and should be used if the necessary equipment is available.

5 On XR200R-1986 and 1987 models the rear suspension unit damping is adjustable over both the compression and rebound strokes. To alter the compression damping, first find the standard position by rotating fully anticlockwise the black plastic knurled knob which is to be found at the hose union end of the unit's remote reservoir. When the knob comes to a stop, check that the yellow dots on the knob and the reservoir body are aligned; if not turn the knock clockwise until the dots align. From this softest setting turn the knob clockwise through 8 positions (2 full turns of the knob) to achieve the standard setting. This is exactly the mid-point of the unit's 16-position compression damping adjustment range; one full turn of the knob in either direction will soften or stiffen the damping, as appropriate, by 4 positions. Only careful experimentation will reveal which setting is best.

6 To adjust the rebound damping, turn the knurled knob at the unit's bottom mounting eye to the desired position. This facility is adjustable through 4 positions; be careful that the adjuster knob locks securely into a detent position whenever the setting is altered.

7 On XR200R 1981, 1982 and 1983 models the rebound damping only is adjustable exactly as described above. For these models position number 1 is softest, number 2 is standard and number 4 provides the stiffest damping. Align the appropriate mark on the adjuster knob with the reference mark on the unit to achieve the desired setting; again, ensure that the knob locks securely into a detent position.

8 The rear suspension unit fitted to XR80R and XR100R models is not adjustable.

7.3 Rear suspension preload adjuster and locknuts

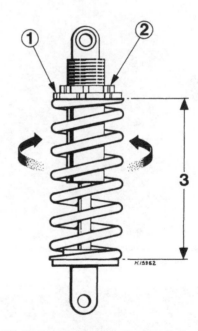

Fig. 7.1 Rear spring preload – Pro-link models

1 Adjuster
2 Locknut
3 Spring length

8 Chain adjustment: general

1 XL185S-B and XL125S and XL185S models from 1981 onwards were fitted with XR-style chain adjusters which use two bolts to push the spindle backwards instead of the previous drawbolt arrangement. The specified chain free play and the method of measuring it remains as described in Routine Maintenance.

2 On XL125R and all XL200R models chain free play must be 30 – 40 mm ($1\frac{1}{4}$ – $1\frac{5}{8}$ in) measured with the tightest spot in the

chain midway between the sprockets on the chain lower run when the machine is on its side stand.

3   On all XR200R models chain free play must be 35 – 45 mm (1$\frac{3}{8}$ – 1$\frac{3}{4}$ in) measured with the tightest spot in the chain midway between the sprockets on the chain upper run when the machine is supported securely on a stand with the rear wheel clear of the ground.

4   On XL125R and all XL/XR200R models the chain is adjusted by slackening the spindle nut and rotating the snail cam adjusters until the required free play is reached. Ensure that the same index mark number stamped in each snail cam aligns with the stopper pin in each swinging arm fork end to preserve accurate wheel alignment. Use the specified torque setting when tightening the spindle nut, and do not forget to recheck the chain and rear brake adjustment before using the machine.

5   On XL200R models the chain must be renewed when the red zone of the snail cam label aligns with the centre of the stopper pin when the chain is correctly adjusted.

6   Chain adjustment on XR80R and XR100R models is carried out as described for XL100S models in Routine Maintenance.

8.2 Chain tension is measured on bottom run on XL models and top run on XR models

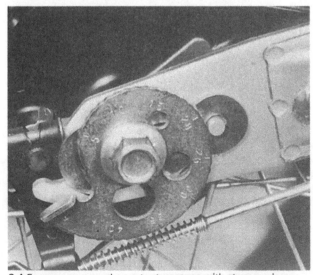

8.4 Ensure same number cutout engages with stopper pin on each side

9 O-ring final drive chain: general

1   Although fitted as standard only to the XL125R-F and later XR200R models, O-ring drive chains are increasingly popular and may be encountered on other models. Grease is sealed into the chain rollers by small O-rings, giving greatly increased life especially during off-road use.

2   These chains may be lubricated only using SAE 80 or 90 gear oil; most aerosol lubricants contain chemicals which will damage the O-rings causing premature and excessive wear. There are however a few aerosol products now sold in a form suitable for O-ring chains; ensure that this is the case before using any aerosol.

3   The chain can be disconnected and removed in the usual way, taking care not to lose the O-rings when the connecting link is dismantled, and can be cleaned in kerosene (paraffin) only. Do not use any strong solvents as these may damage the O-rings and never use a steam cleaner or pressure washer. Lubricate the chain thoroughly with gear oil after cleaning and drying. On reassembly, ensure that the four O-rings are positioned correctly and ensure that there is no space between the connecting link sideplate and spring clip.

10 Final drive chain guides: examination and renewal

1   The chain tensioner slipper block or swinging arm pivot guide block fitted to all XL185S-B and XL125S, XL185S models from 1981 onwards must be renewed when it has worn to a thickness of 10 mm (0.4 in) or less, measured at each end of the block.

2   On XR200R-1981 models the chain guide rollers should be renewed if worn to 20 mm (0.8 in) or less in diameter. Check also that each is free to rotate and that the rear chain guide is not bent.

3   On all later XR200R models and XL200R models, renew the front guide block if worn to 15 mm (0.6 in) or less, as described in paragraph 1, and the rear guide if it is worn to 17 mm (0.7 in) or less or if the chain can be seen through the guide's wear limit opening. Check carefully that the guides are correctly secured and that neither is bent.

4   On XR80R and XR100R models, renew the swinging arm pivot guide block if it is worn at any point to a thickness of 6.0 mm (0.24 in) or less.

11 Front disc brake: checking the fluid level and pad wear – XL125R-F

1   The brake should be checked for correct free play and operation, and for signs of fluid leakage, during every pre-ride check or at the very least once weekly.

2   The fluid level should be checked once every month of every 500 miles (800 km).

3   The complete brake system should be thoroughly checked every four months or 2500 miles (4000 km). Proceed as described below.

4   The hydraulic front brake requires no regular adjustments; pad wear is compensated for by the automatic entry of more fluid into the system from the handlebar reservoir. All that is necessary is to maintain a regular check on the fluid level and the degree of pad wear.

5   To check the fluid level, turn the handlebars until the reservoir is horizontal and check that the fluid level, as seen through the sight glass in the rear face of the reservoir body, is not below the lower level mark on the body. Remember that while the fluid level will fall steadily as the pad friction material is used up, if the level falls below the lower level mark there is a risk of air entering the system; it is therefore sufficient to maintain the fluid level above the lower level mark, by topping-

up if necessary. Do not top up to the higher level mark (formed by a cast line on the inside of the reservoir) unless this is necessary after new pads have been fitted. If topping up is necessary, wipe any dirt off the reservoir, remove the retaining screws and lift away the reservoir cover and diaphragm. Use only good quality brake fluid which meets or exceeds the DOT 4 specification and ensure that it comes from a freshly opened sealed container; brake fluid is hygroscopic, which means that it absorbs moisture from the air, therefore old fluid may have become contaminated to such an extent that its boiling point has been lowered to an unsafe level. Remember also that brake fluid is an excellent paint stripper and will attack plastic components; wash away any spilled fluid immediately with copious quantities of water. When the level is correct, clean and dry the diaphragm, fold it into its compressed state and fit it to the reservoir. Refit the reservoir cover (and plate) and tighten securely, but do not overtighten, the retaining screws.

6   If the brake is in good condition, the lever should have a small amount of free play, followed immediately by firm pressure as the pads come into contact with the disc. If the brake feels weak or spongy it is probably due to the caliper body being locked in place with dirt or corrosion so that it cannot slide across correctly on its axle bolts. This can be cured only by the dismantling and thorough cleaning of the affected components. However, these symptoms can also be produced by the presence of air in the system; this must be removed immediately by bleeding the system. If the brake system is in good condition, but lever travel is thought to be excessive, take the machine to a good Honda Service Agent who will be able to check it; a modified lever is available to reduce travel which can be fitted to earlier machines.

7   To check the degree of pad wear, look closely at the pads from above or below the caliper. Wear limit marks are provided in the form of deep notches cut in the top and bottom edges of the friction material or as red painted lines cut around the outside of the material. If either pad is worn at any point so that the inside end of the mark (next to the metal backing) is in contact with the disc, or if the wear limit marks have been removed completely, both pads must be renewed as a set. If the pads are so fouled with dirt that the marks cannot be seen, or if any oil or grease is seen on them, they must be removed for cleaning and examination.

8   To remove the pads, use an Allen key of suitable size to unscrew the threaded plugs which seal the pad retaining pins, then slacken the retaining pins. Remove the two mounting bolts and withdraw the caliper, complete with its mounting bracket, from the fork leg, taking care not to twist the brake hose. Remove both pad retaining pins and withdraw the pads, noting carefully the presence and exact positions of the anti-rattle spring and the retainer.

9   Thoroughly clean all components, removing all traces of dirt, grease and old friction material then use with care fine abrasive paper to polish clean any corroded items. Check carefully that the caliper body slides easily on its two axle bolts and that there is no damage to any of the caliper components, especially the rubber seals. If it is stiff, remove the mounting bracket from the caliper, clean the axle bolts and check them for wear or damage (which can be cured only by the renewal of the bolts or mounting bracket, as applicable) then smear a good quantity of silicone or PBC (Poly Butyl Cuprysil) based caliper grease over the bolts and caliper bores before refitting the mounting bracket to the caliper. It is essential that the caliper body can move smoothly and easily on the mounting bracket for the brake to be effective. **Warning:** do not use ordinary high-melting point grease; this will melt and foul the pads, rendering the brake ineffective.

10   If the pads are worn to the limit marks, fouled with oil or grease, or heavily scored or damaged by dirt and debris, they must be renewed as a set; there is no satisfactory way of degreasing friction material. If the pads can be used again, clean them carefully using a fine wire brush that is completely free of oil or grease. Remove all traces of road dirt and corrosion, then

use a pointed instrument to clean out the groove(s) in the friction material and to dig out any embedded particles of foreign matter. Any areas of glazing may be removed using emery cloth.

11   On reassembly, if new pads are to be fitted, the caliper pistons must now be pushed back as far as possible into the caliper bores to provide the clearance necessary to accommodate the unworn pads. It should be possible to do this with hand pressure alone. If any undue stiffness is encountered the caliper assembly should be dismantled for examination as described in Section 36. While pushing the pistons back, maintain a careful watch on the fluid level in the handlebar reservoir. If the reservoir has been overfilled, the surplus fluid will prevent the pistons returning fully and must be removed by soaking it up with a clean cloth. Take care to prevent fluid spillage. Apply a thin smear of caliper grease to the outer edge and rear surface of the moving pad and to the pad retaining pins. Take care to apply caliper grease to the metal backing of the pad only and not to allow any grease to contaminate the friction material. Carefully fit the pad anti-rattle spring, ensuring that it is correctly located, and insert the moving pad into its aperture in the caliper mounting bracket. Check that the pad is free to slide in the mounting bracket then refit the second pad and insert the two pad retaining pins, tightening them securely to the specified torque setting. Do not forget to place the retainer correctly in position as the pads are refitted.

12   Replace the caliper assembly on the machine ensuring that the pads engage correctly on the disc and that the hose is not twisted, then refit the mounting bolts and tighten them securely, to the specified torque setting if possible. Check that the pad pins are securely tightened and refit the threaded plugs. Apply the brake lever gently and repeatedly to bring the pads firmly into contact with the disc until full brake pressure is restored. Be careful to watch the fluid level in the reservoir; if the pads have been re-used it will suffice to keep the level above the lower level mark, by topping-up if necessary, but if new pads have been fitted, the level must be restored to the upper level line described above by topping up or removing surplus fluid as necessary. Refit the reservoir cover, and diaphragm as described above.

13   Before taking the machine out on the road, be careful to check for fluid leaks from the system, and that the front brake is working correctly. Remember also that new pads, and to a lesser extent, cleaned pads, will require a bedding-in period before they will function at peak efficiency. Where new pads are fitted, use the brake gently but firmly for the first 50 – 100 miles to enable the pads to bed in fully.

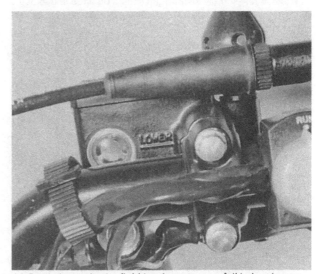

11.5a Hydraulic brake fluid level must never fall below lower level mark on reservoir body

11.5b If topping up is necessary, wipe away any dirt before removing reservoir body cover

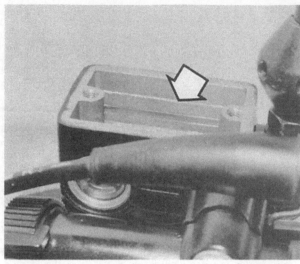

11.5c Upper level mark is formed by cast line on body inside face (arrowed)

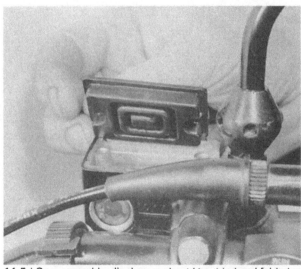

11.5d On reassembly, diaphragm should be dried and folded as shown ...

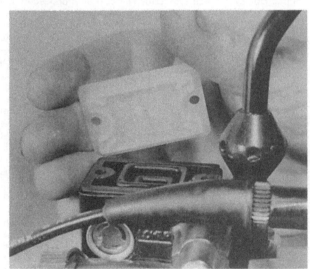

11.5e ... do not forget diaphragm plate

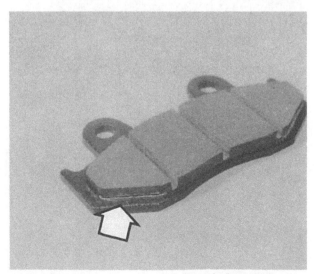

11.7 If either pad is worn to wear limit marks (arrowed) or beyond, both must be renewed

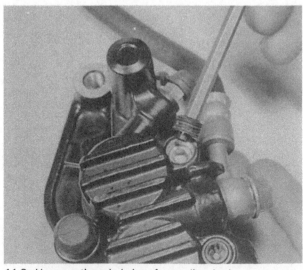

11.8a Unscrew threaded plugs from caliper body ...

11.8b ... to expose the pad retaining pins

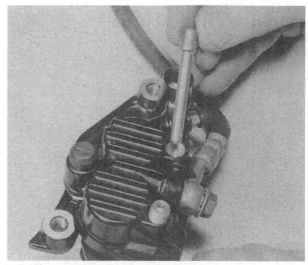

11.8c Pins must be unscrewed to release brake pads

11.9 Check that caliper axle bolts are clean and greased on reassembly

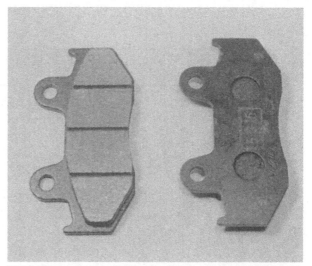

11.10 Both pads must be completely clean before refitting – do not forget to clean out grooves in friction material

11.11a Anti-rattle spring is fitted as shown. Note insulating caps on caliper pistons – do not omit

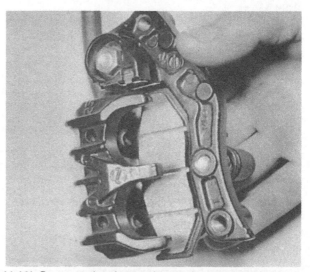

11.11b Ensure retainer is seated correctly on mounting bracket before refitting first (moving) pad

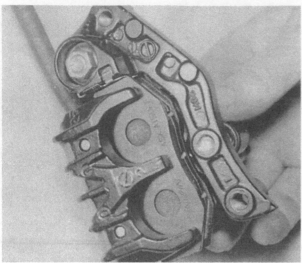

11.11c Position fixed pad and refit retaining pins to secure both pads

12 Front disc brakes: renewing the brake fluid, hydraulic seals and hoses – XL125R-F

1   At intervals of every two years, the master cylinder and caliper of the hydraulic front brake should be dismantled and checked. This is because their seals will deteriorate with age whether the machine is ridden a great deal or hardly at all. The seals should be renewed regardless of their apparent condition and the master cylinder and caliper bores, as well as the caliper pistons, should be checked very carefully for wear or damage. The brake fluid should be changed as described below. Check the brake hose very carefully indeed and renew it if any signs of splitting, cracking or chafing are seen. While Honda make no specific recommendation to this effect, it is usual to renew the hose at the same time as the seals and for the same reason. Refer to the relevant Sections of this Chapter for full information.
2   It is necessary to renew the brake fluid at this interval to preserve maximum brake efficiency by ensuring that the fluid has not been contaminated and deteriorated to an unsafe degree.
3   Before starting work, obtain a new, full can of DOT4 hydraulic fluid and read carefully Section 37 on brake bleeding. Prepare the clear plastic tube and glass jar in the same way as for bleeding the hydraulic system, open the bleed nipple by unscrewing it $\frac{1}{4} - \frac{1}{2}$ a turn with a spanner and apply the front brake lever gently and repeatedly. This will pump out the old fluid. **Keep the master cylinder reservoir topped up at all times**, otherwise air may enter the system and greatly lengthen the operation. The old brake fluid is invariably much darker in colour than the new, making it easier to see when it is pumped out and the new fluid has completely replaced it.
4   When the new fluid appears in the clear plastic tubing completely uncontaminated by traces of old fluid, close the bleed nipple, remove the plastic tubing and replace the rubber cap on the nipple. Top the master cylinder reservoir up to above the lower level mark, unless the brake pads have been renewed, in which case the reservoir should be topped up to its higher level. Clean and dry the rubber diaphragm, fold it into its compressed state and refit the diaphragm and reservoir cover, tightening securely the retaining screws.
5   Wash off any surplus fluid and check that the brake is operating correctly before taking the machine out on the road.

13 Engine mountings: modifications

XR80R and XR100R
1   Due to changes in frame design the engine mountings have been altered slightly from those described in Chapter 1.
2   When removing the engine/gearbox unit, first withdraw the sump guard, which is retained by two bolts at the rear and by one at the front; note that this latter bolt also passes through the engine front mounting bracket and the crankcase.
3   Enlist the aid of an assistant to hold the machine upright while the side stand/left-hand footrest assembly is removed (two bolts), the engine bottom rear mounting bolt and nut are withdrawn, and the side stand assembly is refitted to support the machine.
4   Remove the remaining three bolts which still secure the front mounting bracket, then withdraw the upper rear mounting bolt and lift the engine out of the frame.
5   Reverse the removal procedure to refit the engine, but be careful to ensure that all mounting bolts and nuts are in place before finally tightening any of them. Check that the engine is seated securely but without stress on its mountings, then tighten all fasteners to their specified torque wrench settings.

XR200R – 1986 and 1987 models
6   The full cradle frame employed on these models means that the procedure for removing and refitting the engine/gearbox unit is slightly modified.
7   With all ancillary components detached or disconnected, remove the rubber guard and dismantle the cylinder head steady assembly, then remove the engine protector bar and the front mounting bracket. Remove the nuts from the engine rear lower and upper mounting bolts, tap out each bolt in turn and lift the engine out of the frame.
8   On reassembly, insert all the mounting bolts from left to right; do not forget the spacer fitted between the engine and the frame at the left-hand end of the upper rear mounting bolt and at the right-hand end of the lower front mounting bolt. In addition to these, there may be two separate collars fitted between the front mounting bracket left-hand plate and the frame front downtube.
9   Ensure that the engine is seated securely and without stress on its mountings and that all mounting components are in place before tightening the various nuts and bolts to their specified torque settings.

14 Fuel tank: removal and refitting – XR80R, XR100R and XR200R

Note that the fuel tank is retained by a rubber strap at the rear, and by two bolts passing through rubber mountings at the front. On refitting, ensure that the tank is seated correctly on its mountings and that the bolts are tightened to the specified torque setting.

15 Carburettor adjustment and exhaust emission: general note

1   In some countries legal provision is made for describing and controlling the types and levels of toxic emissions from motor vehicles.
2   In the USA exhaust emission legislation is administered by the Environmental Protection Agency (EPA) which has introduced stringent regulations relating to motor vehicles. The Federal law entitled The Clean Air Act, specifically prohibits the removal (other than temporary) or modification of any component incorporated by the vehicle manufacturer to comply with the requirements of the law. The law extends the prohibition to any tampering which includes the addition of components, use

of unsuitable replacement parts or maladjustment of components which allows the exhaust emissions to exceed the prescribed levels. Violations of the provision of this law may result in penalties of up to $10,000 for each violation. It is strongly recommended that appropriate requirements are determined and understood prior to making any change to or adjustments of components in the fuel, ignition, crankcase breather or exhaust systems.

3   To help ensure compliance with the emission standards some manufacturers have fitted to the relevant systems fixed or pre-set adjustment screws as anti-tamper devices. In most cases this is restricted to a plastic or metal limiter cap fitted to the carburettor pilot adjustment screw, which allows normal adjustment only within narrow limits. Occasionally the pilot screw may be recessed and sealed behind a small metal blanking plug, or locked in position with a thread-locking compound, which prevents normal adjustment.

4   It should be understood that none of the various methods of discouraging tampering actually prevents adjustment, nor, in itself, is re-adjustment an infringement of the current regulations. Maladjustment, however, which results in the emission levels exceeding those laid down, is a violation. It follows that no adjustments should be made unless the owner feels confident that he can make those adjustments in such a way that the resulting emissions comply with the limits. For all practical purposes a gas analyser will be required to monitor the exhaust gases during adjustment, together with EPA data of the permissible Hydrocarbon and CO levels. Obviously, the home mechanic is unlikely to have access to this type of equipment or the expertise required for its use, and therefore, it will be necessary to place the machine in the hands of a competent motorcycle dealer who has the equipment and skill to check the exhaust gas content.

5   XL80S models are fitted with a small plug over the pilot screw to prevent adjustment as described above. This can be deformed and prised out with a sharp pointed instrument. On completion of adjustment a new plug must be drifted in until it seats on the shoulder provided at a depth of 1 mm (0.04 in). All other models subject to the above requirements are fitted with a limiter cap which is stuck to the protruding end of the pilot screw using Loctite 601. It can be prised or cut off if care is taken but must be renewed on completion of adjustment.

16 Carburettor slow (pilot jet: fitting

1   Most later models (see Specifications) are fitted with a removable slow-running (pilot) jet; if cleaning is necessary it can be unscrewed and checked as described in Chapter 2.

2   On refitting, however, great care is necessary with the type of jet fitted to certain models. While Honda only give specific instructions for the XL/XR100-1982 on models, XL125S-1982 models and XL185S-1982 on models, this jet would also appear to be fitted to other models, particularly XR80S-1983 on models, XR100R models, XR125R models, XR200R models and XR200R-1986 on models. It is therefore recommended that great care is taken when refitting the pilot jet of any model described in this Manual.

3   The exact procedure necessary to fit this type of jet is as follows. It must be screwed in carefully until resistance is encountered, then tightened by $\frac{3}{4}$ turn only. Do not overtighten the jet as it is easy to shear it off.

17 Evaporative Emission Control system: general – California models

1   As described in the introduction to this Chapter, certain 1983 and later models sold in California are fitted with equipment to prevent the escape into the atmosphere of fuel vapour ejected as a result of evaporation.

2   The system is simple in operation, consisting of vent tubes connecting the fuel tank filler cap and carburettor float chamber to a canister which contains activated charcoal. While the engine is stopped, the canister filters out fuel vapour, the surplus being fed back into the engine via the air filter as soon as the engine is started. A connection to the air filter allows the tank and carburettor to breathe normally while the engine is running, to replace the fuel consumed.

3   On XL80S and XL100S models the system must be checked every 7500 miles (12 000 km), but at intervals of 12 000 miles (19 200 km) on all other models. Check that the canister is securely mounted and undamaged and that all hoses and connections are undamaged and securely fastened. If any component is found to be worn or damaged, it must be renewed using only genuine Honda replacement parts. While the canister will normally last the life of the machine, its service life will be drastically reduced if it is flooded with fuel or water at any time.

18 Condenser: location – XR80R and all later 100 models

1   On all XR100 models, the XL100S-B and the XL100S-1981 and 1982 models, the condenser was relocated. It was mounted on the ignition HT coil and is not available as a separate component as described for XR80 models in Chapter 3, Section 5.

2   From 1983 on, all XR100S models reverted to having the condenser a separate unit bolted to the frame top tubes above the engine. Removal and refitting is therefore as described in Chapter 3 for all early XL100S models.

3   On all XR80R and XL100R models the condenser is clamped by a single screw to the ignition HT coil mounting bracket. It can be removed and refitted as an individual component, if required.

19 Pulser generator coil: general – all 125, 185 and 200 models

1   The pulser generator backplate can be rotated to adjust the ignition timing by slackening the two mounting screws. Rotate the plate clockwise to advance the ignition timing and anti-clockwise to retard it.

2   If the generator backplate is disturbed for any reason, the air gap between the coil and the rotor must be checked, Turn the engine over until the timing marks on rotor and coil align, then use feeler gauges to measure the gap between the two. The specified air gap is 0.3 – 0.4 mm (0.012 – 0.016 in); the setting is altered by slackening its two mounting screws and moving the coil towards or away from the rotor, as required. Tighten the screws securely and recheck the gap. It is essential that the gap is correct if the pulser generator is to function correctly.

20 CDI unit: testing – all 125, 185 and 200 models

1   Information is now available to permit the testing of the CDI unit, as shown in the accompanying illustrations. All XL125S, XL185S, XR185, XR200 and XR200R-1981, 1982 and 1983 models described in this Manual use a similar CDI unit which is tested as follows.

2   Remove the side panels, the seat and the fuel tank to gain access to the unit and its connections. The unit must be completely disconnected but can remain in place. Honda recommend the use only of either the Sanwa SP.10D Electric Tester (Honda part number 07308-0020000) or the Kowa Electric Tester (TH-5H-1) to give results of sufficient accuracy. Use the accompanying table to make the meter connections, noting that results should be obtained in kilo ohms. To

distinguish between the two black/white wires, that marked (A) is the individual wire ending in a female spade terminal, while that marked (D) is in the main group and ends in a male bullet connector. Of the two green wires, that marked (C) ends in a female bullet connector, while that marked (B) ends in a male bullet connector.

3   The units fitted to XL125R-C, XL200R, and XR200R 1986, 1987 models are tested as shown in the separate table, using the accompanying illustration to identify the terminals. Tests should be conducted using either one of the two testers described above, or the Kowa Digital tester (Honda part number 07411-0020000) or KS-AWM-32-003 (US only). Results should be obtained in kilo ohms on the Sanwa unit, or in the x100 ohm range on either of the Kowa units. If any results are not as given, the unit is faulty and must be renewed.

4   If a CDI unit is suspected of being faulty, an ordinary multimeter may be used provided that care is taken in making the connections exactly as shown; the resistance range may be determined by trial and error. In spite of inevitable inaccuracies in the readings, a reasonable indication of the unit's condition should be obtainable. This should be confirmed, however, by a Honda Service Agent testing the unit on the correct equipment, especially if tracing an elusive ignition system fault.

5   Note that while specific information is not available for the unit fitted to XL125R-F models, test results should be similar to those given for the XL125R-C/XL200R unit.

20.2 Peel back rubber cover to expose CDI unit – XL125/200R models

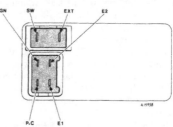

TEST RANGE: SANWA: x kΩ   KOWA: x 100 Ω

UNIT: KΩ

(+) \\ (−)	BLACK/WHITE (D)	GREEN (B)	BLACK/RED	GREEN (C)	BLUE/YELLOW	BLACK/WHITE (A)
BLACK/WHITE (D)		∞	∞	∞	∞	∞
GREEN (B)	2−50		0.5−10	—	∞	∞
BLACK/RED	0.5−10	∞			∞	∞
GREEN (C)	2−50	—	0.5−10		∞	∞
BLUE/YELLOW	2−50	0.5−10	2−50	0.5−10		∞
BLACK/WHITE (A)	∞	∞	∞	∞	∞	

H.15966

Fig. 7.2 CDI unit test table – all XL125S, XL185S, XR185, XR200 and 1981 to 83 XR200R models

+PROBE \\ −PROBE	SW	EXT	P●C	E1●E2	IGN
SW		∞	∞	∞	∞
EXT	0.3−20		∞	∞	∞
P●C	10−150	2−50		2−50	∞
E1●E2	10−100	1−30	∞		∞
IGN	∞	∞	∞	∞	

H.15967

Fig. 7.3 CDI unit test – XL125R-C and all XL200R models

21 Front forks: modifications

1   The modifications to the front forks which require a change in working practice are described below. Check carefully the Specifications Section at the beginning of this Chapter for any further information relevant to the machine being worked on.

XL/XR80 – 1983 on models
2   These models are fitted with simpler forks which are the same in layout as those shown in Fig. 4.2 of Chapter 4. The only differences from those shown are that the damper rod seat is omitted and a threaded retaining plug is fitted in the top of the stanchion to retain the fork spring; on XL80S – 1985 models a separate O-ring is fitted inside this plug to prevent oil leaks. Note the different fork oil quantity and spring free length given for the different models, also that a lock washer is now fitted to the steering stem nut. On reassembly, which is otherwise as described for XL100S and XR200 models, note that the springs are refitted with the closer-spaced coils to the top.

XL100S-B and XL/XR100 – 1981 on models
3   The forks fitted to XL100S-B and XL/XR100 – 1981 models differ in layout from those described in Chapter 4 only

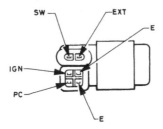

RANGE SANWA: x kΩ   KOWA: x 100 Ω

+PROBE \\ −PROBE	SW	EXT	PC	E	IGN
SW		∞	∞	∞	∞
EXT	0.5−100		∞	∞	∞
PC	1−200	0.5−100		0.5−100	∞
E	0.5−100	0.5−100	∞		∞
IGN	∞	∞	∞	∞	

H.20042

Fig. 7.4 CDI unit test – 1986 to 87 XR200R models

in that the damper rod seat is omitted and a threaded plug is fitted inside the top of the stanchion to retain the fork spring. On reassembly the springs are refitted with the closer-spaced coils to the top and the threaded plug is tightened to the specified torque setting using an 8 mm Allen key. Note the different fork oil quantity and spring free lengths, also that the top yoke has no pinch bolts.

4   For XL/XR100 – 1982 models the top yoke is again fitted with pinch bolts but the forks themselves remain the same as for 1981 models except for a further change in oil capacity and spring free length.

5   For XL/XR 100 models – 1983 onwards, although the forks remain similar in layout, there is a further change in oil quantity and spring free length. Also the spring retaining threaded plug is omitted, the spring being retained by the stanchion cap bolt.

XR80R and XR100R
6   The forks fitted to these models are the same in layout as those shown in Fig. 4.2 of Chapter 4. Refer to the Specifications Section of this Chapter for all information necessary. On reassembly note that the fork spring must be refitted with its tapered end downwards.

XL125S-B, C and XL125S – 1981 on models
7   The XL125S-B and C models were fitted with gaiters and a two-piece fork spring as shown for the XR200 model in Fig. 4.3 of Chapter 4 (but omitting the oil seal seat); all US models also receiving the latter modification. Note the new spring free lengths and revised oil quantity given.

XL125R-C
8   The forks fitted to this model are the same in layout as those shown for the XR200 model in Fig. 4.3 of Chapter 4, except that a tubular spacer is fitted instead of the top spring. The forks are also fitted with air caps. Refer to the Specifications Section of this Chapter for the relevant information. On reassembly, note that the fork springs are installed with the tapered ends downwards. Instructions on checking the fork oil level and air pressure are given earlier in this Chapter.

XL185S-B and XL185S – 1981 on models
9   The XL185S-B model was fitted with gaiters and a two piece fork spring as shown for the XR200 model in Fig. 4.3 of Chapter 4 (but omitting the oil seal seat); the XL185S – 1981 model also receiving the latter modification. Note the new spring free lengths and revised oil quantity given.

10   On the XL185S – 1982 and 1983 models are fitted with new forks whose layout was as described above but now reverting to a single fork spring in each leg. Also, since air valves are fitted, the oil seal seat shown in Fig. 4.3 of Chapter 4 is now fitted. Again, note the new spring free lengths and oil quantity; instructions on setting the air pressure are given earlier in this Chapter.

XL200R
11   The forks fitted to this model are similar in layout to those shown from the XR200 model in Fig. 4.3 of Chapter 4 except that air valves are fitted to the stanchion top bolts. Dismantling and reassembly are as described for the XR200 model noting the differences given in the Specifications Section of this Chapter and that the springs are refitted with their tapered ends downwards. Instructions on setting the fork oil level and air pressure are given earlier in this Chapter.

XR200 – 1982 on models
12   Apart from the addition of air valves to the stanchion top caps and the use of a single spring instead of the previous two, the forks remain as described in Chapter 4. Note the different spring free lengths and fork oil quantity given, and refer to the relevant Sections of this Chapter for setting the fork oil level and air pressure.

XL125R-F and XR200R
13   The layout of the forks fitted to these models is shown in the accompanying illustration. Refer to the Specifications Section of this Chapter for all information necessary. Refer to the relevant Sections of this Chapter for information on setting the fork oil level and air pressure.

14   To dismantle the forks, release the air pressure and remove the top cap, spacer and washer or spring guide (where fitted),

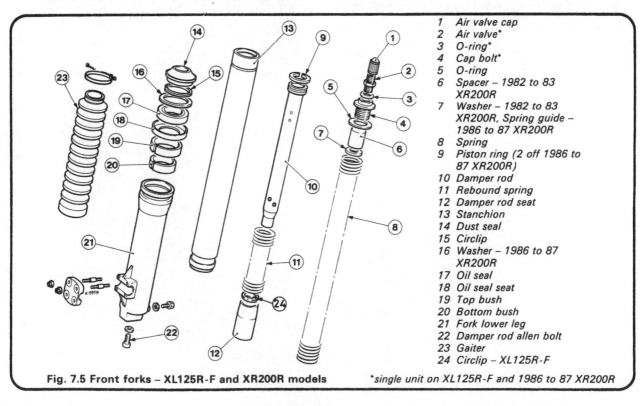

1   Air valve cap
2   Air valve*
3   O-ring*
4   Cap bolt*
5   O-ring
6   Spacer – 1982 to 83 XR200R
7   Washer – 1982 to 83 XR200R, Spring guide – 1986 to 87 XR200R
8   Spring
9   Piston ring (2 off 1986 to 87 XR200R)
10   Damper rod
11   Rebound spring
12   Damper rod seat
13   Stanchion
14   Dust seal
15   Circlip
16   Washer – 1986 to 87 XR200R
17   Oil seal
18   Oil seal seat
19   Top bush
20   Bottom bush
21   Fork lower leg
22   Damper rod allen bolt
23   Gaiter
24   Circlip – XL125R-F

Fig. 7.5 Front forks – XL125R-F and XR200R models   *single unit on XL125R-F and 1986 to 87 XR200R

fork spring and damper rod Allen bolt as described in Chapter 4. Remove the gaiter, dust seal and oil seal retaining circlip and washer (where fitted) from the fork lower leg, then drain the oil. Clamp the fork lower leg by its spindle lug in a vice fitted with soft jaw covers, push the stanchion fully into the lower leg and pull it out as sharply as possible. Repeat the operation several times, using the slide hammer action of the bottom bush against the top bush to dislodge the oil seal. When the seal is released, the stanchion can be withdrawn complete with the seal, back-up ring, and bushes.

15 On XL125R-F models note that a small wire circlip prevents the damper rod from being pulled upwards through the stanchion; remove the circlip to release the rod. Note also the damper components inside the stanchion lower end. Swill out the stanchion with a suitable solvent to remove all traces of old oil and dirt, but note that the complete stanchion must be renewed if these components are found to be damaged or severely worn; they cannot be removed.

16 Check all other fork components as described in Chapter 4; if free play can be felt when the stanchion and bushes are temporarily refitted into the lower leg, or if any bush is excessively worn, the bushes must be renewed. If the Teflon is worn away so that the copper material appears over more than $\frac{3}{4}$ of the whole bearing surface, that bush must be considered worn out. It is best to renew all the bushes together to preserve equal fork performance. The top bush can be slid off the stanchion; the bottom bush is split so that it can be sprung apart and eased out of its recess and off the stanchion lower end.

17 On reassembly, insert the damper rod, complete with piston ring and rebound spring, into the stanchion and press it down so that it projects fully from the stanchion lower end, then refit the damper rod seat, and, on XL125R-F models, the circlip. Check that the bottom bush is correctly seated, smear the stanchion with fork oil and insert the assembly into the fork lower leg. Press the stanchion fully into the lower leg to centralise the damper rod seat, apply a few drops of thread locking compound to the threads of the Allen bolt and refit it, not omitting its sealing washer. Refit the spring, with the washer and spacer (where fitted), and the top cap to retain the damper rod and tighten the Allen bolt to the specified torque setting. Remove the top cap and fork spring.

18 Smearing it with fork oil, slide the top bush down over the stanchion followed by the seal seat (back-up ring) and press the bush fully into its recess in the fork lower leg. Use a hammer and suitable drift to tap the seal seat and bush fully into place. On XR200R 1986 and 1987 models wrap a single layer of

insulating tape or similar around the top of the stanchion to protect the delicate seal lips from the raised edges of the groove. On all models, smear the stanchion with fork oil to protect the seal lips and slide down the oil seal with its marked surface upwards, followed by (where fitted), its plain washer. Press the seal squarely into the fork lower leg as far as possible by hand only, then use a length of tubing of the same inside and outside diameters as the seal or washer to tap the seal into place until the circlip groove is exposed. Do not risk distorting the seal by forcing it any further into the leg. Refit the circlip, dust seal and gaiter.

19 Refill the fork to the correct level with oil, refit the spring (and washer, spring guide and spacer, where fitted), then refit the top cap and pressurise the fork leg with air before refitting it to the machine. Note that the closer-pitched coils should be facing downwards when refitting the fork spring.

20 On XR200R 1986 and 1987 models note that the fork legs must be refitted so that the groove just below the stanchion upper end is aligned with the top yoke upper surface. On all models fitted with gaiters, rotate the gaiters so that their breather holes face to the rear, then raise each gaiter into contact with the bottom yoke before tightening its clamp securely.

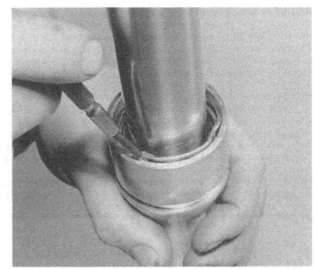

21.14 Use pointed instrument to displace fork seal retaining circlip

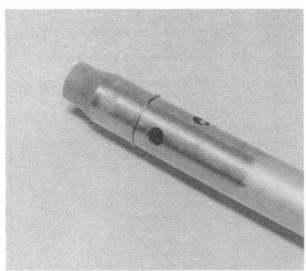

21.15a XL125R-F – damper rod is retained in stanchion by a wire circlip ...

21.15b ... and various damper components are sealed inside stanchion lower end

21.16a Examine sliding surfaces of bushes for wear (ie top bush inside, bottom bush outside)

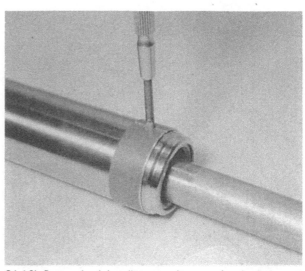

21.16b Bottom bush is split to permit removal and refitting – do not distort bush by opening gap too far

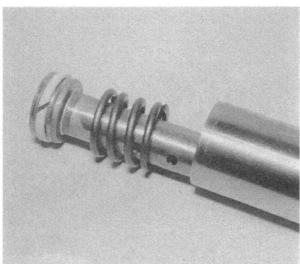

21.17a Fit piston ring and rebound spring to damper rod and insert assembly into stanchion

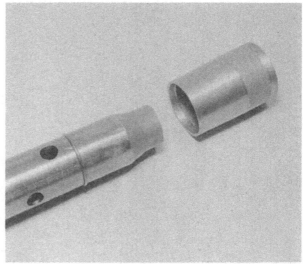

21.17b Fit wire circlip to damper rod lower end (XL125R-F) and refit damper rod seat ...

21.17c ... then insert stanchion assembly into lower leg

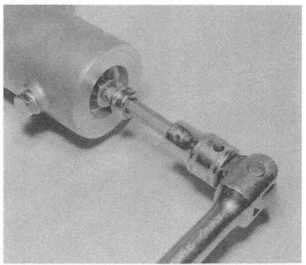

21.17d Tighten damper rod Allen bolt to specified torque setting

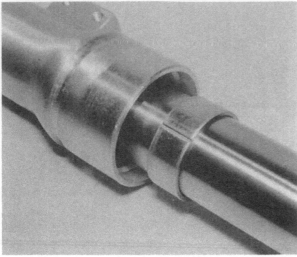

21.18a Smear oil over bush and stanchion before refitting top bush

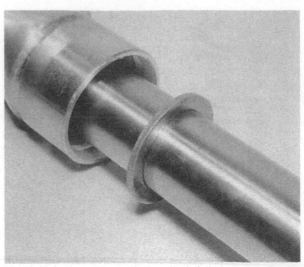

21.18b Fit back-up ring and press bush fully into lower leg ...

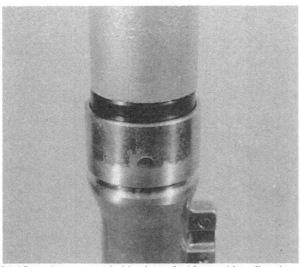

21.18c ... then use a suitable piece of tubing to drive oil seal into place ...

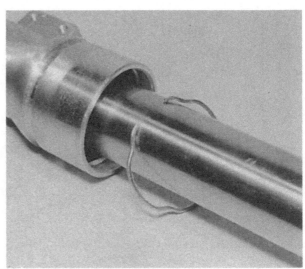

21.18d ... before refitting circlip ...

21.18e ... followed by dust seal ...

21.18f ... and fork gaiter

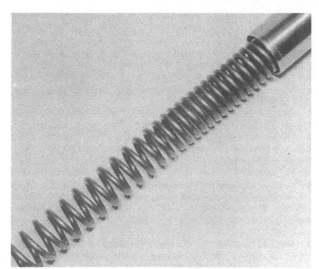

21.19 Fork spring is refitted with closer-pitched coils downwards

22 Steering head bearings: removal and refitting – XL125R and all XL/XR200R models

1   These machines are fitted with taper roller steering head bearings. The procedure for checking and adjusting them is as described in Routine Maintenance and Chapter 4. To gain access to them for greasing or renewal, remove from the machine the front wheel, the forks and the top yoke as described in Chapter 4, Sections 2, 3 and 4. Withdraw the dust cover.
2   If the special Honda socket spanner, Part Number 07916-3710100 (XR200R – 1986 and 1987 models) or 07916-KA50100 (all other models) is not available, use a C-spanner or pin spanner to unscrew the slotted adjusting nut from the top of the steering stem, supporting the bottom yoke so that it does not fall clear. With the nut unscrewed, withdraw the bottom yoke and lift the top bearing out of the steering head.
3   The bottom bearing is a tight interference fit around the bottom of the steering stem. Hold the yoke securely in a vice with padded jaws and use two large levers to displace the bearing. If this fails, a knife-edged bearing puller must be applied. On XR200R – 1981, 1982 and 1983 models, an alternative is to remove the steering stem pinch bolt from the bottom yoke and press the steering stem down through the bottom yoke using an hydraulic press, taking great care that the yoke is fully supported around the stem.
4   The bearing outer races can be drifted from their locations by passing a long drift through the steering head from the opposite end. Tap evenly all around each race so that it does not tilt and stick in the steering head.
5   Renew the dust seal under the bottom bearing as a matter of course; its good condition is vital to exclude dirt and water from the bearings.
6   Wash the bearings in a suitable solvent, removing all traces of old grease and road dirt, then examine the rollers, cages and races for signs of wear or damage. This is revealed by any form of pitting or scratches, by scuff marks or discolouration on any component or by actual damage such as cracked rollers, cages or races. If any signs of such wear or damage is seen, the bearing must be renewed; this will of course, mean the renewal of the inner bearing, complete with rollers, and the outer race as a set.
7   The outer races are refitted using a fabricated drawbolt arrangement as shown in the accompanying illustration. This is the safest method of inserting the races squarely into the steering head until they seat against their locating shoulders.

8   To refit the bottom bearing, first refit the dust seal, then place the bearing on the steering stem and slide it as far down as possible. Obtain a length of heavy tubing longer than the steering stem; the tube must fit closely over the stem and must fit against the bearing inner race without touching the rollers or cage. This can then be used as a drift, a few sharp blows from a hammer being sufficient to seat the bearing and seal on the bottom yoke. Take great care not to damage the seal or bearing and ensure that the bottom yoke is fully supported. On XR200R – 1981, 1982 and 1983 models place the seal and bearing on the bottom yoke, and push the steering stem up through them as far as possible then fit the Honda tool Part Number 07946-4300100 (basically a length of tubing carefully shaped, as described above) over the stem upper end, against the bearing. The assembly can then be placed in an hydraulic press so that the bottom yoke seal and bearing can be pressed down onto the steering stem shoulder. When the assembly is complete, refit the steering stem pinch bolt and tighten it to a torque setting of 4.0 – 5.0 kgf m (29 – 36 lbf ft). On XR200R – 1986 and 1987 models the same equipment can be used to press the dust seal and bearing on to the one-piece steering stem/bottom yoke.
9   Pack both bearings with grease and smear grease over the steering stem and inside the steering head, particularly over the outer races. Insert the bottom yoke into the steering head, refit the top bearing and screw down the adjusting nut.
10 To bed the bearings in, tighten the nut firmly using hand pressure only, then rotate the bottom yoke through its full movement. The steering head should turn smoothly with no signs of rough spots or notchiness. If all is well, unscrew the nut until pressure is released, then tighten it again until resistance is **just** evident. If a suitable method can be devised, the nut should be tightened to the specified torque setting.
11 The object is to set the adjuster nut so that the bearings are under a **very light** loading, just enough to remove any free play. If the nut is overtightened, the bearings will be damaged and the steering will be stiff, causing the machine to roll from side to side at low speeds. When the initial setting is correct, refit the dust cover (where applicable) and reassemble the front forks as described in Chapter 4, but remember to check the bearing adjustment before taking the machine out on the road.

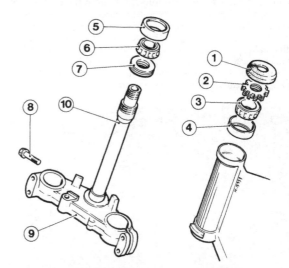

Fig. 7.6 Steering head bearings – XL125R and all XL/XR200R models – typical

1	Dust cover	6	Bottom inner bearing
2	Adjuster nut	7	Dust seal
3	Top inner bearing	8	Pinch bolt
4	Outer race	9	Bottom yoke
5	Outer race	10	Steering stem

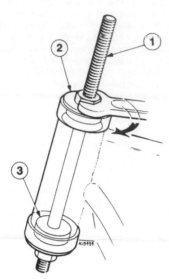

Fig. 7.7 Drawbolt arrangement for fitting bearing outer races

1   High tensile bolt   2   Heavy washer   3   Guide

23 Pro-Link rear suspension: removal and refitting

1   Remove the rear wheel as described in Section 31 of this Chapter, then remove the seat and both side panels. On XR200R models it may be necessary to remove the air filter casing; slacken the clamp and pull the connecting hose off the carburettor, remove the three casing mounting bolts and manoeuvre the casing out to the left. On all XL125R, XL200R and XR200R 1981, 1982 and 1983 models, slacken the upper rear (10 mm) engine mounting bolt. On XL and later XR models disconnect the brake pedal return spring from the swinging arm.
2   While it is possible, with a little care, to remove the pivot bolts and separate the suspension unit, linkage and swinging arm from each other so that all three can be removed individually, it is recommended that the complete suspension be removed as a single assembly, as described below.
3   Remove the suspension unit top mounting bolt. On XR200R models, remove the remote reservoir mounting bolts

or screws and release the reservoir from its mountings. On all models remove the linkage front arm/frame pivot bolt and the swinging arm pivot bolt. If any bolt is difficult to remove because of dirt or corrosion, apply a liberal quantity of penetrating fluid and allow time for it to work before tapping out the bolt with a hammer and drift. Manoeuvring it carefully to avoid damage, lift out the swinging arm and rear suspension, disengaging it from the chain.
4   On reassembly, grease all bearings thoroughly and check that all components are in place, then manoeuvre the assembly into position, remembering to pass the swinging arm left-hand fork end through the chain. Refit the swinging arm pivot bolt, greasing it thoroughly and tighten its retaining nut to the torque setting specified. Check that it moves easily, with no trace of stiffness or free play, then align the linkage front arm with its frame mounting, grease the pivot bolt and refit it, followed by refitting the suspension unit top mounting bolt, suitably greased. Tighten their retaining nuts to the specified torque settings.
5   Refit all components that were removed on dismantling and check for correct chain adjustment, suspension movement and rear brake operation before taking the machine out on the road.

23.1a Slacken rear upper engine mounting bolt ...

23.1b ... and disengage brake pedal return spring

23.3a Remove suspension unit top mounting bolt ...

23.3b ... and linkage front arm/frame pivot bolt

23.4a Check all bearings are greased and all components refitted ...

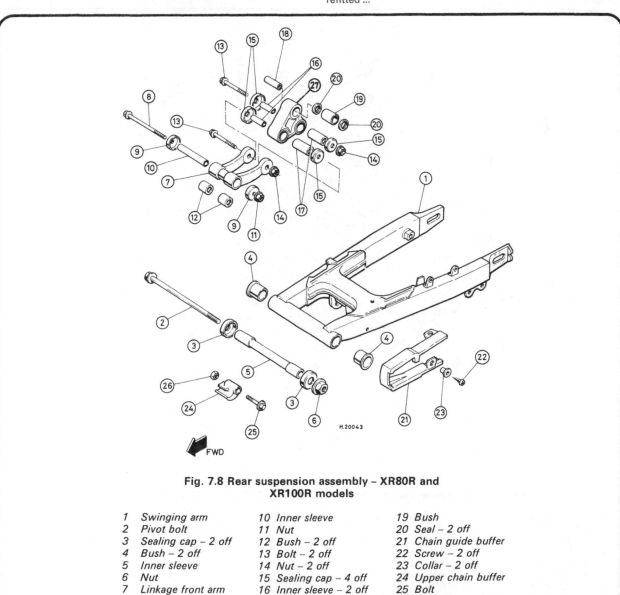

Fig. 7.8 Rear suspension assembly – XR80R and XR100R models

1	Swinging arm	10	Inner sleeve
2	Pivot bolt	11	Nut
3	Sealing cap – 2 off	12	Bush – 2 off
4	Bush – 2 off	13	Bolt – 2 off
5	Inner sleeve	14	Nut – 2 off
6	Nut	15	Sealing cap – 4 off
7	Linkage front arm	16	Inner sleeve – 2 off
8	Bolt	17	Bush – 2 off
9	Sealing cap – 2 off	18	Inner sleeve

19	Bush
20	Seal – 2 off
21	Chain guide buffer
22	Screw – 2 off
23	Collar – 2 off
24	Upper chain buffer
25	Bolt
26	Washer
27	Linkage rear arm

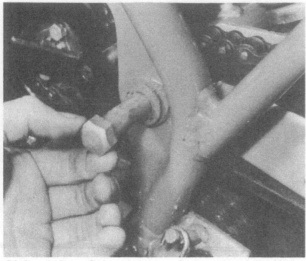

23.4b ... before refitting rear suspension – grease pivot bolts to prevent corrosion

23.4c Tighten all retaining nuts to the specified torque settings

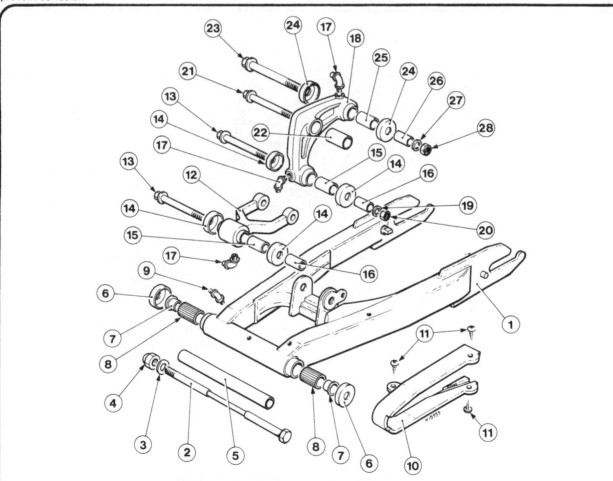

Fig. 7.9 Rear suspension assembly – XL125R, XL200R and 1981 to 83 XR200R models

1	Swinging arm	9	Grease nipple
2	Pivot bolt	10	Chain guide buffer
3	Washer	11	Screw – 3 off
4	Nut	12	Linkage front arm
5	Inner sleeve	13	Bolt – 2 off
6	Sealing cap – 2 off	14	Sealing cap – 4 off
7	Bush – 2 off	15	Bush – 2 off
8	Needle roller bearing – 2 off – XR only		

16 Inner sleeve – 2 off
17 Grease nipple – 2 off (where fitted)
18 Linkage rear arm
19 Spring washer – XR only
20 Nut
21 Bolt

22 Bush
23 Bolt
24 Sealing cap – 2 off
25 Bush – 2 off
26 Inner sleeve
27 Spring washer – XR only
28 Nut

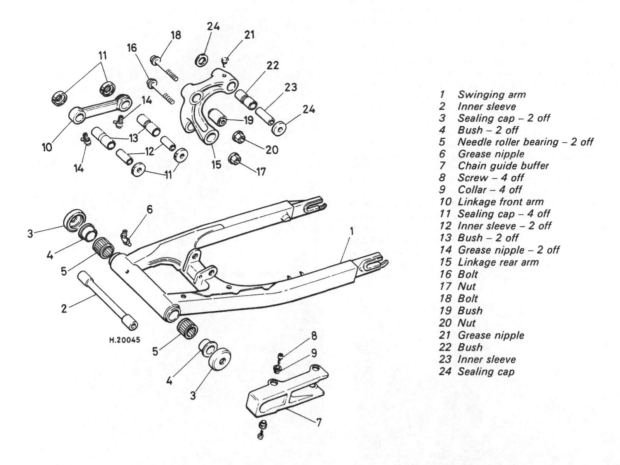

1   Swinging arm
2   Inner sleeve
3   Sealing cap – 2 off
4   Bush – 2 off
5   Needle roller bearing – 2 off
6   Grease nipple
7   Chain guide buffer
8   Screw – 4 off
9   Collar – 4 off
10 Linkage front arm
11 Sealing cap – 4 off
12 Inner sleeve – 2 off
13 Bush – 2 off
14 Grease nipple – 2 off
15 Linkage rear arm
16 Bolt
17 Nut
18 Bolt
19 Bush
20 Nut
21 Grease nipple
22 Bush
23 Inner sleeve
24 Sealing cap

Fig. 7.10 Rear suspension assembly – 1986 to 87 XR200R models

24 Pro-link rear suspension: examination and renovation

1   Before commencing work, check with a local Honda Service Agent exactly what is available for the machine being worked on. In some cases bushes are available only as part of a larger component.
2   Referring to the photographs and illustration accompanying the text, dismantle the assembly into its component parts and wash carefully each one in solvent to remove all traces of dirt and old grease. Components such as the chainguard, pillion footrests and linkage shields need not be removed unless necessary.
3   All pivot bearings are formed by a bush of metal or synthetic material, or in some cases bonded rubber bushes or needle roller bearings, pressed into the bearing housing. A steel inner sleeve is fitted in the centre of this bearing and the pivot bolt passes through the sleeve. Sealing caps are fitted at each end of each bearing.
4   Wear will only be found, therefore, between the pivot bolt and inner sleeve and between the inner sleeve and bearing. After thorough cleaning, reassemble each bearing and feel for free play between the components. If excessive free play is felt, examine each component, renewing any that shows signs of deep scoring or scuffing. Where dimensions are given, check the amount of wear by direct measurement. Also check all components for wear due to the presence of dirt or corrosion, and check the swinging arm and linkage arms for signs of cracks, distortion or other damage, renewing any worn items.
5   Check that all pivot bolts are straight and unworn and that their threads are undamaged. Use emery paper to polish off all traces of corrosion and smear grease over them on reassembly.
6   The bushes or needle roller bearings can be removed either by tapping them out or by using a variation of a drawbolt arrangement as shown in the accompanying illustration. Note that since this will almost certainly damage the bush or bearing, they should not be disturbed unless renewal is necessary. On refitting, the drawbolt arrangement shown must be used to avoid the risk of damage. Clean the housing thoroughly and smear grease over the bearing and its housing to aid refitting. Needle roller bearings are refitted with their marked surfaces facing outwards.
7   The sealing caps should be examined closely and renewed if worn; they must be in good condition to exclude dirt and water from the bearings.
8   On reassembly, pack molybdenum disulphide grease (containing at least 45% molybdenum disulphide) into each bearing and sealing cap, and smear it over all inner sleeves and pivot bolts. On XR200R models check that all grease nipples are clear by pumping grease into the bearings.
9   Refer to the accompanying photograph and illustrations on reassembly to ensure that all components are correctly refitted. On XL125R, XL200R and early XR200R models, note that the linkage front arm has the word 'UP' cast on one surface of its single larger mounting boss to show which way round it is fitted. Do not forget to refit the sealing caps to each bearing, ensuring that their lips seat correctly. All linkage pivot bolts are fitted from right to left, particularly the suspension unit bottom mounting bolt on XL125R, XL200R and early XR200R models; note that the unit bottom mountings threaded side must be on the left. Tighten all retaining nuts to the torque settings given in the Specifications Section of this Chapter.

25 Pro-link rear suspension: removing and refitting the suspension unit

1 Support the machine securely on a stand with the rear wheel clear of the ground, then remove the seat, the side panels and, on 200 models only, the air filter casing (see Section 23). On XR200R models, remove its mounting bolts and release the remote reservoir from its mountings. Remove its retaining nut and tap out the suspension unit top mounting bolt.

2 Exposing the unit bottom mounting bolt may prove to be awkward; remove if necessary the mudguards and chainguard to gain extra working space, then either lift the wheel and support it on blocks as soon as the mounting bolt is accessible, or remove the linkage pivot bolts and lower the unit to the ground. Be careful not to damage the unit.

3 Reassembly is a reversal of the dismantling procedure. If the mounting bolts are to be renewed, use only genuine Honda parts to be sure of mountings of sufficient strength. Smear grease over the bolts and refit them; on XL125R, XL200R and early XR200R models the bolts are fitted from right to left so that the unit's bottom mounting threaded side must, therefore, be on the left. Tighten all bolts to the torque settings specified.

24.2a Separate linkage into individual components ...

24.2b ... but swinging arm need only be stripped if necessary for repairs

24.3a Check for wear between inner sleeve and pivot bush ...

24.3b ... but note that different types of bearing or bush are used

24.8a Liberally grease inner sleeves on refitting ...

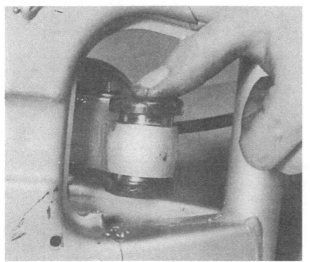

24.8b ... and do not omit sealing caps

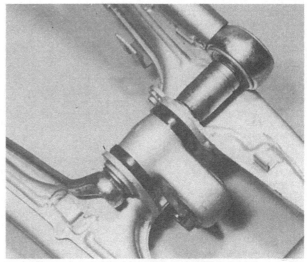

24.8c Tighten all bolts to specified torque setting

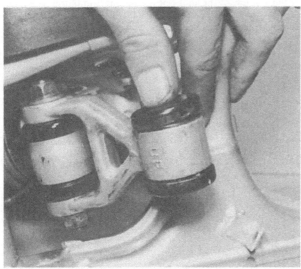

24.9a Note 'Up' mark on linkage front arm

24.9b Suspension unit bottom mounting bolt must fit from right to left

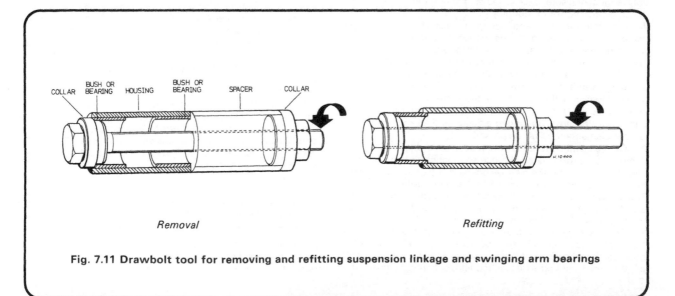

Removal

Refitting

Fig. 7.11 Drawbolt tool for removing and refitting suspension linkage and swinging arm bearings

26 Pro link rear suspension unit: dismantling, examination and reassembly

XR80R and XR100R

1　Clamp the unit by its bottom mounting in a vice with padded jaws and devise a method of compressing the spring safely so that the retaining collar can be withdrawn. This is best achieved using a pair of the spring compressors that are widely available for use on the MacPherson strut front suspension of many modern cars.

2　With the spring removed the unit is fully dismantled and should be checked for faults as described in Chapter 4. Note that the top mounting eye can be unscrewed if the damper is to be renewed.

XL125R and XL200R

3　Clamp the unit top mounting in a vice with padded jaws and use two C-spanners to slacken the spring adjuster locknut. Unscrew the locknut and slowly unscrew the adjusting nut until spring pressure is released. Withdraw the nuts and spring.

4　Holding the unit lower mounting in a vice, slacken the locknut and unscrew the mounting from the damper rod, followed by the seat stop, spring guide, spring seat and dust seal. The damper unit is sealed and cannot be dismantled further. Renew the unit if it is worn, dented, damaged or showing signs of oil leakage. To test it, pull out the rod to its fullest extent and place the bottom end on a set of scales. Measure the force necessary to compress the rod by 10 mm (0.4 in); if less than the limits given in the Specifications Section of this Chapter the unit is worn out, if more than the limit, the rod is probably bent or the unit is damaged in such a way as to produce the extra friction. In either case the unit must be renewed. Measure the spring length and renew it if it has settled to the service limit specified, or shorter. Check the top mounting as described in Section 24.

5　Reassembly is the reverse of the dismantling sequence. On refitting the lower mounting, align its cutout with the pin on the seat stop, then apply thread locking compound to the damper rod threads and tighten the locknut to a torque setting of 6.0 – 7.5 kgf m (43 – 54 lbf ft). Do not forget to set the spring's installed length as described in Section 7 of this Chapter before refitting the unit to the machine.

XR200R-C and XR200R – 1981 and 1982 models

6　Clamp the unit top mounting in a vice with padded jaws, taking care to avoid damage to the hose connection, then use C-spanners to slacken fully the spring lock and adjuster nuts until spring pressure is released and the spring stop can be pulled off the bottom mounting.

7　Examine the unit and spring as described above. The damper unit is sealed and must be renewed if faulty.

8　To dismantle the reservoir, wrap tape around its bottom cap and remove it, using pliers. Depress the valve core to release the nitrogen gas in the reservoir, ensuring that the valve is pointing away from your eyes to avoid the risk of dirt or debris being blown into them. Remove the valve core and hold the assembly over a drip tray while the hose connections are unscrewed. Remove the spring adjuster and locknuts. The unit is now dismantled as far as possible.

9　On reassembly, refit the spring adjuster and locknuts to the damper body, then clamp the unit top mounting in a vice with padded jaws so that the oil hose orifice is uppermost. Compress fully the damper rod and fill the unit with Automatic Transmission Fluid or the chosen grade of fork oil. Slowly pull out the damper rod while filling to ensure that the unit is filled to capacity with no trapped air pockets.

10　Pump the damper rod slowly several times to remove all traces of trapped air. This must be done very carefully indeed if good performance is to be obtained. Use a rag to catch the spilt oil, then refill the unit as described above. Prime the hose with oil, renew the sealing O-ring at each end of the hose and screw it into the damper body. Pull the damper rod fully out and refill the hose with oil.

11　Fill the reservoir with oil and screw it on to the hose. Holding the reservoir uppermost, tighten securely the hose connections and locknut, to a torque setting of 2.0 – 3.5 kgf m (14.5 – 25 lbf ft) if possible. Move the damper rod slowly through its full stroke; if the movement is smooth and even, all is well, but if it is jerky and irregular, air is present in the system and must be removed by careful bleeding, as described above.

12　When all air is removed, refit the valve to the reservoir and have it pressurised to 284 psi (20 kg/cm^2) with nitrogen gas. Honda recommend only the use of nitrogen. Check the damper compression force as described in paragraphs 4 above. If all is well, refit the spring, adjust its installed length as required and refit the unit to the machine.

XR200R – 1983 model

13　Remove the unit from the machine and remove the spring and remote reservoir as described above to leave the bare damper. The unit on this model can be dismantled and rebuilt as follows.

14　Using a hammer and a 3 mm ($\frac{1}{8}$ in) punch, unscrew the cap at the bottom of the unit body, then lift it until the wire circlip or stop ring can be displaced with a sharp instrument. The complete damper rod assembly can then be withdrawn.

15　Cut the piston ring and pull it off the piston, then displace all O-rings; the piston ring and O-rings should be renewed as a matter of course whenever the unit is dismantled. Examine the damper rod and body, looking for scratches, score marks and other signs of wear or for bending and distortion due to accidental impact. Renew any component found to be worn or damaged.

16　Two special tools, a guide and an installing sleeve (Honda Part Numbers 07947 – KA30100 and 07974-KA30200 respectively), are available to assist the delicate operation of refitting the piston ring to the piston. If these are not available, use grease or insulating tape to minimise the risk of damage as the piston ring is very carefully eased into place. If in doubt, take the assembly to a Honda Service Agent for the work to be done.

17　Refit the O-rings to the damper rod, coat the assembly in oil and refit it into the body. Refit the stop ring to secure the rod, ensuring it seats correctly in its groove, then refit the cap and tighten it securely with the hammer and punch. Complete reassembly as described above.

XR200R – 1986 and 1987 models

18　As previously noted, the unit fitted to later XR200R models has a more sophisticated damping system which incorporates a number of adjustments. This does, however, mean that the unit is much more complicated and therefore more difficult to strip and reassemble. Also, not all the internal components are available separately, thus making it doubtful whether it is worthwhile repairing a badly worn or damaged suspension unit. Note that the unit contains pressurised nitrogen gas, as described above.

19　In view of the above and of the need for a number of special tools and workshop facilities during the course of work, it is recommended that these units be regarded as sealed for all practical purposes. The average owner would be well advised to take the complete unit to a Honda Service Agent or to a suspension expert for attention.

27 Footrests: general – XR80R and XR100R

These models are now fitted with two separate hinged, footrest assemblies which are each secured by two bolts to the bottom of the frame rear downtubes.

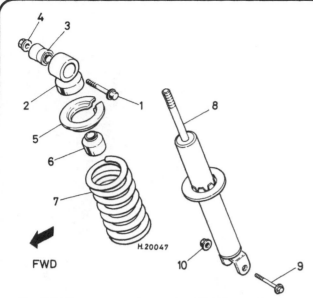

Fig. 7.12 Rear suspension unit – XR80R and XR100R models

1	Bolt	6	Stop rubber
2	Top mounting eye	7	Spring
3	Bush	8	Damper unit
4	Nut	9	Bolt
5	Retaining collar	10	Nut

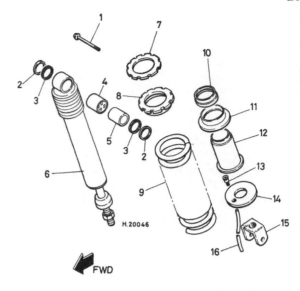

Fig. 7.13 Rear suspension unit – XL125R and XL200R models

1	Bolt	9	Spring
2	Sealing cap – 2 off	10	Dust seal
3	Dust seal – 2 off	11	Spring seat
4	Bush	12	Spring guide
5	Inner sleeve	13	Pipe joint
6	Damper unit	14	Seat stop
7	Adjuster locknut	15	Lower mounting
8	Spring adjuster	16	Drain pipe

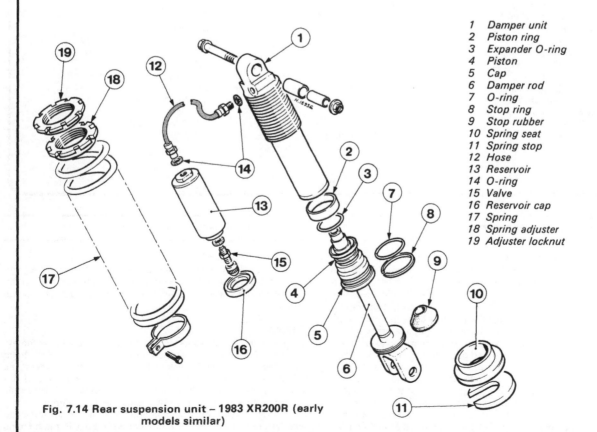

1 Damper unit
2 Piston ring
3 Expander O-ring
4 Piston
5 Cap
6 Damper rod
7 O-ring
8 Stop ring
9 Stop rubber
10 Spring seat
11 Spring stop
12 Hose
13 Reservoir
14 O-ring
15 Valve
16 Reservoir cap
17 Spring
18 Spring adjuster
19 Adjuster locknut

Fig. 7.14 Rear suspension unit – 1983 XR200R (early models similar)

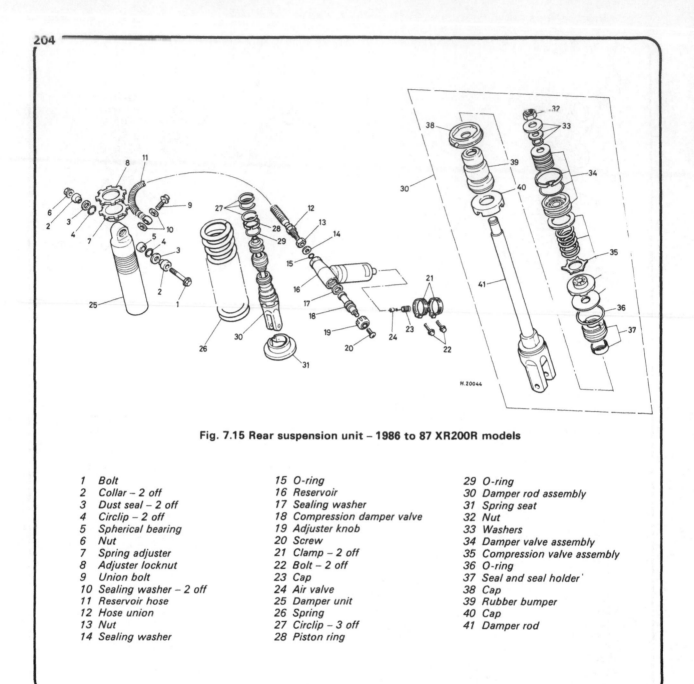

Fig. 7.15 Rear suspension unit – 1986 to 87 XR200R models

1 Bolt	15 O-ring	29 O-ring
2 Collar – 2 off	16 Reservoir	30 Damper rod assembly
3 Dust seal – 2 off	17 Sealing washer	31 Spring seat
4 Circlip – 2 off	18 Compression damper valve	32 Nut
5 Spherical bearing	19 Adjuster knob	33 Washers
6 Nut	20 Screw	34 Damper valve assembly
7 Spring adjuster	21 Clamp – 2 off	35 Compression valve assembly
8 Adjuster locknut	22 Bolt – 2 off	36 O-ring
9 Union bolt	23 Cap	37 Seal and seal holder
10 Sealing washer – 2 off	24 Air valve	38 Cap
11 Reservoir hose	25 Damper unit	39 Rubber bumper
12 Hose union	26 Spring	40 Cap
13 Nut	27 Circlip – 3 off	41 Damper rod
14 Sealing washer	28 Piston ring	

28 Speedometer drive: general – XL125R-F

1   The speedometer drive gearbox is mounted on the front wheel, on the hub right-hand side. The gearbox must be regarded as a sealed unit which cannot be repaired and requires no maintenance except for packing with grease whenever the wheel bearings are lubricated; note that the only component that can be removed if worn is the gearbox drive gear although the thrust washer(s) behind it are also listed separately.
2   The drive ring fitted in the hub can be removed as soon as the hub dust seal has been levered out. On refitting, ensure that the drive ring tabs are located correctly in the slots in the hub and that the dust seal is renewed. Smear a small amount of grease over the drive ring and seal lips before refitting the wheel to the machine. Note that the commonest fault with this type of speedometer drive is that the drive ring's raised ears are flattened by careless refitting of the gearbox; the ears can be bent back into position if care is exercised.

29 Front wheel: removal and refitting

XL125R-C and XL200R
1   The spindle is threaded into the left-hand fork lower leg and must be unscrewed on wheel removal. On refitting, tighten the spindle securely, then tighten the nut to the specified torque setting.

XR200R
2   The spindle is threaded into the left-hand fork lower leg, no nut being fitted, and is clamped by four nuts to the right-hand lower leg. On removal, slacken the four nuts then slacken the spindle itself before removing the clamp completely and un-screwing the spindle. If suitable care is exercised, it is only necessary to slacken the clamp instead of removing it.
3   On refitting, refit the spindle and screw it into the left-hand leg, then refit the clamp (if removed) with the 'UP' mark facing upwards. Tighten the spindle to a torque setting of 5.0 – 8.0 kgf

m (36 − 58 lbf ft), reconnect the front brake cable, push the machine off its stand and pump the forks vigorously up and down to align the right-hand leg on the spindle. Tighten the top clamp nuts first, then the bottom nuts; all clamp nuts should be tightened to a torque setting of 1.0 − 1.4 kgf m (7 − 10 lbf ft)..

All drum brake models
4   If the brake appears spongy after the wheel has been refitted, slacken the spindle nut (or spindle) and apply the brake firmly to centralise the shoes and backplate on the drum. Maintain brake pressure while tightening the spindle nut (or spindle) to the specified torque setting.

XL125R-F
5   Slacken the four nuts securing the spindle clamp to the fork right-hand lower leg, then slacken the spindle itself; there is no need to remove the clamp completely.
6   Lift the machine on to a stand so that it is supported securely upright with the front wheel clear of the ground, then unscrew the spindle and withdraw it while holding the wheel in position. Noting carefully how it is aligned, withdraw the speedometer gearbox from the wheel and lower the wheel away from the machine. There is no need to disconnect the speedometer cable; the wheel can be rotated about the fork left-hand lower leg so that the gearbox can be detached easily. Once the wheel has been removed, the hub left-hand spacer can be withdrawn. It is advisable to wedge a piece of wood between the pads to prevent the risk of the pistons being displaced, and allowing fluid to leak, should the brake lever be applied accidentally while the wheel is removed.
7   On refitting, insert the spacer into the hub left-hand side and check that the thread in the fork left-hand lower leg and the spindle clamp on the right-hand lower leg are completely clean. If disturbed, the spindle clamp is refitted with the 'UP' mark facing upwards. Thoroughly clean the spindle, removing all traces of dirt, corrosion and old grease, then smear clean grease along its length.
8   Lift the wheel into position so that the disc passes between the brake pads and ensuring that the spacer is not displaced, then refit the speedometer gearbox ensuring that the raised ears of the drive ring engage with the slots in the gearbox drive gear. Align the wheel with both fork legs and refit the spindle, passing it through the loosely assembled clamp.
9   Lightly tighten the two upper clamp nuts then rotate the speedometer gearbox so that the cable runs straight to its

clamp on the lower leg; a lug on the gearbox will butt against the rear of the lug on the lower leg to ensure this. Tighten the wheel spindle to a torque setting of 5.0 − 8.0 kgf m (36 − 58 lbf ft) and apply the brake lever repeatedly until the pads are pushed back into contact with the disc and full brake pressure is restored.
10   Push the machine off its stand, apply the front brake and pump the forks vigorously up and down to align the right-hand lower leg on the spindle. Tighten securely first the upper two clamp nuts, then the lower two nuts, all to a torque setting of 1.0 − 1.4 kgf m (7 − 10 lbf ft).
11   Before taking the machine out on the road, check that the forks, brake and speedometer are functioning correctly and that the wheel will rotate freely.

---

30 Front wheel: modifications

XL80S − 1983 on, XR80 − 1983 on, XL100S − 1983 on and XR100 − 1983 on
1   Note that the wheel layout is changed slightly as the wheel hub is of a similar type to that shown in Fig. 5.2.

XL125R-F
2   While the wheel is very different in appearance, working procedures are substantially the same as for drum brake models. Note that the hub right-hand oil seal must be carefully prised out to reduce the speedometer drive ring and the hub left-hand cover and oil seal must be removed before the bearings can be extracted. While the brake disc does not prevent wheel bearing removal and can therefore remain in position while the bearings are overhauled, it is recommended that the disc is first removed whenever such work is to be undertaken to prevent the risk of distorting or damaging it. Refer to the accompanying illustration for details.

XR80R, XR100R, XL125R-C, XL200R and XR200R − 1986 on
3   While the hub is slightly different in appearance, no longer having the separate spoke plate fitted to earlier models, working procedures remain unchanged. Note that the hub right-hand spacer has an integral dust shield on the later XR200R models, and the change in brake drum diameter on XR80R and XR100R models.

29.5 Wheel removal, XL125R-F − slacken spindle clamp nuts and unscrew spindle

29.6 Speedometer drive cable is disconnected by removing retaining screw − not necessary for wheel removal

29.7a Insert spacer into hub left-hand side before refitting

29.7b Note 'Up' mark and arrow cast into spindle clamp to show which way up it must be fitted

29.8 Raised ears of speedometer drive ring must engage with slots in gearbox drive gear

29.9 Rotate speedometer gearbox until its lug butts against fork lower leg locating lug

29.10 Tighten spindle clamp nuts exactly as described – do not try to close gap between clamp lower end and fork leg

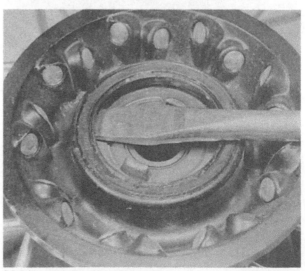

30.2 XL125R-F – oil seal must be removed to withdraw speedometer drive ring

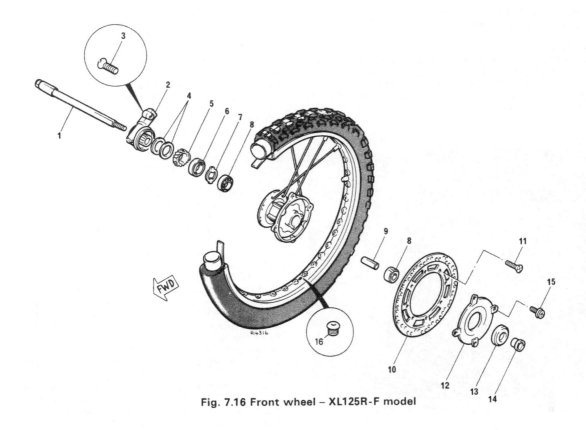

Fig. 7.16 Front wheel – XL125R-F model

1	Spindle	5	Speedometer gear
2	Speedometer gearbox	6	Dust seal
3	Screw	7	Drive ring
4	Thrust washer – 2 off	8	Bearing – 2 off

9	Spacer	13	Oil seal
10	Brake disc	14	Spacer
11	Bolt – 4 off	15	Screw – 4 off
12	Cover	16	Plug

31 Rear wheel: removal and refitting

XR80R and XR100R

1 Apart from the fact that there is no split pin securing the wheel spindle nut and that there is no longer a brake torque arm to be disconnected, rear wheel removal on these models is as described in Chapter 5.

2 On refitting, ensure that the brake backplate slot engages correctly with the lug in the swinging arm fork end.

XL125R and XL/XR200R

3 Support the machine securely on a stand so that the rear wheel is raised clear of the ground. On XL models unscrew the brake adjusting nut and disconnect the brake rod from the operating arm; on XR models press forward against spring pressure the retaining plate and disengage the cable from the operating arm.

4 On all models slacken the spindle nut sufficiently to enable both snail cams to be rotated forward until they can be hooked over the stopper pins, disengage the chain from the sprocket and hang it over the swinging arm fork end, then slacken the spindle nut further until the stopper plate on the right-hand side can be unhooked from the stopper pin. The wheel assembly can then be removed from the machine.

5 On reassembly, smear grease over the length of the spindle and refit it from right to left, passing through the right-hand snail cam, the stopper plate, the brake backplate, the wheel hub, the hub left-hand spacer and the left-hand snail cam before the spindle nut is refitted. Note that the snail cams are fitted with their marked surfaces outwards.

6 Offer up the wheel assembly to the machine ensuring that the lug on the swinging arm right-hand fork inside end engages correctly with the groove in the brake backplate and that the small hole in the stopper plate fits over the stopper pin. Hook the snail cams over the stopper pins and tighten the wheel spindle nut until the cams are just free to move.

7 Adjust the chain as described earlier in this Chapter, then connect the rear brake again; on XR200R models be careful to ensure that the retaining plate is securely engaged around the operating arm clevis pin and that the latter is securely fastened by its split pin or R-clip.

8 On all models, tighten carefully the brake adjusting nut until a rubbing sound as the wheel is spun indicates that the brake shoes are coming into contact with the drum, then unscrew the nut until the sound disappears. Free play at the pedal tip should then be 20 – 30 mm ($\frac{3}{4}$ – $1\frac{1}{4}$ in) for XL and late XR models and 15 – 20 mm ($\frac{5}{8}$ – $\frac{3}{4}$ in) for early XR models. Apply the brake firmly to centralise the shoes and backplate on the drum and tighten the spindle nut to a torque setting of 8.0 – 11.0 kgf m (50 – 80 lbf ft).

9 Check that the chain is correctly adjusted and lubricated, that the rear wheel is free to revolve smoothly and easily, that the rear brake and stop lamp switch (where fitted) are adjusted

correctly and operating properly and that all disturbed components are securely fastened.

All models

10 If the brake appears spongy after chain adjustment or rear wheel removal, slacken the spindle nut and apply the brake firmly to centralise the shoes and backplate on the drum. Maintain brake pressure while tightening the spindle nut to the specified torque setting.

31.4 Hook snail cams over stopper pins to allow chain removal

32 Rear wheel: modifications

XL100S and XR100

1   Although the wheel itself is unchanged, the sprocket mounting and cush-drive have been modified and are now of the type shown in Fig. 5.5 of Chapter 5. Proceed as described in Chapter 5 for the XL/XR80 models.

XR80R and XR100R

2   Again, the wheel itself is the same in layout but the cush drive has been deleted completely, the sprocket being attached directly to the hub by means of four bolts and nuts. Refer to the accompanying illustration for details.

XL125R and all XL/XR200R models

3   The hub left-hand wheel bearing and oil seal are secured by a threaded retainer which is screwed into the hub and staked at four points to prevent it from unscrewing.

4   The retainer is removed and refitted using a special tool. Part Number 07710-0010401 in conjunction with an adaptor. Part Number 07710-0010100 (for 200 models) or Part Number 07710-010300 (for 125 models). If these are not available, and a peg spanner of suitable size cannot be found, obtain a steel strip (an old tyre lever would be ideal) and drill two holes in it to correspond with two diagonally opposite holes in the retainer. Pass a small bolt through each hole and secure them with nuts to complete a fabricated peg spanner of the type shown in the accompanying illustration.

5   On refitting, apply grease to the retainer threads and screw it into the hub. Tighten it securely and use a hammer and punch to stake it to the hub at four points around its outer edge.

6   On XR200R – 1986 and 1987 models, the cush drive has been deleted, the sprocket being attached directly to the hub by means of six bolts and nuts. Refer to the accompanying illustration for details.

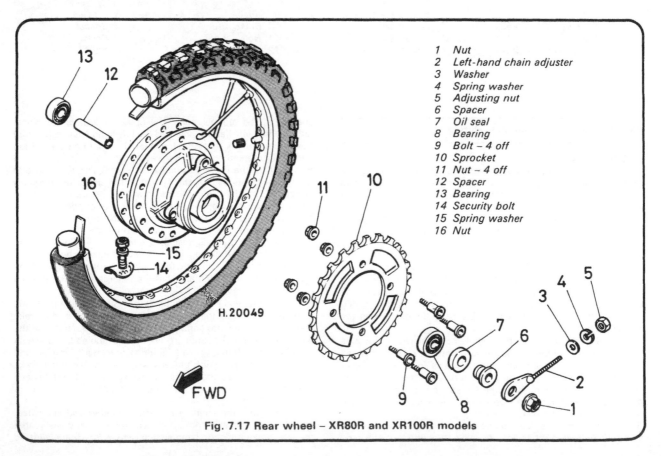

1   Nut
2   Left-hand chain adjuster
3   Washer
4   Spring washer
5   Adjusting nut
6   Spacer
7   Oil seal
8   Bearing
9   Bolt – 4 off
10 Sprocket
11 Nut – 4 off
12 Spacer
13 Bearing
14 Security bolt
15 Spring washer
16 Nut

H.20049

FWD

Fig. 7.17 Rear wheel – XR80R and XR100R models

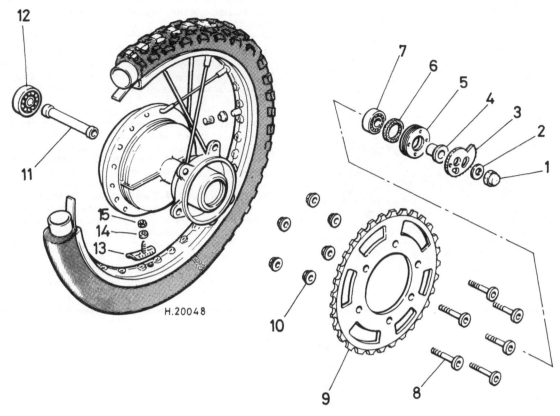

Fig. 7.18 Rear wheel – 1986 to 87 XR200R models

1 Nut	5 Bearing retainer	9 Sprocket	13 Securing bolt
2 Washer	6 Oil seal	10 Nut – 6 off	14 Spring washer
3 Left-hand chain adjuster	7 Bearing	11 Spacer	15 Nut
4 Spacer	8 Bolt – 6 off	12 Bearing	

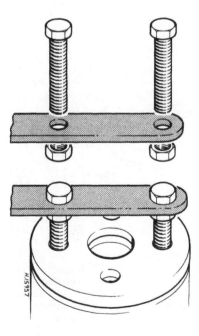

Fig. 7.19 Fabricated peg spanner for removing and refitting the wheel bearing retainer

33 Front brake disc: examination, removal and refitting – XL125R-F

1   The brake disc can be checked for wear and for warpage whilst the front wheel is still in the machine. Using a micrometer, measure the thickness of the disc at the point of greatest wear. If the measurement is less than the recommended service limit, the disc should be renewed. Check the warpage (runout) of the disc by setting up a suitable pointer close to the outer periphery of the disc and spinning the front wheel slowly. If the total warpage is more than 0.30 mm (0.012 in), the disc should be renewed. A warped disc, apart from reducing the braking efficiency, is likely to cause juddering during braking and will cause the brake to bind when it is not in use.

2   The brake disc should also be checked for bad scoring on its contact area with the brake pads. If any of the above mentioned faults are found, the disc should be removed from the wheel for renewal or for repair by skimming. Such repairs should be entrusted only to a reputable engineering firm. A local motorcycle dealer may be able to assist in having the work carried out.

3   Remove the front wheel from the machine as described in Section 29 of this Chapter. Unscrew the four screws and lift away the cover from the centre of the disc, then unscrew the four mounting bolts and withdraw the disc.

4   Refitting is the reverse of the above, but when tightening the retaining nuts do so in an even and diagonal sequence to avoid stress on the disc. Secure to the recommended torque setting.

33.3 XL125R-F brake disc is retained by dome-headed hexagon socket bolts

34 Master cylinder: examination and renovation – XL125R-F

1   If the regular checks described in Section 11 reveal the presence of a fluid leak, or if a loss of brake pressure is encountered at any time, the handlebar-mounted master cylinder assembly must be removed and dismantled for checking.

2   Disconnect the stop lamp switch wires at the switch. The switch need not be disturbed unless the master cylinder is to be renewed. Place a clean container beneath the caliper unit and run a clear plastic tube from the caliper bleed nipple to the container. Unscrew the bleed nipple by one full turn and drain the system by operating the brake lever repeatedly until no more fluid can be seen issuing from the nipple.

3   Position a pad of clean cloth beneath the point where the brake hose joins the master cylinder to prevent brake fluid from dripping. Unscrew the union gland nut, then unscrew the union itself. Once any excess fluid has drained from the union connection, wrap the end of the hose in rag or polythene and then attach it to a point on the handlebars.

4   Remove the brake lever by unscrewing its shouldered pivot screw and locknut, remove the reservoir cover and diaphragm,

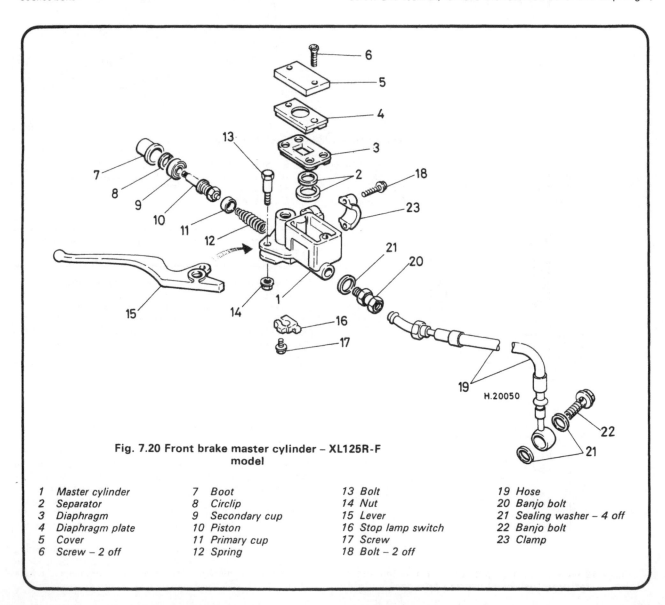

Fig. 7.20 Front brake master cylinder – XL125R-F model

1	Master cylinder	7	Boot	13	Bolt	19 Hose
2	Separator	8	Circlip	14	Nut	20 Banjo bolt
3	Diaphragm	9	Secondary cup	15	Lever	21 Sealing washer – 4 off
4	Diaphragm plate	10	Piston	16	Stop lamp switch	22 Banjo bolt
5	Cover	11	Primary cup	17	Screw	23 Clamp
6	Screw – 2 off	12	Spring	18	Bolt – 2 off	

then remove the two clamp bolts and withdraw the master cylinder assembly.

5   Use the flat of a small screwdriver to prise out the rubber dust seal boot. This will expose a retaining circlip which must be removed using a pair of circlip pliers which have long, straight jaws. With the circlip removed, the piston and cup assembly can be pulled out. Be very careful to note the exact order in which these components are fitted.

6   Note that if a vice is used to hold the master cylinder at any time during dismantling and reassembly, its jaws must be padded with soft alloy or wooden clamps and the master cylinder must be wrapped in soft cloth to obviate any risk of the assembly being marked or distorted.

7   Place all the master cylinder component parts in a clean container and wash each part thoroughly in new brake fluid. Lay the parts out on a sheet of clean paper and examine each one as follows.

8   Inspect the unit body for signs of stress failure around both the brake lever pivot lugs and the handlebar mounting points. Carry out a similar inspection around the hose union boss. Examine the cylinder bore for signs of scoring or pitting. If any of these faults are found, the unit body must be renewed.

9   Inspect the surface of the piston for signs of scoring or pitting and renew it if necessary. It is advisable to discard all the components of the piston assembly as a matter of course as the replacement cost is relatively small and does not warrant re-use of components vital to safety. If measuring equipment is available, compare the dimensions of the master cylinder bore and the piston with those given in the Specifications Section of this Chapter and renew any component that is worn beyond the set wear limit. Inspect the threads of the brake hose union for any signs of failure and renew the union if in the slightest doubt. Renew each of the gasket washers located one either side of the hose union.

10   Check before reassembly that any traces of contamination remaining within the reservoir body have been removed. Inspect the diaphragm to see that it is not perished or split. It must be noted that any reassembly work must be undertaken in ultra-clean conditions. Particles of dirt entering the component will only serve to score the working points of the cylinder and thereby cause early failure of the system.

11   When reassembling and fitting the master cylinder, follow the removal and dismantling procedures in the reverse order whilst paying particular attention to the following points. Make sure that the piston components are fitted the correct way round and in the correct order. Immerse all of these components in new brake fluid prior to reassembly and refer to the figure accompanying this text when in doubt as to their fitted positions. When refitting the master cylinder assembly to the handlebar, position the assembly so that the reservoir will be exactly horizontal when the machine is in use. Tighten the clamp top bolt first, and then the bottom bolt, to a torque setting of 0.8 – 1.2 kgf m (6 – 9 lbf ft). Connect the brake hose to the master cylinder, ensuring that a new sealing washer is fitted and tightening the hose union to the specified torque setting.

12   Bleed the brake system after refilling the reservoir with new hydraulic fluid, then check for leakage of fluid whilst applying the brake lever. Push the machine forward and bring it to a halt by applying the brake. Do this several times to ensure that the brake is operating correctly before taking the machine for a test run. During the run, use the brakes as often as possible and on completion, recheck for signs of fluid loss.

35 Brake hose: examination and renovation: XL125R-F

1   A flexible brake hose is used as a means of transmitting hydraulic pressure to the caliper unit once the front brake lever is applied.

2   When the brake assembly is being overhauled, or at any time during a routine maintenance or cleaning procedure, check the condition of the hose for signs of leakage, damage, deterioration or scuffing against any cycle components. Any such damage will mean that the hose must be renewed immediately. The union connections at either end of the hose must also be in good condition, with no stripped threads or damaged sealing washers. Do not tighten these union bolts over the recommended torque setting as they are easily sheared if overtightened.

36 Brake caliper: examination and renovation – XL125R-F

1   If the regular checks described in Section 11 reveal the presence of any fluid leaks or of any wear, damage or corrosion which will impair the caliper's efficiency, the unit must be removed and dismantled for checking. Start by removing the brake pads as described in Section 11.

2   Pull the mounting bracket out of the caliper body. If they are damaged or worn, the two axle bolts can be unscrewed from the bracket and renewed individually. Detach the rubber dust seal from the upper axle bolt and pull out of the caliper body the rubber dust seal/bush which fits around the lower axle bolt; these should be renewed if cracked, split or worn in any way.

3   The simplest way of ejecting the pistons from the caliper bore is to place the caliper in a plastic bag, tying the neck tightly around the brake hose to catch the inevitable shower of brake fluid. Apply the brake lever repeatedly, using normal hydraulic pressure to force out the pistons. Ensure that the pistons leave the bores at the same time; if one sticks at any point the other piston must be restrained by firm hand pressure so that the full hydraulic pressure can overcome the resistance. It would be very difficult to extract one piston alone from this type of caliper without risking damage. When the pistons have been ejected, apply the brake lever until all remaining fluid is pumped into the bag, then unscrew the union bolt to disconnect the brake hose and remove the caliper from the machine, taking care to prevent the spillage of any brake fluid.

4   An alternative to the above method will require a source of compressed air. Attach a length of clear plastic tubing to the bleed nipple, placing the tube lower end in a suitable container. Open the bleed nipple by one full turn and pump gently on the front brake lever to drain as much fluid as possible from the hydraulic system. When no more fluid can be seen issuing from the bleed nipple, tighten it down again, withdraw the plastic tube and disconnect the hydraulic hose at the caliper union. Wrap a large piece of cloth lcosely around the caliper and apply a jet of compressed air to the union orifice. Be careful not to use too high an air pressure or the pistons may be damaged.

5   If either of the above methods fails to dislodge the pistons, no further time should be wasted and the complete caliper assembly should be renewed. If the pistons and caliper bores are so badly damaged or so corroded that hydraulic or air pressure cannot move the pistons, the caliper will not be safe for further use even if a method can be devised of removing the pistons and cleaning the bearing surfaces.

6   When each piston has been removed, it should be placed in a clean container to ensure that it cannot be damaged and the seals should be picked out of the caliper bores, pressing each inwards to release it, and discarded. Wrap the exposed end of the hydraulic hose in clean rag or polythene to prevent the entry of dirt. Carefully wash off all spilt brake fluid from the machine using fresh water; brake fluid is an excellent paint stripper and will attack any painted metal or plastic component.

7   Clean the caliper components thoroughly in hydraulic fluid. **Never** use petrol or cleaning solvent for cleaning hydraulic brake parts otherwise the rubber components will be damaged. Discard all the rubber components as a matter of course. The replacement cost is relatively small and does not warrant re-use of components vital to safety. Check the pistons and caliper cylinder bores for scoring, rusting or pitting. If any of these defects are evident it is unlikely that a good fluid seal can be maintained and for this reason the components should be

renewed. If measuring equipment is available, compare the dimensions of the caliper bores and the pistons with those given in the Specifications Section of this Chapter, renewing any component that is worn to beyond the set wear limit.

8   Inspect the shank of each axle bolt for any signs of damage or corrosion and clean or renew each one as necessary; if the matching bores in the caliper body are worn or damaged, the body must be renewed. It is essential that the body can move smoothly and easily on the axle bolts. Check also that the bleed nipple is clear and its threads undamaged, and that the dust cap is in good condition; renew either component if faulty.

9   On reassembly, the components must be absolutely clean and dry; cleanliness is essential to prevent the entry of dirt into the system. Soak the new seals in hydraulic fluid before refitting them to the caliper bores; check that each seal is securely located in its groove without being twisted or distorted. Smear a liberal quantity of clean brake fluid over the pistons and caliper bores then very carefully insert the pistons (with their recessed faces outwards) with a twisting motion, as if screwing them in, to avoid damaging or dislodging the seals. When the piston outer ends project 3 – 5 mm above the caliper they are

fully refitted; do not press them fully into the caliper bores. Wipe off any surplus brake fluid.

10 Fit the rubber bush and seal to the caliper body and mounting bracket, check that the axle bolts are tightened securely on the bracket and smear silicone- or PBC-based brake caliper grease over the bolt bearing surfaces. Pack a small amount of grease into the matching caliper passages and refit the mounting bracket, ensuring that the seals are correctly engaged on their locating shoulders. Renew the sealing washers if worn or distorted and connect the brake hose to the caliper. Check that the hose is correctly aligned and tighten the union bolt to the specified torque setting. Refit the pads to the caliper and the caliper assembly to the machine, as described in Section 11.

11 Refill the master cylinder reservoir with new hydraulic fluid and bleed the system by following the procedure given in Section 37 of this Chapter. On completion of bleeding, carry out a check for leakage of fluid whilst applying the brake lever. Push the machine forward and bring it to a halt by applying the brake. Do this several times to ensure that the brake is operating correctly before taking the machine out on the road.

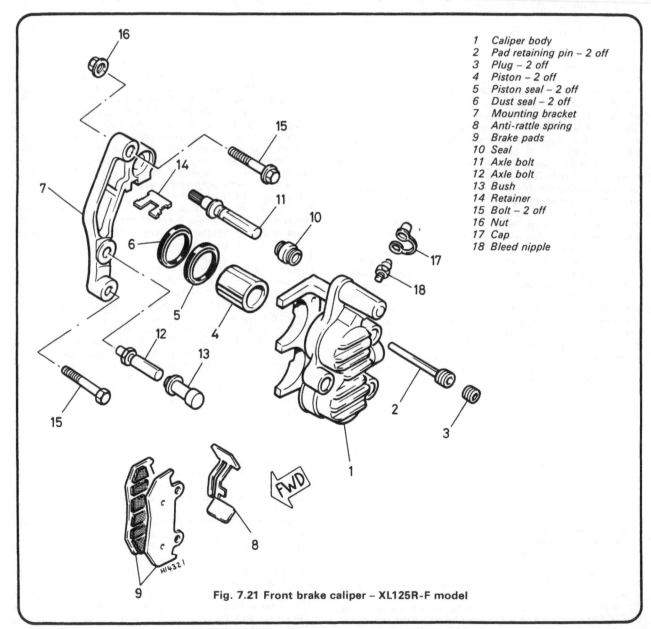

1   Caliper body
2   Pad retaining pin – 2 off
3   Plug – 2 off
4   Piston – 2 off
5   Piston seal – 2 off
6   Dust seal – 2 off
7   Mounting bracket
8   Anti-rattle spring
9   Brake pads
10 Seal
11 Axle bolt
12 Axle bolt
13 Bush
14 Retainer
15 Bolt – 2 off
16 Nut
17 Cap
18 Bleed nipple

Fig. 7.21 Front brake caliper – XL125R-F model

37 Bleeding the hydraulic brake system – XL125R-F

1   If the brake action becomes spongy, or if any part of the hydraulic system is dismantled, it is necessary to bleed the system in order to remove all traces of air. The procedure for bleeding the hydraulic system is best carried out by two people.
2   Check the fluid level in the reservoir and top up with fluid of the specified type if required. Keep the reservoir at least half full during the bleeding procedure; if the level is allowed to fall too far, air will enter the system requiring that the procedure be started again from scratch. Refit the reservoir cover to prevent the ingress of dust or the ejection of a spout of fluid.
3   Remove the dust cap from the caliper bleed nipple and clean the area with a rag. Place a clean glass jar below the caliper and connect a pipe from the bleed nipple to the jar. A clear plastic tube should be used so that air bubbles can be more easily seen. Place some clean hydraulic fluid in the glass jar so that the pipe is immersed below the fluid surface throughout the operation.
4   If parts of the system have been renewed, and thus the system must be filled, open the bleed nipple about one turn and pump the brake lever until fluid starts to issue from the clear tube. Tighten the bleed nipple and then continue the normal bleeding operation as described in the following paragraphs. Keep a close check on the reservoir level whilst the system is being filled.
5   Operate the brake lever as far as it will go and hold it in this position against the fluid pressure. If spongy brake operation has occurred it may be necessary to pump the brake lever rapidly a number of times until pressure is achieved. With pressure applied, loosen the bleed nipple about half a turn. Tighten the nipple as soon as the lever has reached its full travel and then release the lever. Repeat this operation until no more air bubbles are expelled with the fluid into the glass jar. When this condition is reached, the air bleeding operation should be complete, resulting in a firm feel to the brake operation. If sponginess is still evident, continue the bleeding operation; it may be that an air bubble trapped at the top of the system has yet to work down through the caliper.
6   When all traces of air have been removed from the system, top up the reservoir and refit the diaphragm and cover. Check the entire system for leaks, and check also that the brake system in general is functioning efficiently before using the machine on the road.
7   Brake fluid drained from the system will almost certainly be contaminated, either by foreign matter or more commonly by the absorption of water from the air. All hydraulic fluids are to some degree hygroscopic, that is, they are capable of drawing water from the atmosphere, and thereby degrading their specification. In view of this, and the relative cheapness of the fluid, old fluid should always be discarded.
8   Great care should be taken not to spill hydraulic fluid on any painted cycle parts; it is very effective paint stripper. Also, the plastic glasses in the instrument heads, and most other plastic parts, will be damaged by contact with this fluid.

38 Rear brake: modifications

1   On all models with Pro-Link rear suspension, note that the rear brake operating arm is fitted so that it points upwards when the rear wheel is in the machine.
2   On XR200R-1986 and 1987 models, note that there is an additional seal and a special washer fitted to the camshaft on the inside of the backplate. If the camshaft is disturbed for any reason, note carefully how the washer is fitted before disturbing it and ensure it is correctly fitted on reassembly.
3   Note also that the brake pedal mounting has been modified on these later XR200R models. To remove the pedal, unscrew the pinch bolt which is to be found inside the frame downtube, disconnect the operating rod from the arm on the brake

backplate and unhook the pedal return spring. Carefully prise the lever off the pivot splines and withdraw the lever and pivot, noting the presence of the two pivot seals; on reassembly these must be refitted one each side of the frame lug and the pivot itself must be well greased.
4   On XR80R and XR100R models the brake pedal mounting is now inside the right-hand footrest mounting bracket, the pedal pivoting on a separate sleeve which is fitted over a pivot bolt. To remove the pedal, disconnect the operating rod from the arm on the brake backplate and unhook the pedal return spring from the swinging arm. Unscrew the pedal pivot bolt, slacken the remaining footrest bracket mounting bolt and withdraw the pedal. The amount of wear of the pedal pivot inside diameter and of the sleeve outside diameter can be checked by direct measurement; renew either component if it is found to be excessively worn or if it shows any other signs of wear or damage. Apply a liberal smear of grease to all bearing surfaces on reassembly.

37.3 Bleeding the hydraulic brake system to remove air bubbles

39 Voltage regulator/rectifier unit: testing

1   All late models (except XR models) are fitted with a combined voltage regulator/rectifier unit to control the generator output. The unit is a sealed metal box easily identifiable by the colour of the five wires leading to it. Tests are as follows, but only the testers specified must be used for accurate results. If an insulation resistance tester using a high voltage source is used, it may damage the unit and give the operator a shock. An ordinary multimeter may be used but may not give exactly the same readings, as described in Section 20 of this Chapter for the CDI unit. A test conducted with a multimeter can be expected to produce a fair indication of the unit's condition but results should be confirmed by a Honda Service Agent using the correct equipment before any definite conclusions are reached.

XL100S-B, XL125S-B/C, XL185S-B, all XL100S, XL125S and XL185S models

2   On XL100S models the unit is mounted behind the right-hand side panel, in front of the air filter casing. On all other models it is mounted as described in Chapter 6 for the rectifier. Using only a Sanwa SP10D Electric Tester (Honda part number 07308-0020000) or a Kowa Electric Tester (TH-5H-1), make the meter connections as shown in the accompanying table and compare the results obtained with those given. Results should

be in kilo ohms on the Sanwa tester, or in the x 100 phm range with the Kowa unit.

3   If any results differ markedly from those given, the unit is proven faulty and must be renewed, although this should always be confirmed by an expert using the correct equipment before the unit is discarded. Repairs are not possible.

XL125R and XL200R models

4   The unit on these models is mounted on the left-hand side of the frame, immediately behind the steering head, and is tested as described above, using either of the two units mentioned or also the Kowa Digital Tester (Honda part number 07411-0020000) or KS-AHM-32-003 (US only). Make the meter connections as shown in the separate table; results should be obtained in kilo ohms on the Sanwa tester, or in the x 100 ohm range on either of the Kowa units.

39.4 Location of electrical components – XL125/200R – Finned AC regulator above voltage regulator/rectifier unit

SANWA: KΩ

+PROBE / -PROBE	YELLOW	PINK	RED	BLACK	GREEN
YELLOW		∞	1—25	∞	∞
PINK	10—250		1—25	1.5—3.5	6—150
RED	∞	∞		∞	∞
BLACK	5—120	5—120	10—250		5—45
GREEN	1—25	1—25	3—80	1.5—40	

KOWA: [x 100]

+PROBE / -PROBE	YELLOW	PINK	RED	BLACK	GREEN
YELLOW		∞	2—50	∞	∞
PINK	100—∞		2—50	6—150	50—∞
RED	∞	∞		∞	∞
BLACK	50—∞	50—∞	100—∞		50—500
GREEN	2—50	2—50	8—200	8—200	

H.1596B

Fig. 7.22 Voltage regulator/rectifier unit test table – XL100S-B, XL125S-B and C, XL185S-B, all XL100S, XL125S and XL185S models 1982 on

40 Resistor: testing – XL185S-B

1   A 30-watt resistor is fitted to this model, being connected to the main lighting switch to soak up the excess power generated by the lighting coil when the switch is in the 'P' position. The unit is mounted next to the ignition HT coil.

2   Maintenance is restricted to ensuring that the unit is clean and that its connection and mounting point are clean and securely fastened.

3   If problems such as persistent bulb-blowing arise which indicate a fault in the resistor, it can be tested by measuring the resistance between its wire terminal and the mounting point or a similar good earth; a value of 1 ohm should be measured. If the result obtained differs widely from this, the unit is faulty and must be renewed.

41 AC regulator: testing – XL125R, XL200R and XR200R-1986 and 1987 models

1   These models are fitted with a (second) regulator unit to control the output of the lighting coil. It is a heavily finned unit mounted just above the voltage regulator/rectifier unit, and requires no maintenance save to ensure that both the unit and its mountings and connections are clean and securely fastened.

2   If the lights are too dim, or if the bulbs keep blowing with melted filaments indicating excess power, the lighting output should be checked. Note that although information on outputs is not available for the XL125R, results should be very similar to those given for the XL200R models.

3   On XL125R models, switch the lights on; on all models switch the dipswitch to the main beam position. Remove the headlamp unit and connect an AC voltmeter between the headlamp blue and white wires then start the engine, increase speed slowly to 5000 rpm, and note the reading obtained. If the system is in good order a reading of 13.5 – 14.5 volts should be measured (XL200R models), or 12.5 – 13.5 volts (XR200R models).

4   If the result differs widely from that given, the generator or regulator must be at fault, but a check should be made that the bulbs are all of the correct type and rating, and that all switches and wiring are in good condition; this is particularly true if the output is low.

5   Repeat the test with the regulator disconnected; a reading of at least 15 volts should be obtained at 2500 rpm (XL200R). **Do not** attempt to measure the maximum output, which increases steadily to 25 volts at 10 000 rpm; an engine should never be run unloaded at high speed as serious engine damage will occur.

6   If the unregulated output is different from that listed, the generator is at fault; this can be confirmed by measuring the

RANGE: SANWA: kΩ
KOWA: x100Ω

+ PROBE / – PROBE	YELLOW	PINK	GREEN	RED	BLACK
YELLOW		∞	∞	1—20	∞
PINK	∞		∞	1—20	∞
GREEN	1—20	1—20		3—100	0.2—20
RED	∞	∞	∞		∞
BLACK	1—50	1—50	0.2—10	3—100	

H.1596.9

Fig. 7.23 Voltage regulator/rectifier unit test table – XL125R and XL200R models

resistance of the lighting coil which should be 0.2 – 1.0 ohm (XL200R), or 0.2 – 1.2 ohm (XR200R).

7   If the unregulated output is correct, the regulator must be at fault; this can only be tested using the equipment listed in Sections 20 and 39 of this Chapter for the CDI unit and regulator/rectifier unit of the XL125/200R models. If alternative equipment is used, the result should not be considered conclusive until confirmed by a Honda Service Agent using the correct equipment.

8   Setting the Sanwa unit to the kilo ohms scale and either of the Kowa units to the x 100 ohm scale, disconnect the regulator and connect the meter positive (+) probe to its white/yellow wire terminal, with the meter negative (–) probe to the green wire terminal; a reading of 10 – 900 ohms should be recorded for all XL models, 100 ohms to infinite resistance for the XR models. On reversing the meter connections exactly the same reading should be obtained. If the results obtained do not correspond with those given, the unit is faulty and must be renewed.

42 Headlamp bulb failure: general – all UK XL125S and XL185S models

1   If persistant bulb failure is encountered on any of these models, first check that the bulbs used are of the correct type and wattage and that they are of good quality. Incorrect bulbs are usually the cause of frequent bulb failures.

2   If the fault persists, check carefully the operation of the lighting switch and, particularly, the dip switch; use a multimeter or an ohmmeter to check that there is continuity between the various terminals as indicated in the relevant wiring diagram at the back of this Manual. A switch that is stiff through corrosion or sloppy due to excessive wear can cause excessive voltage surges as it is operated. For safety's sake, renew any switch that is found to be defective.

3   If the switches are in good order, check the headlamp mountings for security against vibration and then check all connections in the lighting circuit, ensuring that all are clean and securely fastened. Check with particular care the bulb holder and all earth connections (green wires).

4   One other possible cause of bulb failure is the effect of constant very high engine speed; high frequency vibration or fluctuations in the power supply at high engine speeds could well reduce bulb life noticeably. The only way of checking this is to ensure that the machine is used in a more relaxed manner for a long enough period so that firm conclusions can be reached.

5   If all of the above checks fail to cure bulb-blowing, the problem is most likely to be caused by voltage surges; the filament will have a characteristic melted appearance if this is the case. Honda have made available a regulator unit to be fitted into the lighting circuit, if necessary, to control the excess power. A local Honda Service Agent should be able to check whether or not the fault is due to a defect and may be able to supply details of the modification.

43 Bulbs: renewal

1   To renew the headlamp or pilot lamp bulb on XL125R models, first remove the two mounting bolts and withdraw the headlamp nacelle, then unscrew the two bolts which retain the headlamp assembly. Partially withdraw the assembly from its bracket and peel back the rubber cover over the headlamp bulb holder. The bulb holders and bulbs can then be removed and refitted as described in Chapter 6.

2   To renew the turn signal lamp bulbs on XL125R-F models, note that each lamp lens is retained by a single screw which passes through the rear (ie through the body) of the lamp unit.

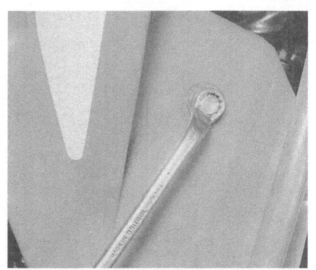

43.1a XL125R – remove mounting bolts to release headlamp nacelle ...

43.1b ... then remove mounting bolts to release headlamp assembly – align stamped reference marks on refitting to provide correct beam alignment

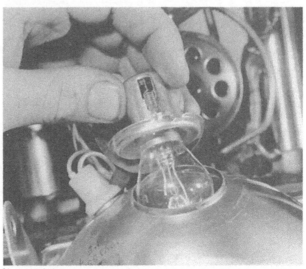

43.1c Peel back rubber cover and unhook headlamp bulbholder

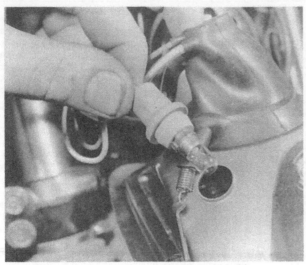

43.1d Pilot lamp bulb is a bayonet fitting in its holder

43.2 XL125R-F – turn signal lamp lens retaining screw is reached from rear of lamp unit

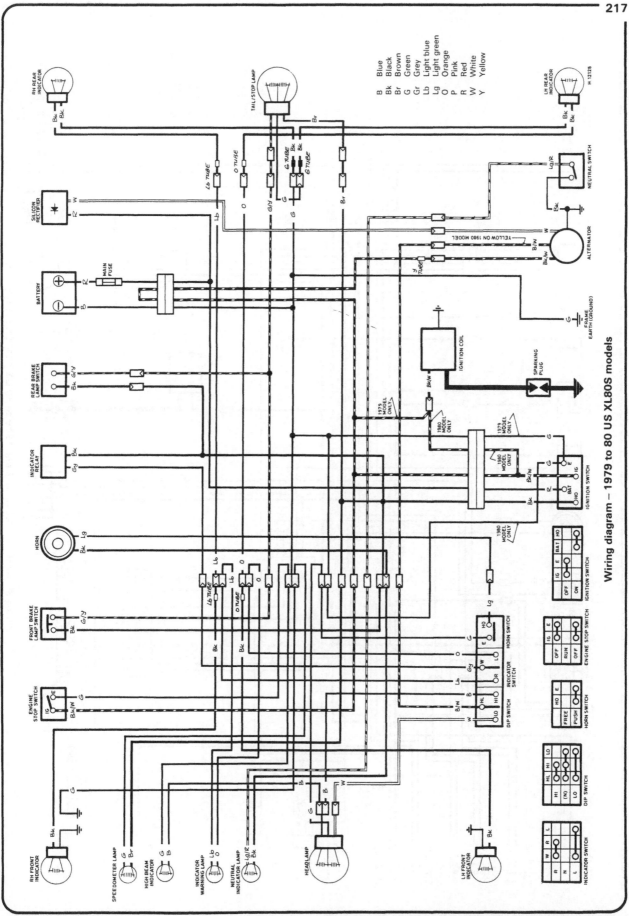

Wiring diagram – 1979 to 80 US XL80S models

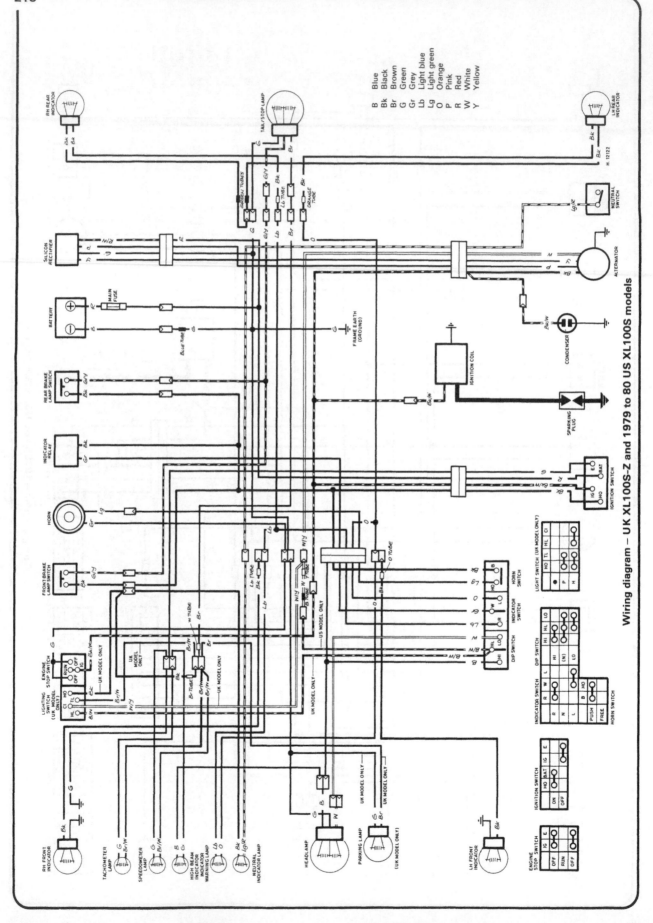

Wiring diagram – UK XL100S-Z and 1979 to 80 US XL100S models

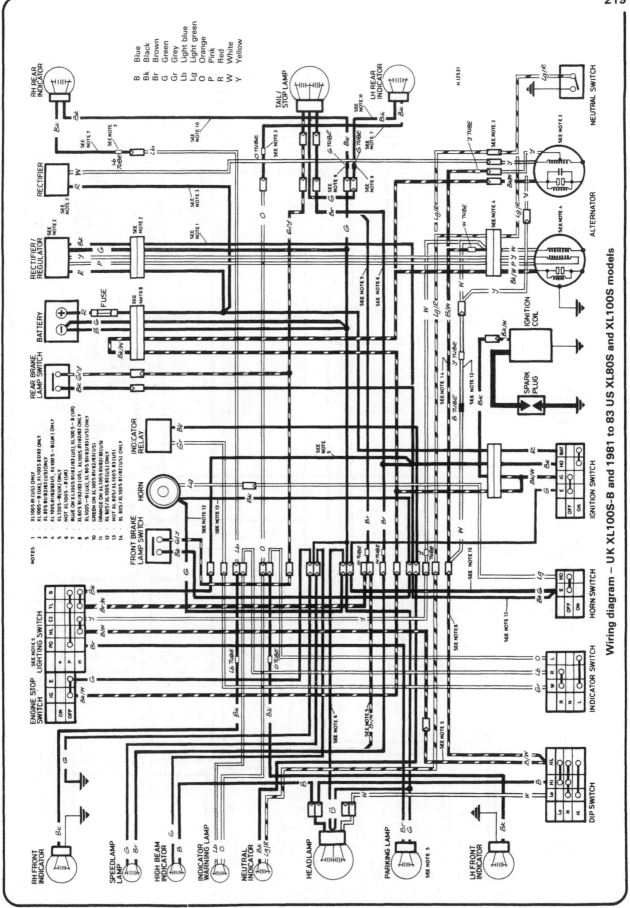

Wiring diagram – UK XL100S-B and 1981 to 83 US XL80S and XL100S models

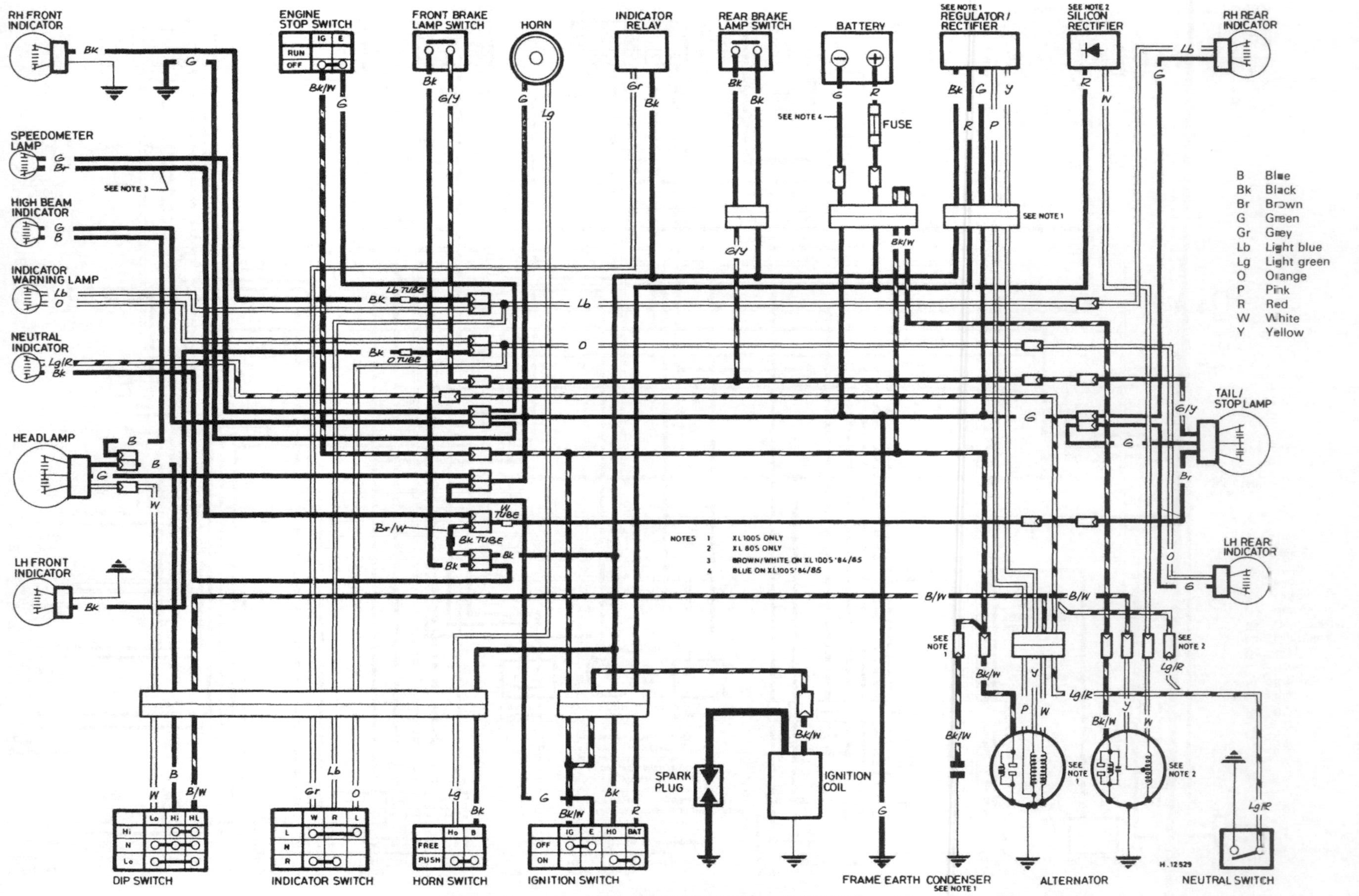

Wiring diagram – 1984 to 85 US XL80S and XL100S models

RH REAR INDICATOR

B	Blue
Bk	Black
Br	Brown
G	Green
Gr	Grey
Lb	Light blue
Lg	Light green
O	Orange
P	Pink
R	Red
W	White
Y	Yellow

TAIL/STOP LAMP

LH REAR INDICATOR

NEUTRAL SWITCH

SEE NOTE 2

ALTERNATOR

H.12323

SEE NOTE 2 SILICON RECTIFIER

SEE NOTE 1

SEE NOTE 1 REGULATOR/RECTIFIER

CONDENSER SEE NOTE 1

FRAME EARTH

FUSE

BATTERY

SEE NOTE 4

IGNITION COIL

SPARK PLUG

REAR BRAKE LAMP SWITCH

NOTES
1. XL100S ONLY
2. XL80S ONLY
3. BROWN/WHITE ON XL100S '84/85
4. BLUE ON XL100S '84/85

INDICATOR RELAY

IGNITION SWITCH

	BAT	HO	E	IG	
OFF					
ON					

HORN

HORN SWITCH

	HO	B
FREE		
PUSH		

FRONT BRAKE LAMP SWITCH

INDICATOR SWITCH

	L	
W	R	
	L	N	R

ENGINE STOP SWITCH

	IG	E
RUN		
OFF		

SEE NOTE 3

DIP SWITCH

	B/W	HL
Hi		
Lo		
	Hi	N	Lo

RH FRONT INDICATOR

SPEEDOMETER LAMP

HIGH BEAM INDICATOR

INDICATOR WARNING LAMP

NEUTRAL INDICATOR

HEADLAMP

LH FRONT INDICATOR

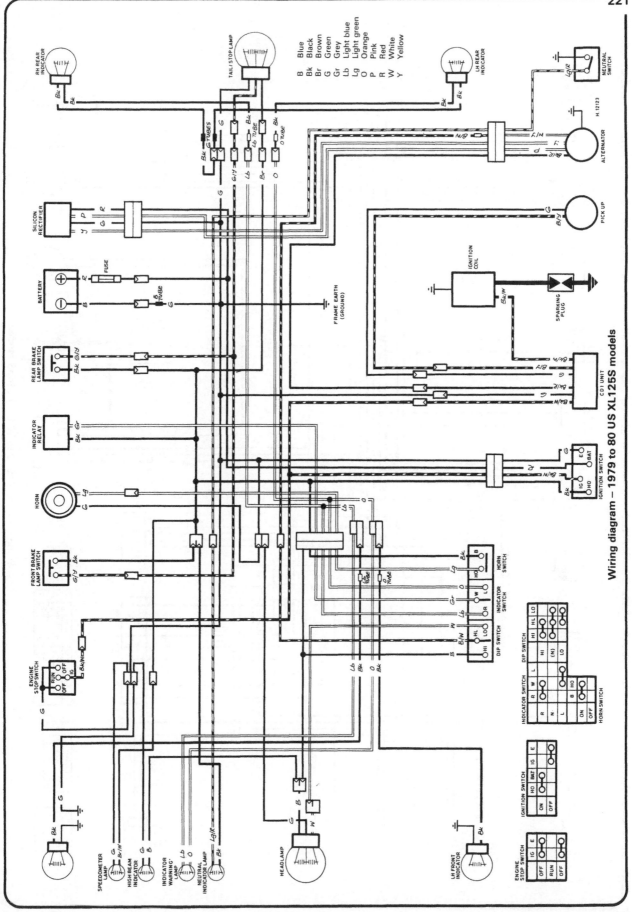

Wiring diagram – 1979 to 80 US XL125S models

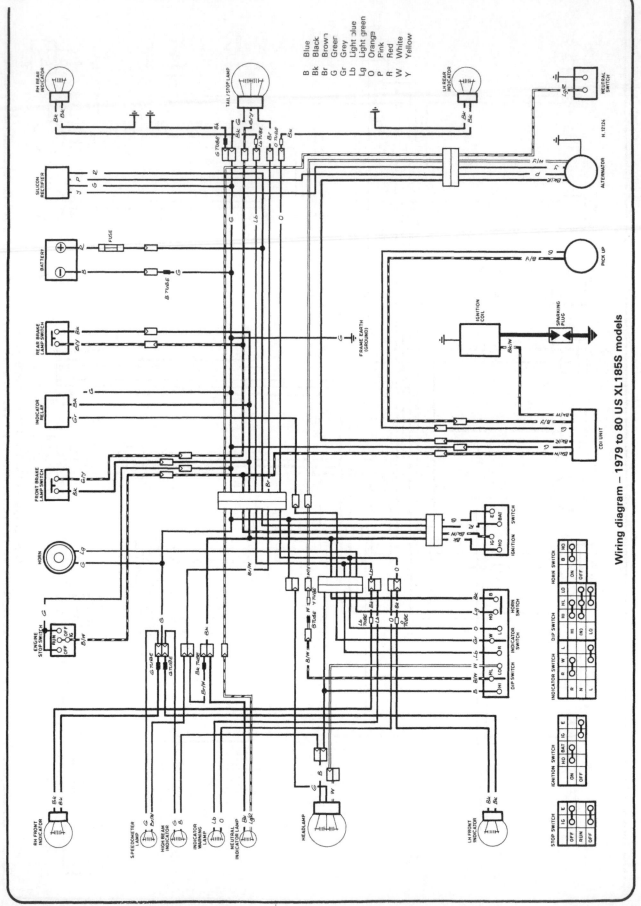

Wiring diagram – 1979 to 80 US XL185S models

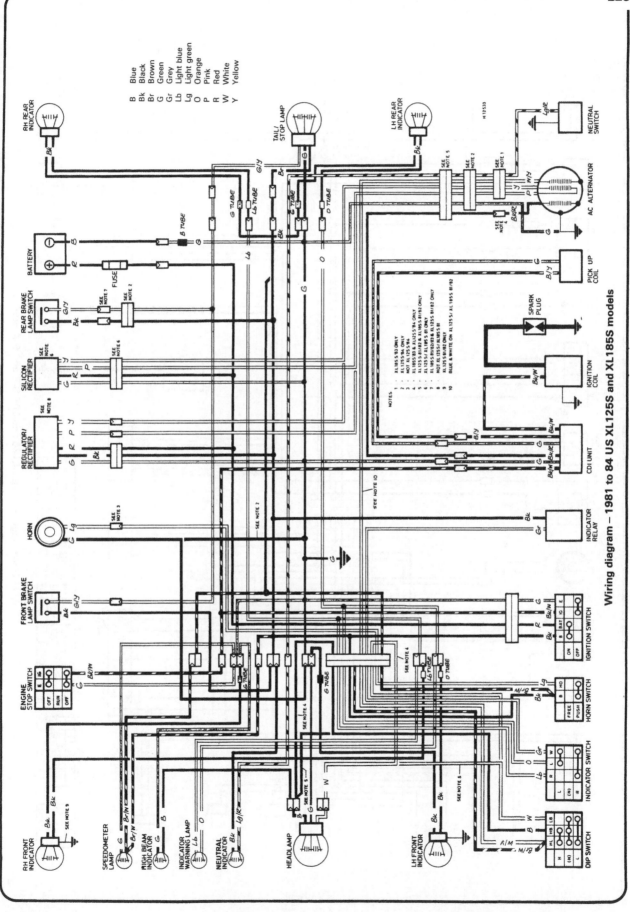

Wiring diagram – 1981 to 84 US XL125S and XL185S models

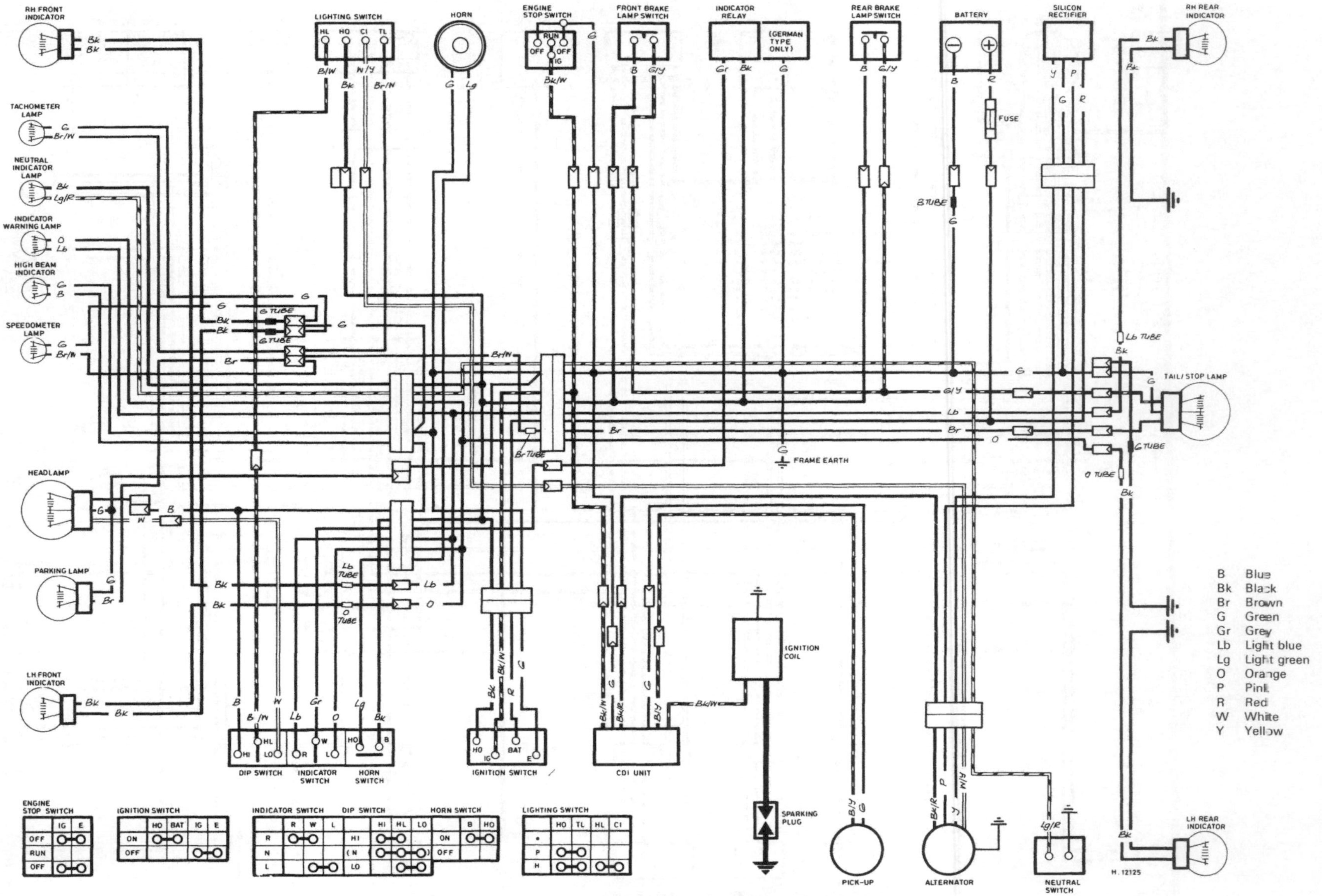

Blue — B
Black — Bk
Brown — Br
Green — G
Grey — Gr
Light blue — Lb
Light green — Lg
Orange — O
Pink — P
Red — R
White — W
Yellow — Y

RH REAR INDICATOR
TAIL/STOP LAMP
LH REAR INDICATOR
H 12125

SILICON RECTIFIER
NEUTRAL SWITCH
BATTERY
FUSE
ALTERNATOR
REAR BRAKE LAMP SWITCH
PICK-UP
INDICATOR RELAY (GERMAN TYPE ONLY)
IGNITION COIL
SPARKING PLUG
FRONT BRAKE LAMP SWITCH
CDI UNIT
ENGINE STOP SWITCH
HORN
IGNITION SWITCH
LIGHTING SWITCH
HORN SWITCH
DIP SWITCH
INDICATOR SWITCH

RH FRONT INDICATOR
TACHOMETER LAMP
NEUTRAL INDICATOR LAMP
INDICATOR WARNING LAMP
HIGH BEAM INDICATOR
SPEEDOMETER LAMP
HEADLAMP
PARKING LAMP
LH FRONT INDICATOR

FRAME EARTH

LIGHTING SWITCH
HL TL HO C1

IGNITION SWITCH
HO BAT IG

ENGINE STOP SWITCH
IG E
OFF
RUN
OFF

Wiring diagram – UK XL125 and 185 S-Z and S-A models

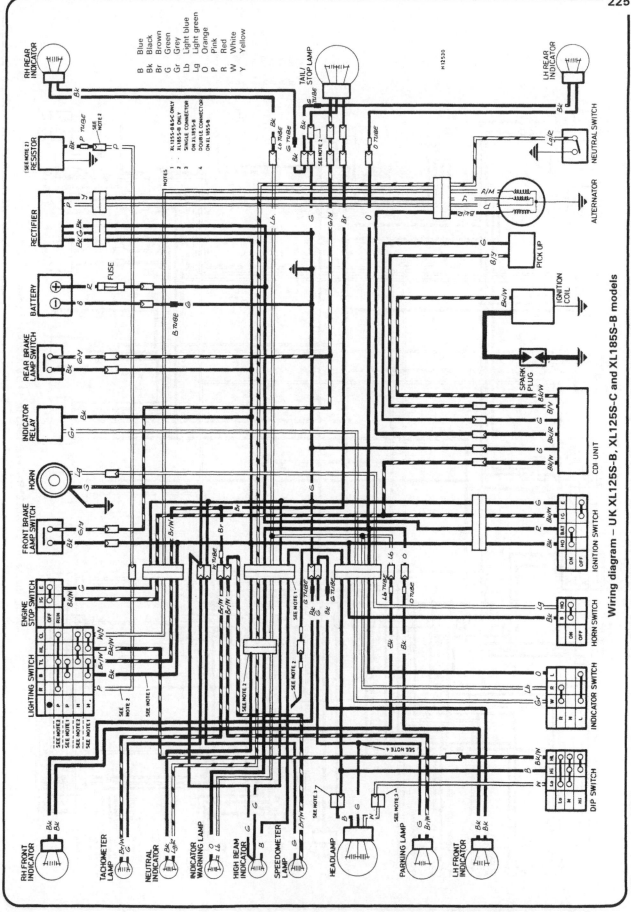

Wiring diagram – UK XL125S-B, XL125S-C and XL185S-B models

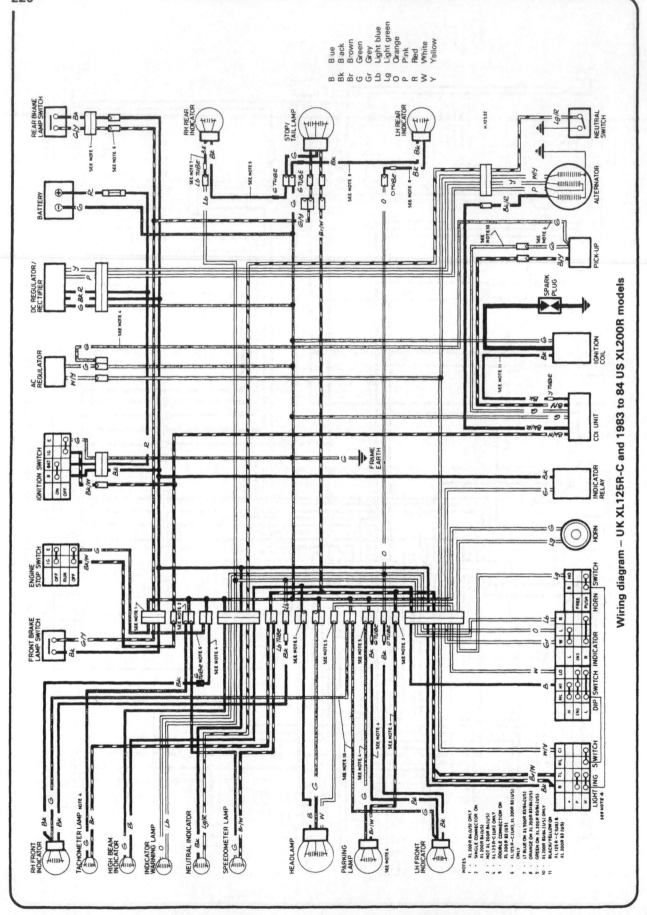

Wiring diagram – UK XL125R-C and 1983 to 84 US XL200R models

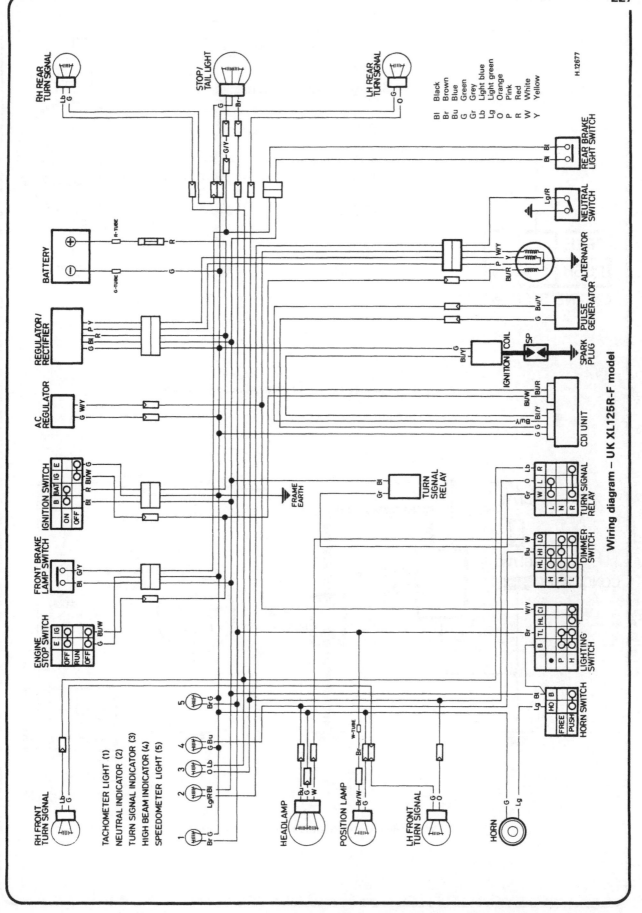

Wiring diagram – UK XL125R-F model

H.12677

RH REAR TURN SIGNAL

STOP/ TAIL LIGHT

LH REAR TURN SIGNAL

Bl — Black
Br — Brown
Bu — Blue
G — Green
Gr — Grey
Lb — Light blue
Lg — Light green
O — Orange
P — Pink
R — Red
W — White
Y — Yellow

REAR BRAKE LIGHT SWITCH

NEUTRAL SWITCH

BATTERY

ALTERNATOR

REGULATOR/ RECTIFIER

PULSE GENERATOR

IGNITION COIL

SPARK PLUG

AC REGULATOR

CDI UNIT

IGNITION SWITCH

TURN SIGNAL RELAY

FRAME EARTH

TURN SIGNAL RELAY

FRONT BRAKE LAMP SWITCH

DIMMER SWITCH

LIGHTING SWITCH

ENGINE STOP SWITCH

HORN SWITCH

RH FRONT TURN SIGNAL

TACHOMETER LIGHT (1)
NEUTRAL INDICATOR (2)
TURN SIGNAL INDICATOR (3)
HIGH BEAM INDICATOR (4)
SPEEDOMETER LIGHT (5)

HEADLAMP

POSITION LAMP

LH FRONT TURN SIGNAL

HORN

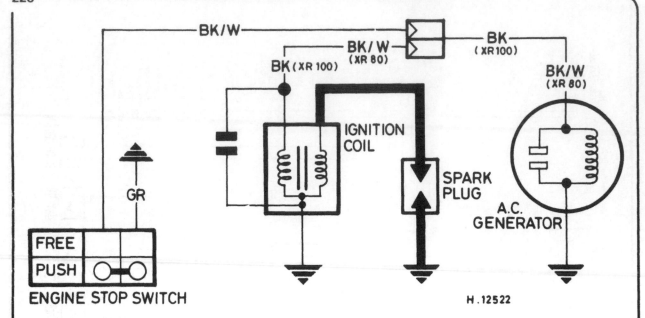

BK Black
GR Grey
W White

Wiring diagram – 1981 to 84 US XR80 and XR100 models

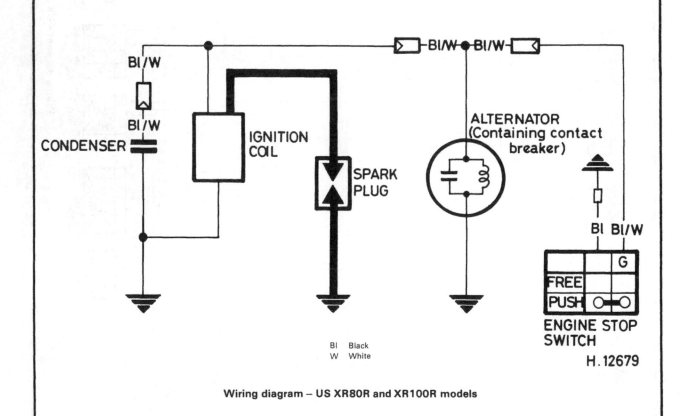

Bl Black
W White

Wiring diagram – US XR80R and XR100R models

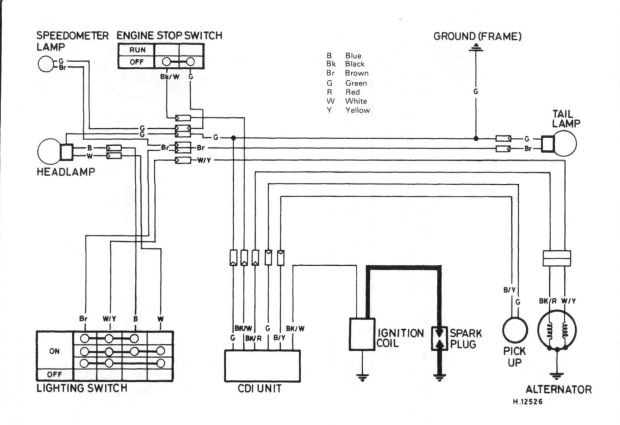

Wiring diagram – UK XR200-A and 1979 to 81 US XR185 and XR200 models

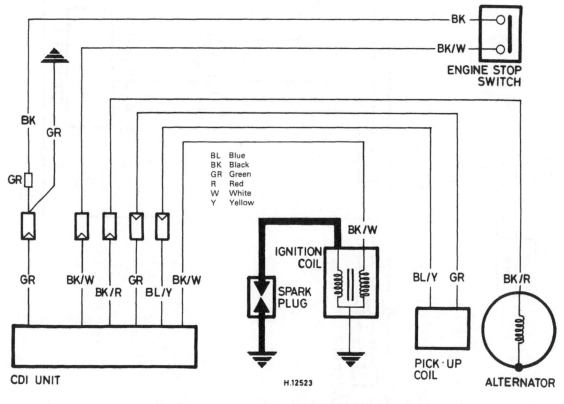

Wiring diagram – 1982 to 84 US XR200 models

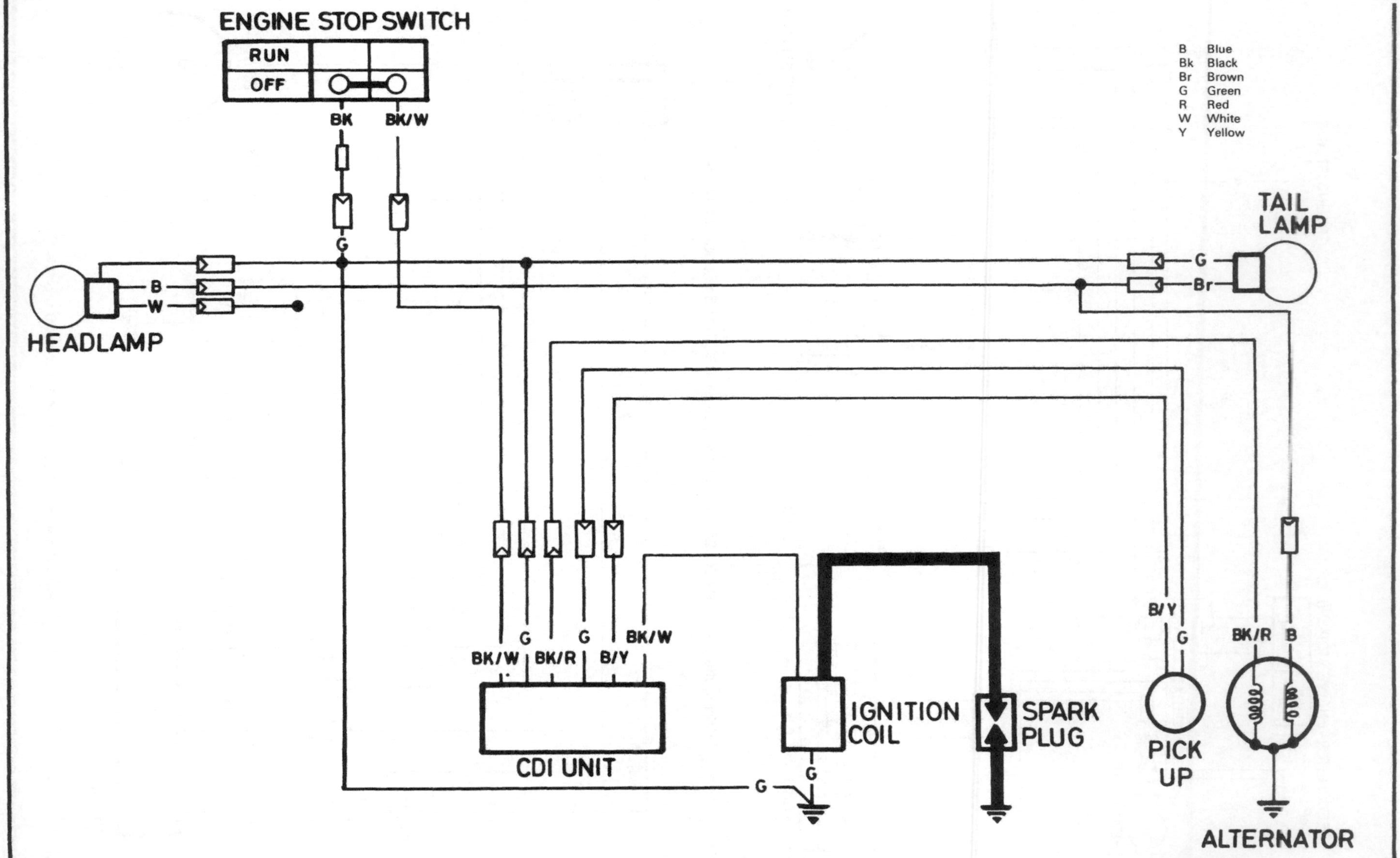
ENGINE STOP SWITCH
RUN
OFF
BK
BK/W
G
B Blue
Bk Black
Br Brown
G Green
R Red
W White
Y Yellow
TAIL LAMP
G
Br
B
W
HEADLAMP
BK/W
G
BK/R
G
B/Y
BK/W
CDI UNIT
IGNITION COIL
G
SPARK PLUG
B/Y
G
PICK UP
BK/R
B
ALTERNATOR
H.12527

Wiring diagram – 1980 to 83 US XR200R models

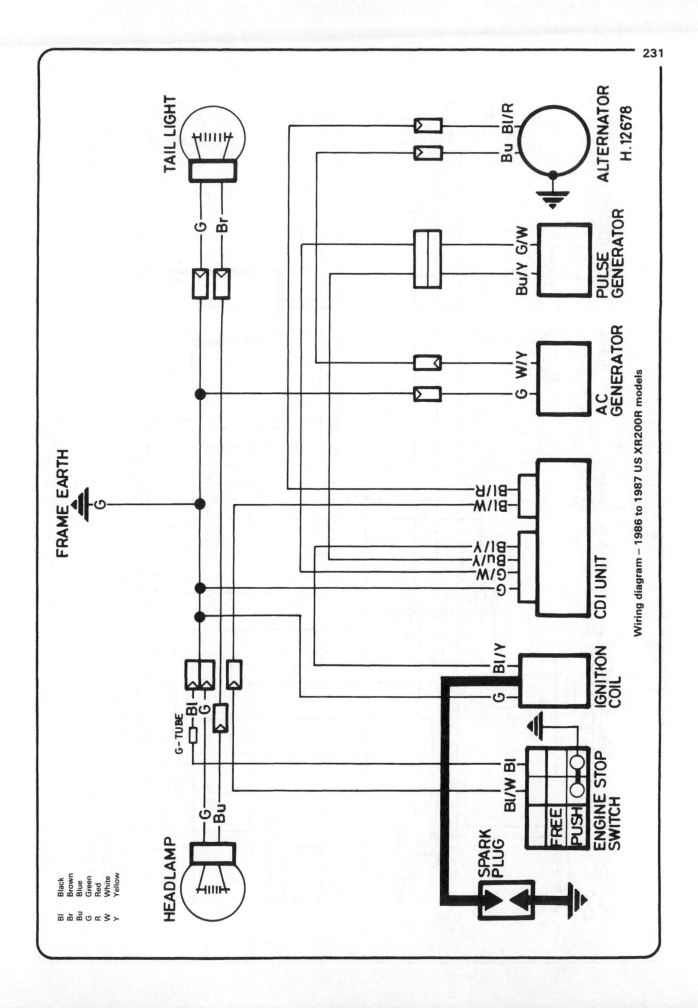

Wiring diagram – 1986 to 1987 US XR200R models

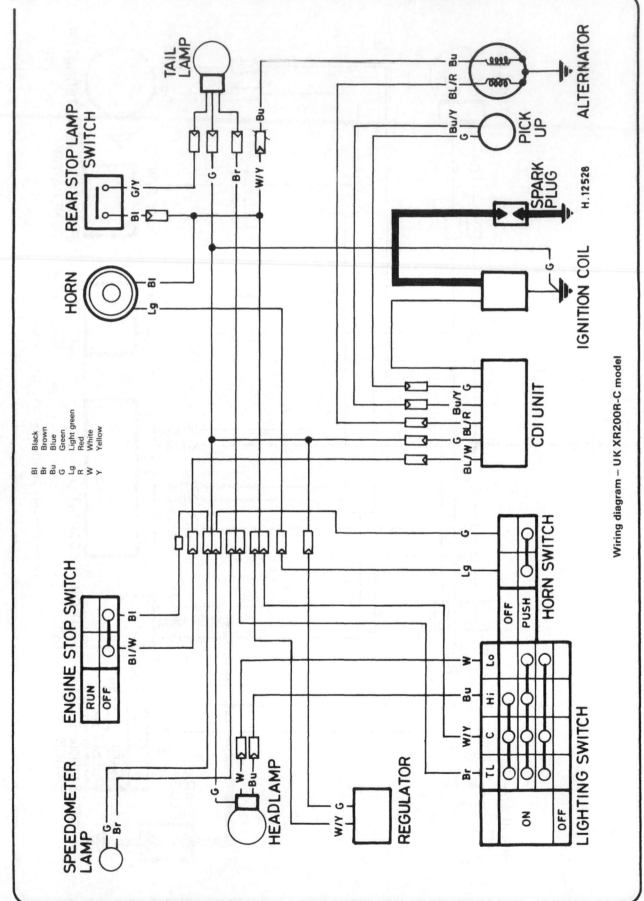

Wiring diagram – UK XR200R-C model

Bl   Black
Br   Brown
Bu   Blue
G   Green
Lg   Light green
R   Red
W   White
Y   Yellow

Conversion factors

Length (distance)
Inches (in)	X	25.4	= Millimetres (mm)	X 0.0394	= Inches (in)
Feet (ft)	X	0.305	= Metres (m)	X 3.281	= Feet (ft)
Miles	X	1.609	= Kilometres (km)	X 0.621	= Miles

Volume (capacity)
Cubic inches (cu in; in^3)	X	16.387	= Cubic centimetres (cc; cm^3)	X 0.061	= Cubic inches (cu in; in^3)
Imperial pints (Imp pt)	X	0.568	= Litres (l)	X 1.76	= Imperial pints (Imp pt)
Imperial quarts (Imp qt)	X	1.137	= Litres (l)	X 0.88	= Imperial quarts (Imp qt)
Imperial quarts (Imp qt)	X	1.201	= US quarts (US qt)	X 0.833	= Imperial quarts (Imp qt)
US quarts (US qt)	X	0.946	= Litres (l)	X 1.057	= US quarts (US qt)
Imperial gallons (Imp gal)	X	4.546	= Litres (l)	X 0.22	= Imperial gallons (Imp gal)
Imperial gallons (Imp gal)	X	1.201	= US gallons (US gal)	X 0.833	= Imperial gallons (Imp gal)
US gallons (US gal)	X	3.785	= Litres (l)	X 0.264	= US gallons (US gal)

Mass (weight)
Ounces (oz)	X	28.35	= Grams (g)	X 0.035	= Ounces (oz)
Pounds (lb)	X	0.454	= Kilograms (kg)	X 2.205	= Pounds (lb)

Force
Ounces-force (ozf; oz)	X	0.278	= Newtons (N)	X 3.6	= Ounces-force (ozf; oz)
Pounds-force (lbf; lb)	X	4.448	= Newtons (N)	X 0.225	= Pounds-force (lbf; lb)
Newtons (N)	X	0.1	= Kilograms-force (kgf; kg)	X 9.81	= Newtons (N)

Pressure
Pounds-force per square inch (psi; lbf/in^2; lb/in^2)	X	0.070	= Kilograms-force per square centimetre (kgf/cm^2; kg/cm^2)	X 14.223	= Pounds-force per square inch (psi; lbf/in^2; lb/in^2)
Pounds-force per square inch (psi; lbf/in^2; lb/in^2)	X	0.068	= Atmospheres (atm)	X 14.696	= Pounds-force per square inch (psi; lbf/in^2; lb/in^2)
Pounds-force per square inch (psi; lbf/in^2; lb/in^2)	X	0.069	= Bars	X 14.5	= Pounds-force per square inch (psi; lbf/in^2; lb/in^2)
Pounds-force per square inch (psi; lbf/in^2; lb/in^2)	X	6.895	= Kilopascals (kPa)	X 0.145	= Pounds-force per square inch (psi; lbf/in^2; lb/in^2)
Kilopascals (kPa)	X	0.01	= Kilograms-force per square centimetre (kgf/cm^2; kg/cm^2)	X 98.1	= Kilopascals (kPa)

Torque (moment of force)
Pounds-force inches (lbf in; lb in)	X	1.152	= Kilograms-force centimetre (kgf cm; kg cm)	X 0.868	= Pounds-force inches (lbf in; lb in)
Pounds-force inches (lbf in; lb in)	X	0.113	= Newton metres (Nm)	X 8.85	= Pounds-force inches (lbf in; lb in)
Pounds-force inches (lbf in; lb in)	X	0.083	= Pounds-force feet (lbf ft; lb ft)	X 12	= Pounds-force inches (lbf in; lb in)
Pounds-force feet (lbf ft; lb ft)	X	0.138	= Kilograms-force metres (kgf m; kg m)	X 7.233	= Pounds-force feet (lbf ft; lb ft)
Pounds-force feet (lbf ft; lb ft)	X	1.356	= Newton metres (Nm)	X 0.738	= Pounds-force feet (lbf ft; lb ft)
Newton metres (Nm)	X	0.102	= Kilograms-force metres (kgf m; kg m)	X 9.804	= Newton metres (Nm)

Power
Horsepower (hp)	X	745.7	= Watts (W)	X 0.0013	= Horsepower (hp)

Velocity (speed)
Miles per hour (miles/hr; mph)	X	1.609	= Kilometres per hour (km/hr; kph)	X 0.621	= Miles per hour (miles/hr; mph)

Fuel consumption*
Miles per gallon, Imperial (mpg)	X	0.354	= Kilometres per litre (km/l)	X 2.825	= Miles per gallon, Imperial (mpg)
Miles per gallon, US (mpg)	X	0.425	= Kilometres per litre (km/l)	X 2.352	= Miles per gallon, US (mpg)

Temperature
Degrees Fahrenheit = (°C x 1.8) + 32   Degrees Celsius (Degrees Centigrade; °C) = (°F - 32) x 0.56

*It is common practice to convert from miles per gallon (mpg) to litres/100 kilometres (l/100km),
where mpg (Imperial) x l/100 km = 282 and mpg (US) x l/100 km = 235

English/American terminology

Because this book has been written in England, British English component names, phrases and spellings have been used throughout. American English usage is quite often different and whereas normally no confusion should occur, a list of equivalent terminology is given below.

English	American	English	American
Air filter	Air cleaner	Number plate	License plate
Alignment (headlamp)	Aim	Output or layshaft	Countershaft
Allen screw/key	Socket screw/wrench	Panniers	Side cases
Anticlockwise	Counterclockwise	Paraffin	Kerosene
Bottom/top gear	Low/high gear	Petrol	Gasoline
Bottom/top yoke	Bottom/top triple clamp	Petrol/fuel tank	Gas tank
Bush	Bushing	Pinking	Pinging
Carburettor	Carburetor	Rear suspension unit	Rear shock absorber
Catch	Latch	Rocker cover	Valve cover
Circlip	Snap ring	Selector	Shifter
Clutch drum	Clutch housing	Self-locking pliers	Vise-grips
Dip switch	Dimmer switch	Side or parking lamp	Parking or auxiliary light
Disulphide	Disulfide	Side or prop stand	Kick stand
Dynamo	DC generator	Silencer	Muffler
Earth	Ground	Spanner	Wrench
End float	End play	Split pin	Cotter pin
Engineer's blue	Machinist's dye	Stanchion	Tube
Exhaust pipe	Header	Sulphuric	Sulfuric
Fault diagnosis	Trouble shooting	Sump	Oil pan
Float chamber	Float bowl	Swinging arm	Swingarm
Footrest	Footpeg	Tab washer	Lock washer
Fuel/petrol tap	Petcock	Top box	Trunk
Gaiter	Boot	Torch	Flashlight
Gearbox	Transmission	Two/four stroke	Two/four cycle
Gearchange	Shift	Tyre	Tire
Gudgeon pin	Wrist/piston pin	Valve collar	Valve retainer
Indicator	Turn signal	Valve collets	Valve cotters
Inlet	Intake	Vice	Vise
Input shaft or mainshaft	Mainshaft	Wheel spindle	Axle
Kickstart	Kickstarter	White spirit	Stoddard solvent
Lower leg	Slider	Windscreen	Windshield
Mudguard	Fender		

Index

Haynes Motorcycle Manuals – The Complete List

Title	Book No
APRILIA RS50 (99 - 06) & RS125 (93 - 06)	4298
Aprilia RSV1000 Mille (98 - 03) ◆	4255
BMW 2-valve Twins (70 - 96) ◆	0249
BMW K100 & 75 2-valve Models (83 - 96) ◆	1373
BMW R850, 1100 & 1150 4-valve Twins (93 - 04) ◆	3466
BMW R1200 (04 - 06) ◆	4598
BSA Bantam (48 - 71)	0117
BSA Unit Singles (58 - 72)	0127
BSA Pre-unit Singles (54 - 61)	0326
BSA A7 & A10 Twins (47 - 62)	0121
BSA A50 & A65 Twins (62 - 73)	0155
DUCATI 600, 620, 750 and 900 2-valve V-Twins (91 - 05) ◆	3290
Ducati MK III & Desmo Singles (69 - 76) ◇	0445
Ducati 748, 916 & 996 4-valve V-Twins (94 - 01) ◆	3756
GILERA Runner, DNA, Ice & SKP/Stalker (97 - 07)	4163
HARLEY-DAVIDSON Sportsters (70 - 03) ◆	2534
Harley-Davidson Shovelhead and Evolution Big Twins (70 - 99) ◆	2536
Harley-Davidson Twin Cam 88 (99 - 03) ◆	2478
HONDA NB, ND, NP & NS50 Melody (81 - 85) ◇	0622
Honda NE/NB50 Vision & SA50 Vision Met-in (85 - 95) ◇	1278
Honda MB, MBX, MT & MTX50 (80 - 93)	0731
Honda C50, C70 & C90 (67 - 03)	0324
Honda XR80/100R & CRF80/100F (85 - 04)	2218
Honda XL/XR 80, 100, 125, 185 & 200 2-valve Models (78 - 87)	0566
Honda H100 & H100S Singles (80 - 92) ◇	0734
Honda CB/CD125T & CM125C Twins (77 - 88) ◇	0571
Honda CG125 (76 - 07) ◇	0433
Honda NS125 (86 - 93) ◇	3056
Honda CBR125R (04 - 07)	4620
Honda MBX/MTX125 & MTX200 (83 - 93) ◇	1132
Honda CD/CM185 200T & CM250C 2-valve Twins (77 - 85)	0572
Honda XL/XR 250 & 500 (78 - 84)	0567
Honda XR250L, XR250R & XR400R (86 - 03)	2219
Honda CB250 & CB400N Super Dreams (78 - 84) ◇	0540
Honda CR Motocross Bikes (86 - 01)	2222
Honda CRF250 & CRF450 (02 - 06)	2630
Honda CBR400RR Fours (88 - 99) ◇ ◆	3552
Honda VFR400 (NC30) & RVF400 (NC35) V-Fours (89 - 98) ◇ ◆	3496
Honda CB500 (93 - 01) ◇	3753
Honda CB400 & CB550 Fours (73 - 77)	0262
Honda CX/GL500 & 650 V-Twins (78 - 86)	0442
Honda CBX550 Four (82 - 86) ◇	0940
Honda XL600R & XR600R (83 - 00)	2183
Honda XL600/650V Transalp & XRV750 Africa Twin (87 to 07) ◆	3919
Honda CBR600F1 & 1000F Fours (87 - 96) ◆	1730
Honda CBR600F2 & F3 Fours (91 - 98) ◆	2070
Honda CBR600F4 (99 - 06) ◆	3911
Honda CB600F Hornet & CBF600 (98 - 06) ◇ ◆	3915
Honda CBR600RR (03 - 06) ◆	4590
Honda CB650 sohc Fours (78 - 84)	0665
Honda NTV600 Revere, NTV650 and NT650V Deauville (88 - 05) ◇ ◆	3243
Honda Shadow VT600 & 750 (USA) (88 - 03)	2312
Honda CB750 sohc Four (69 - 79)	0131
Honda V45/65 Sabre & Magna (82 - 88)	0820
Honda VFR750 & 700 V-Fours (86 - 97) ◆	2101
Honda VFR800 V-Fours (97 - 01) ◆	3703
Honda VFR800 V-Tec V-Fours (02 - 05) ◆	4196
Honda CB750 & CB900 dohc Fours (78 - 84)	0535
Honda VTR1000 (FireStorm, Super Hawk) & XL1000V (Varadero) (97 - 00) ◆	3744
Honda CBR900RR FireBlade (92 - 99) ◆	2161
Honda CBR900RR FireBlade (00 - 03) ◆	4060
Honda CBR1000RR Fireblade (04 - 07) ◆	4604
Honda CBR1100XX Super Blackbird (97 - 07) ◆	3901
Honda ST1100 Pan European V-Fours (90 - 02) ◆	3384
Honda Shadow VT1100 (USA) (85 - 98)	2313
Honda GL1000 Gold Wing (75 - 79)	0309
Honda GL1100 Gold Wing (79 - 81)	0669

Title	Book No
Honda Gold Wing 1200 (USA) (84 - 87)	2199
Honda Gold Wing 1500 (USA) (88 - 00)	2225
KAWASAKI AE/AR 50 & 80 (81 - 95)	1007
Kawasaki KC, KE & KH100 (75 - 99)	1371
Kawasaki KMX125 & 200 (86 - 02) ◇	3046
Kawasaki 250, 350 & 400 Triples (72 - 79)	0134
Kawasaki 400 & 440 Twins (74 - 81)	0281
Kawasaki 400, 500 & 550 Fours (79 - 91)	0910
Kawasaki EN450 & 500 Twins (Ltd/Vulcan) (85 - 04)	2053
Kawasaki EX500 (GPZ500S) & ER500 (ER-5) (87 - 05) ◆	2052
Kawasaki ZX600 (ZZ-R600 & Ninja ZX-6) (90 - 06) ◆	2146
Kawasaki ZX-6R Ninja Fours (95 - 02) ◆	3541
Kawasaki ZX-6R (03 - 06) ◆	4742
Kawasaki ZX600 (GPZ600R, GPX600R, Ninja 600R & RX) & ZX750 (GPX750R, Ninja 750R) ◆	1780
Kawasaki 650 Four (76 - 78)	0373
Kawasaki Vulcan 700/750 & 800 (85 - 04) ◆	2457
Kawasaki 750 Air-cooled Fours (80 - 91)	0574
Kawasaki ZR550 & 750 Zephyr Fours (90 - 97) ◆	3382
Kawasaki Z750 & Z1000 (03 - 08) ◆	4762
Kawasaki ZX750 (Ninja ZX-7 & ZXR750) Fours (89 - 96) ◆	2054
Kawasaki Ninja ZX-7R & ZX-9R (94 - 04) ◆	3721
Kawasaki 900 & 1000 Fours (73 - 77)	0222
Kawasaki ZX900, 1000 & 1100 Liquid-cooled Fours (83 - 97) ◆	1681
KTM EXC Enduro & SX Motocross (00 - 07) ◆	4629
MOTO GUZZI 750, 850 & 1000 V-Twins (74 - 78)	0339
MZ ETZ Models (81 - 95) ◇	1680
NORTON 500, 600, 650 & 750 Twins (57 - 70)	0187
Norton Commando (68 - 77)	0125
PEUGEOT Speedfight, Trekker & Vivacity Scooters (96 - 05) ◇	3920
PIAGGIO (Vespa) Scooters (91 - 06)	3492
SUZUKI GT, ZR & TS50 (77 - 90) ◇	0799
Suzuki TS50X (84 - 00) ◇	1599
Suzuki 100, 125, 185 & 250 Air-cooled Trail bikes (79 - 89) ◇	0797
Suzuki GP100 & 125 Singles (78 - 93) ◇	0576
Suzuki GS, GN, GZ & DR125 Singles (82 - 05) ◇	0888
Suzuki 250 & 350 Twins (68 - 78)	0120
Suzuki GT250X7, GT200X5 & SB200 Twins (78 - 83) ◇	0469
Suzuki GS/GSX250, 400 & 450 Twins (79 - 85)	0736
Suzuki GS500 Twin (89 - 06) ◆	3238
Suzuki GS550 (77 - 82) & GS750 Fours (76 - 79)	0363
Suzuki GS/GSX550 4-valve Fours (83 - 88)	1133
Suzuki SV650 & SV650S (99 - 05) ◆	3912
Suzuki GSX-R600 & 750 (96 - 00) ◆	3553
Suzuki GSX-R600 (01 - 03), GSX-R750 (00 - 03) & GSX-R1000 (01 - 02) ◆	3986
Suzuki GSX-R600/750 (04 - 05) & GSX-R1000 (03 - 06) ◆	4382
Suzuki GSF600, 650 & 1200 Bandit Fours (95 - 06) ◆	3367
Suzuki Intruder, Marauder, Volusia & Boulevard (85 - 06) ◆	2618
Suzuki GS850 Fours (78 - 88)	0536
Suzuki GS1000 Four (77 - 79)	0484
Suzuki GSX-R750, GSX-R1100 (85 - 92), GSX600F, GSX750F, GSX1100F (Katana) Fours ◆	2055
Suzuki GSX600/750F & GSX750 (98 - 02) ◆	3987
Suzuki GS/GSX1000, 1100 & 1150 4-valve Fours (79 - 88)	0737
Suzuki TL1000S/R & DL1000 V-Strom (97 - 04) ◆	4083
Suzuki GSX1300R Hayabusa (99 - 04) ◆	4184
Suzuki GSX1400 (02 - 07) ◆	4758
TRIUMPH Tiger Cub & Terrier (52 - 68)	0414
Triumph 350 & 500 Unit Twins (58 - 73)	0137
Triumph Pre-Unit Twins (47 - 62)	0251
Triumph 650 & 750 2-valve Unit Twins (63 - 83)	0122
Triumph Trident & BSA Rocket 3 (69 - 75)	0136
Triumph Bonneville (01 - 07) ◆	4364
Triumph Daytona, Speed Triple, Sprint & Tiger (97 - 05) ◆	3755
Triumph Triples and Fours (carburettor engines) (91 - 04) ◆	2162
VESPA P/PX125, 150 & 200 Scooters (78 - 06)	0707
Vespa Scooters (59 - 78)	0126
YAMAHA DT50 & 80 Trail Bikes (78 - 95) ◇	0800
Yamaha T50 & 80 Townmate (83 - 95) ◇	1247
Yamaha YB100 Singles (73 - 91) ◇	0474

Title	Book No
Yamaha RS/RXS100 & 125 Singles (74 - 95)	0331
Yamaha RD & DT125LC (82 - 87) ◇	0887
Yamaha TZR125 (87 - 93) & DT125R (88 - 02) ◇	1655
Yamaha TY50, 80, 125 & 175 (74 - 84) ◇	0464
Yamaha XT & SR125 (82 - 03) ◇	1021
Yamaha Trail Bikes (81 - 00)	2350
Yamaha 2-stroke Motocross Bikes 1986 - 2006	2662
Yamaha YZ & WR 4-stroke Motocross Bikes (98 - 07)	2689
Yamaha 250 & 350 Twins (70 - 79)	0040
Yamaha XS250, 360 & 400 sohc Twins (75 - 84)	0378
Yamaha RD250 & 350LC Twins (80 - 82)	0803
Yamaha RD350 YPVS Twins (83 - 95)	1158
Yamaha RD400 Twin (75 - 79)	0333
Yamaha XT, TT & SR500 Singles (75 - 83)	0342
Yamaha XZ550 Vision V-Twins (82 - 85)	0821
Yamaha FJ, FZ, XJ & YX600 Radian (84 - 92)	2100
Yamaha XJ600S (Diversion, Seca II) & XJ600N Fours (92 - 03) ◆	2145
Yamaha YZF600R Thundercat & FZS600 Fazer (96 - 03) ◆	3702
Yamaha FZ-6 Fazer (04 - 07) ◆	4751
Yamaha YZF-R6 (99 - 02) ◆	3900
Yamaha YZF-R6 (03 - 05) ◆	4601
Yamaha 650 Twins (70 - 83)	0341
Yamaha XJ650 & 750 Fours (80 - 84)	0738
Yamaha XS750 & 850 Triples (76 - 85)	0340
Yamaha TDM850, TRX850 & XTZ750 (89 - 99) ◇ ◆	3540
Yamaha YZF750R & YZF1000R Thunderace (93 - 00) ◆	3720
Yamaha FZR600, 750 & 1000 Fours (87 - 96) ◆	2056
Yamaha XV (Virago) V-Twins (81 - 03) ◆	0802
Yamaha XVS650 & 1100 Drag Star/V-Star (97 - 05) ◆	4195
Yamaha XJ900F Fours (83 - 94) ◆	3239
Yamaha XJ900S Diversion (94 - 01) ◆	3739
Yamaha YZF-R1 (98 - 03) ◆	3754
Yamaha YZF-R1 (04 - 06) ◆	4605
Yamaha FZS1000 Fazer (01 - 05) ◆	4287
Yamaha FJ1100 & 1200 Fours (84 - 96) ◆	2057
Yamaha XJR1200 & 1300 (95 - 06) ◆	3981
Yamaha V-Max (85 - 03) ◆	4072

ATVs

Title	Book No
Honda ATC70, 90, 110, 185 & 200 (71 - 85)	0565
Honda Rancher, Recon & TRX250EX ATVs	2553
Honda TRX300 Shaft Drive ATVs (88 - 00)	2125
Honda TRX300EX, TRX400EX & TRX450R/ER ATVs (93 - 06)	2318
Kawasaki Bayou 220/250/300 & Prairie 300 ATVs (86 - 03)	2351
Polaris ATVs (85 - 97)	2302
Polaris ATVs (98 - 06)	2508
Yamaha YFS200 Blaster ATV (88 - 02)	2317
Yamaha YFB250 Timberwolf ATVs (92 - 00)	2217
Yamaha YFM350 & YFM400 (ER and Big Bear) ATVs (87 - 03)	2126
Yamaha Banshee and Warrior ATVs (87 - 03)	2314
Yamaha Kodiak and Grizzly ATVs (93 - 05)	2567
ATV Basics	10450

TECHBOOK SERIES

Title	Book No
Twist and Go (automatic transmission) Scooters Service and Repair Manual	4082
Motorcycle Basics TechBook (2nd Edition)	3515
Motorcycle Electrical TechBook (3rd Edition)	3471
Motorcycle Fuel Systems TechBook	3514
Motorcycle Maintenance TechBook	4071
Motorcycle Modifying	4272
Motorcycle Workshop Practice TechBook (2nd Edition)	3470

◇ = not available in the USA ◆ = Superbike

The manuals on this page are available through good motorcycle dealers and accessory shops.
In case of difficulty, contact: **Haynes Publishing**
(UK) +44 1963 442030 (USA) +1 805 498 6703
(SV) +46 18 124016
(Australia/New Zealand) +61 3 9763 8100

MCL23.12/07